Biogeography

Biogeography

Paul Müller

University of Saabrüchen

Edited by

J. R. Flenley

The University of Hull

S. A. Burgess

D. Beeson

1817

HARPER & ROW, PUBLISHERS, New York

Cambridge, Philadelphia, San Francisco,

London, Mexico City, São Paulo, Singapore, Sydney

Sponsoring Editor: Claudia Wilson
Cover Design: Diane Lufrano
Printer and Binder: The Maple Press

Biogeography

Distributed by Harper & Row, Publishers, Inc.

Library of Congress Cataloging in Publication Data

Muller, Paul, 1940 Oct. 11–
Biogeography.

Translation of: Biogeographie.
Bibliography: p.
Includes index.
1. Biogeography. I. Flenley, John. II. Burgess, S. A. III. Beeson, D. IV. Title.
QH84.M8213 1986 574.9 86–306
ISBN 0–06–044634–X

86 87 88 89 9 8 7 6 5 4 3 2 1

Contents

Foreword

Biogeography is the science of propagation areas. Its research topics cover the geographic demands of living systems, the space and time requirements of populations and individuals. Biogeography is consequently concerned with the biological significance of geographic regions. It pursues geographic knowledge with the tools of the natural historian. Knowledge about plants and animals is a simple prerequisite to further research in biogeography. The history and characteristics of particular species are merely one factor, as fundamental thermodynamic law is another, in the investigation of ecosystems. The propagation area provides the unit within which all these different concepts, which determine the nature of life in the Biosphere, can be brought together and studied. In this way biogeography provides the basis for a deeper understanding of our own existence, as well as suggesting possible solutions to the most urgent of our environmental difficulties.

This book is a first attempt to provide students of biology and the Earth sciences with an overall survey of biogeography as a field. The interested reader will at once see that biogeography is no mere supplement to conventional zoogeography, or geography of vegetation: though it has the same roots, it is a new science in its own right. Its foundations were laid by my teachers Professor de Lattin (Zoology, Saarbrücken) and Professor J. Schmithüsen (Geography, Saarbrücken) and by my scientific 'godfathers' Professors C. Kosswig (Hamburg) and W.F. Reinig (Nürtingen).

My warmest gratitude is due to a great many friends and colleagues for their comments and criticisms on the text. I should like particularly to thank Professors P. Banarescu (Bucharest), L. Brundin (Stockholm), T. Dick (Porto Allegre), Guderian (Essen), J. Haffer (Essen), Kloft (Bonn). J. Leclerq (Gembloux), R. Lewis (Washington), E. Mayr (Harvard), D. Mora (Bogotá), Miyawaki (Tokyo), Økland (Oslo), J.M. Pelt (Metz), D. Povolny (Brünn), F. Salomonsen (Copenhagen), P. Sawaya (São Paulo), V. Soçava (Irkutsk), Z. Varga (Debrecen), F. Vuilleumier (New York) and Walden (Stockholm).

My thanks also go to my undergraduate and graduate students in Saarbrücken (University of the Saarland) and Porto Allegre (Federal University of Rio Grande do Sul). Miss Kreis saw to the notes and the index in the original edition, and D. Schwang was responsible for the programming work for the computer-drawn maps. I am deeply indebted to my colleagues H. Ellenberg, J. Goergen, P. Nagel, A. Schäfer, H. Schreiber, H. Steiniger, G. Wagner and E. Zimen for countless discussions and many suggestions without which this text could never have been composed.

My secretary, Mrs Kunzmann, who has devoted herself selflessly to the development of biogeography since it appeared at Saarbrücken, often working on Sundays and holidays, once more deserves my thanks for the work she did on retyping each and every new version of the German language manuscript.

This book could never have been written without my wife's help. She read the proofs, acted as my frankest critic and provided me with the kind of working conditions essential for the composition of any book.

Finally I should like to thank the publishers of the German edition, R. Ulmer, and my colleague Professor Kloft (Bonn) who has made so important a contribution to the development of biogeography in Germany, and who first gave me the idea of writing this book. It was out of my dialogue with him, enriched by his many suggestions and critical comments, that this *Biogeography* was born.

Paul Müller

Biogeography

Chapter 1

Biogeography — Research Objective and History

1.1 RESEARCH OBJECTIVE OF BIOGEOGRAPHY

Biogeography is the evaluation of distribution areas (ranges) by means of clarification of the structure, function, history, and indicator significance of propagation areas.[1] The subject is a quantitative attempt at deciphering the information available from propagation areas concerning the ecological potential, genetic viability, and phylogeny of populations and biotic communities as well as the spatially and temporally varying behaviour of environmental factors. Biogeography also includes utilizing the study of propagation areas to achieve a better understanding of our habitat (Fig. 1). The basic material for biogeographic investigations consists of the organisms, populations, and communities within the countries of the world. Thus, biogeography must attempt to understand the environment, animals and plants which constitute the subsystems of the global ecosystem. It must examine, by means of experimental analyses, the relationship between complex ecological systems (ecosystems) and propagation areas, i.e., in this case, the spatial and temporal affinities between individual organisms and populations. Inevitably, when making such an examination, the boundaries between biogeography and ecology become blurred, so much so that North American ecologists, such as MacArthur and Wilson (1971), etc., described themselves as biogeographers, without recognizing a "real difference between biogeography and ecology." The actual research objective which remains uniquely the province of biogeography, however, is the propagation area. The apparently simple question: "Why does a certain species exist in area *y* (or why is it missing in area *y*)?" (cf. Müller 1977), can only be answered by an evaluation of the propagation area of that species with respect to space and time.

[1]The German term is "Die Arealsystem" which has no exact English equivalent. For a definition of the term, see Chapter 3.

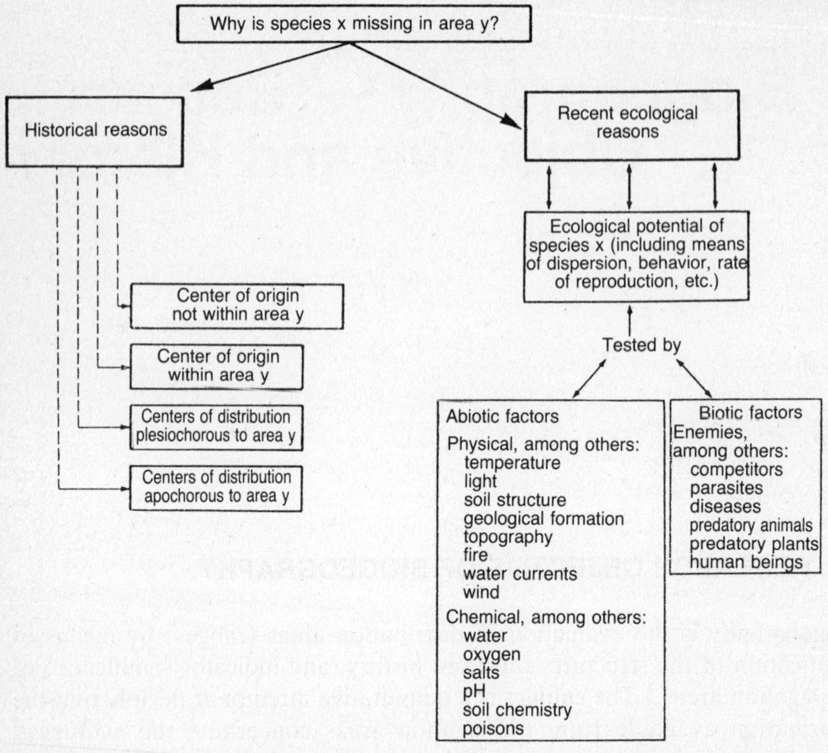

Figure 1 One of the principal questions of biogeography, i.e. the reasons for the presence or the absence of a certain species within a given area can only be answered if a simultaneous analysis of recent ecological and historical factors is performed. The classification of the informational content of the total propagation area in which the species exists as one of the elements constitutes the center point of this analysis.

The expectations of other sciences from biogeography can only be fulfilled by concentrating on this research objective. While biologists expect biogeography to make contributions to the theory of the mechanics of evolution, the geographic objective of biogeography is the elucidation of the properties of geographical regions. This is only possible by continually clarifying the information provided by organisms living within the geographical regions.

1.2 HISTORY OF BIOGEOGRAPHY

A history dealing with the "theory of propagation areas" has not yet been written. But then, it can only be written once the relative importance of phytogeographical and zoogeographical discoveries, findings from theoretical systems analyses, further development of scientific methods, formation of phylogenetic theories and the results of ecosystem research have been assessed more clearly. Still, it may be appropriate at least to point out some outlines of the history of biogeography, particularly, however, that of zoo- and phytogeography as separate subjects. The subject began at the same time as the natural sciences, geography and biology. It was only in the 19th century that distribution became a prominent objective of research.

Theophrastus (372 to 288 B.C.), who dealt with the botanical findings during Alexander's March, in his *Natural History of Plants*, distinguished for the first time between "forms of growth" and "forms of life", and, by means of comparative observations, he formed the opinion that fruit matured because of climatic and soil conditions. He proposed that the true scientist should clarify natural processes by explaining the underlying processes causing them; this suggestion was put into effect by his follower, Straton of Lamsakos, who conducted experimental tests. These early attempts are worth mentioning because their ideas were revived in various places during the 12th century (Hays 1972, Mason 1974).

The beginnings of the study of spatial relationships of animals were first recorded by Albertus Magnus (1193 to 1280), Francis of Assisi (1181 to 1216), and Frederick II of Hohenstaufen (1194 to 1250). Francis of Assisi recognized animals as "co-inhabitants" of the earth, whereas Frederick's falcon book, *The Art of hunting with Birds,* reflects observations of birds and areas. Albertus Magnus did not possess this critical attitude. In his work *De animalibus* (26 volumes), he described the animal world as it was known at that time; he assumed, however, the knowledge of Thomas of Cantimpre (*De natura verum*) and let the latter's fables survive in his own works. Chu Hsi (1131 to 1200), perhaps the most important Neo-Confucian in China, recognized at about the same time that fossils must be the remains of organic beings. After the discovery of the "landscape" in painting, it was not until Leonardo da Vinci (1452 to 1519) that we find further ideas representing the bases for the first biogeographic insights. He no longer was "horrified" by the mountains. In the Monte Rosa area, he studied the vertical zonation of a mountain up to the snowline, with its specialized assemblages of organisms. Paracelsus (1493 to 1541) replaced the study of old writers with his own observations and stressed the point: "I only pay heed to what I have discovered myself and what I have seen confirmed in practice and by experience."

Otto Brunfels (1488 to 1534), called "the Father of Botany" by Linnaeus (1707 to 1778), published the first book on herbs in 1532 (*Contrafyt Kräuterbuch*) and Hieronymus Bock (1498 to 1554), who in 1539 published his *Neu-Kreutter-Buch*, for the first time also indicated the places where the plants had actually been found. Physicians, who more and more disassociated themselves from scholastic thinking and from accepting the authority of the Ancients, recognized their own observations and critical thinking as the actual basis of real science, and thus shaped future developments (Schmithüsen 1970). Here, we would like particularly to mention Conrad Gessner (1516 to 1565), the "German Pliny" and one of the first great biogeographers. His *Historia animalium* (7 volumes) was one of the first works dealing with scientific zoology. Cuvier called Gessner's *Thierbuch*, which appeared in 1551, *"la première base de toute la zoologie moderne."* In his bird book, published in 1555, he mentioned 316 bird species, among which were seven local titmouse species and two goldcrests between which he differentiated on the basis of their migratory behaviour. He described for the first time birds of paradise from New Guinea, parrots from Surinam, the South American muscovy duck, and the North American turkey.

During the same period, Charles de Lecluse (1526 to 1609) examined the plant world of Southern France. In contrast to earlier books on herbs, in his works, de Lecluse no longer selected the plants according to their usefulness. A long time before Albrecht Haller (1708 to 1777), he described the physiognomy of alpine plants. In 1535, Leonard Fuchs was appointed to the University of Tübingen, where he conducted botanical excursions, even with medical students. In 1542, he published his illustrated plant book *De historia stirpium*.

Hans Staden deserves special mention. His book, published in 1557, was not only one of the first ethnological monographs of the New World, it was also a very detailed description of the plants and animal species existing there. Among other things he observed that in Brazil, saltwater fish (Mugilidae) moved into fresh waters in order to spawn. At about the same time (1574), Josias Simler (1530 to 1576) published the first monograph of the Alps. His last three chapters also deal with the organisms (trees, herbs, animals). One of the most beautiful German printed works from the Baroque era is the copperplate bible printed in Ulm and belonging to the city physician of Zurich, Johann Jacob Scheuchzer (1672 to 1733); this bible contains one of the most "interesting misinterpretations (Wey 1966) in the history of intellectual development." It pertains to the discovery of a fossil near Lake Constance, in the general area where the present University of Constance is located. This fossil, a giant salamander, *Andrias scheuchzeri*, was interpreted to be a prehistoric human being. In spite of his wrong track, Scheuchzer became one of the founders of palaeontology. Linnaeus (1707 to 1788), with his *Systema naturae*, brought "order" into the busy world of nature; he also recognized the importance of phenology in connection with climatology. In order to be able to show cartographically the climatic differences between the various regions of the world, he recommended that the flowering periods of plants and the migration periods of animals be systematically studied and that climatic maps be derived from the floral and faunal calendars.

Linnaeus's contemporary, Haller (1708 to 1777), is as well known for his poetry about the Alps as for his physiognomic descriptions of Alpine vegetation. His two large works dealing with the vegetation of Switzerland are interesting from the point of view of biogeography, because they show more clearly than ever before the climatically-dependent altitudinal zonation of the vegetation. During the same period, Johann Georg Gmelin (1709 to 1755), who became famous because of his travels in Siberia and who, in 1749, was nominated Professor of Botany and Chemistry in Tübingen, was also engaged in research.

Buffon's (1707 to 1788) *Histoire naturelle* must be mentioned particularly in relation to the ideas for the development of animal distribution on the earth formulated therein. He assumed that the animals had spread from the poles over the earth, and he also considered the change of the species under the influence of climatic changes. His concept includes former land bridges since his conception of the dispersal of the animals required, for instance, a hypothetical "Atlantis" as a land connection to America. Peter Simon Pallas (1811 to 1871), physician and zoologist, for the first time showed that the vegetational character of the Southern Russian Steppes was dependent upon the climatic conditions of the area.

Eberhard Wilhelm Zimmerman (1743 to 1815) is the first significant German zoogeographer. His zoogeographic works were the result of the monographic treatment of individual groups of animals. His contemporaries, Saussure (1740 to 1799), Ramond (1756 to 1827) and Soulavie (1752 to 1813) studied bioclimatic problems. It was particularly Soulavie who was determined to recognize the laws the Creator had applied in the distribution of creatures. From plant fossils, he derived the temperature of earlier eras and concluded that the altitudinal limit of the grapevine had become progressively lower since the 17th century.

Arthur Young (1741 to 1820) in 1790 investigated the northern limits of the olive tree, vine and maize; perhaps inspired by Soulavie, Wildenow (1765 to 1812) went one step further by stating "We understand from the history of plants the influence of climate on vegetation, changes which plants have probably undergone, how Nature cares for the preservation of them, the migration of plants and finally their distribution across the earth." He referred to the fact that in many areas man had already changed a great deal and that certain problems could only be researched "where nature has been allowed to work undisturbed". This was the first time that the distinction between natural land and cultivated land was mentioned. At the head of plant geography is Alexander von Humboldt (1769 to 1859). On his journey to South and Central America he emphasized clearly the value of vegetation for describing areas of land. Vegetation is a deciding factor in the make up of an area. In his *Physiognomik der Gewächse* (*Physiognomy of Plants*) he puts forward principles which determined the course of plant geography in the following period. The concept of "plant association" can be traced back to him. Vegetation as the deciding factor for the basic form of a region and hence of the land was for Humboldt the starting point and most important aim of his interest in the plant world. J.F. Schouw (1787 to 1852) was the first to distinguish the field of research which we know as plant geography and he also makes a distinction between the science of phytogeography and pure botanical geography. His atlas of plant distribution contains the first maps in phytogeography.

In 1835, O. Heer (the originator of plant sociology) published a study: *Conditions of Vegetation in the Southeastern Part of the Canton Glarus*, where the plant world was characterized by communities and by regional types, while in 1872, A. Grisebach, in his *Vegetation of the Earth,* introduced the concept of plant formations which he understood to be plant complexes having a distinct physiognomic character.

Wilhelm Philipp Schimper (1808 to 1880) is considered the founder of "bryogeography". In collaboration with the pharmacist, Philipp Bruch, he started publishing the *Bryologia Europaea* in 1836 (completed in 1866, compare Mägdefrau 1975).

At the time Humboldt ceased to think, there appeared the study by Charles Darwin (1809 to 1882), the "Copernicus of modern Biology" (Haeckel). His work *On the Origin of Species by Means of Natural Selection* (which appeared in 1859) changed the world. It is of significance to the subject of biogeography in that it was the facts of distribution which Darwin collected during his trip around the world that supplied the initiative as well as the proof for his theory of evolution. It is not surprising that at the same time, A.R. Wallace arrived at analogous conclusions in the region of the Sunda islands. Wallace and Darwin had a significant influence on the direction of zoogeography. While Humboldt, among others, was in particular responsible for a more ecological aspect of vegetation geography, Wallace and Darwin determined the historical, phytogenetical aspect of zoogeography.

Around the turn of the century, however, there apeared a great many ecologically-orientated vegetation and zoogeographies. Examples are the vegetation geographies by Schimper (1898) and Warming (1895). Under the influence of the brilliant genius of J.H. Fabre (1823 to 1915), Haeckel (1866), and Möbius (1877), zoogeography began to be studied in terms of ecology. Ekman (1876 to 1964) and Friederichs (1878 to 1969) were active during the same period of time and Hesse recognized in 1924 that "historical zoogeography has been busily expanded and today shows a nice series of well-founded and coherent results. However, in no direction have more sins been committed with respect to zoogeography than here, due to insufficiently founded and superficial hypotheses. Ecological zoogeography, on the other hand, carries the germs for a truly causal science." This historically established contradiction between ecology and history continues in many of the "zoogeographies" up to the middle of the present century, very much in contrast to the vegetation geography (compare among others: Dansereau 1957, Cole 1975, Schmithüsen 1976). It is only deLattin (1967) and Udvardy (1969) who admit that the history of areas is nothing more or less than the ecology of past eras. Areas are the central point. However, they are tied to their environment via the elements constituting these areas which, as genetic structures, possess history and abilities; these thoughts are already at the core of Andrewartha's (1961) and Krebs' (1972) interest. An examination of propagation areas was made possible thereby.

Chapter 2

The Biosphere

We have become used to living on a planet where flowers bloom, birds hatch from eggs, human beings reflect upon the bases of their existence. Sunrise and sunset have become commonplace. We have understood for a long time that the earth is only part, and not the center, of the universe. A wealth of theoretical considerations and experimental findings lead us to suspect that life might at least be thinkable on other planets, within other solar systems. However, the more we delve into the questions regarding the bases of the existence and reaction of living systems on earth, the greater our admiration for the inhabited layer around the earth which is at most 20 km in elevation and which, around the turn of the century was named the "biosphere". This is the fragile stage on which cooperation and antagonism take place among and between the living systems — starting with the individual organism through the biotic communities to complexly structured ecosystems — which finally leads to the development of globally distributed major habitat units such as tropical rain forests, savannahs, or boreal coniferous forests. They form the *ecological* macro-structure of the biosphere, a pattern whose characteristic features are defined by an interplay of earthly and cosmic forces. It is, however, very closely linked to the potentials and specific characteristics of the organisms and biotic communities. They form the genetic *structure of our planet.*

2.1 CONDITIONS AND LIMITS OF LIVING SYSTEMS WITHIN THE BIOSPHERE

2.1.1 The Biosphere and the Sun

The biosphere is the inhabited part of the earth, which is the highest-ranking ecosystem. It comprises the domicile of all living systems. Its most important exterior energy source is the sun. This controls the climate and the gas exchange between the atmosphere and the oceans, as well as the productivity of the vegetation, the blooming seasons of the plants, and the distribution of the animals in space and time. The duration of daylight determines whether the caterpillar of *Araschnia levana* will develop into the light-colored spring form, *levana,* or into the dark-colored summer form, *prorsa,* whether some insects will be forced into pupation due to the reduction of the daylight period, and whether long-day or short-day plants will start to bloom.

On average, the sun's radiation is at 1,350 Watt/m^2 (solar constant) when reaching the upper limit of the atmosphere, the protective envelope of our globe. Upon entering the earth's atmosphere, it diminishes by nearly ⅔ of the cosmic value (to 544 Watt/m^2), and, in Germany, at ground level, it amounts to a yearly mean of only 114 Watt/m^2. This value is dependent on each individual location, on certain climatic factors (cloudiness, among others), and the suspended particulate matter distributed within the atmosphere (Fig. 2). For this reason, a mean value of 220 Watt/m^2 is assumed for

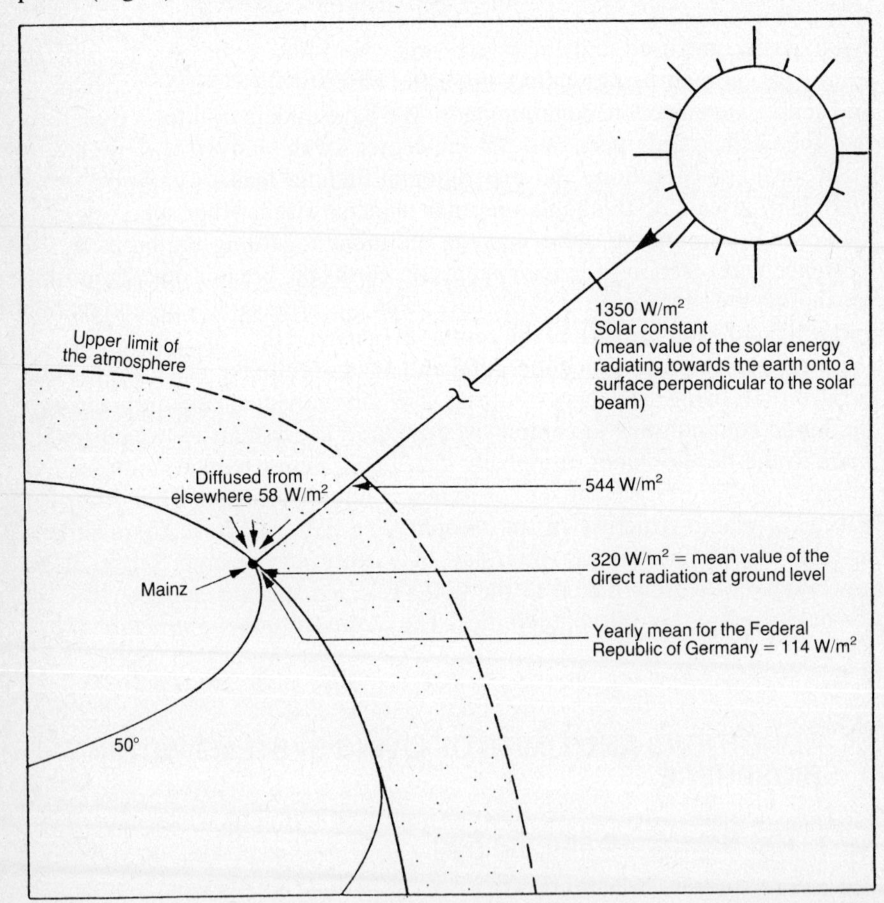

Figure 2 The sun is the decisive driving force behind the biosphere. Its energy which at the point of origin measures 1350 W/m^2, decreases to only 320 W/m^2 when reaching ground level.

the entire biosphere. Although the total energy produced by human beings amounts to less than 1% of the irradiated solar energy, there are significant regional exceptions.

Primary radiation by the sun, astronomical conditions, topography and distribution of land and sea (total land on the earth: 148.9×10^6 km^2 = 29%; ocean: 361.1×10^6 km^2 = 71%. In the northern hemisphere, the ratio land:sea is 39:61; in the southern hemisphere, 19:81), have a deep impact on the climatic processes. The above factors also control the amount of water involved annually in the global hydrological cycle (= 496.1×10^3 km^3).

The oceans account for an average of 82% of the total water circulation in the global hydrological cycle. The evaporation from the oceans (= 1176 mm) exceeds by 110 mm the precipitation falling on those areas. The difference is made up by runoff from the land surfaces where an average of 746 mm of precipitation falls — although evaporation from the land amounts to only 480 mm. Runoff thus compensates for the deficits of the oceans. Half of the total runoff enters the Atlantic, the Pacific Ocean receives 30%, the Indian Ocean 14% and the Arctic Ocean 7%. Both the Pacific and the Atlantic receive 70% of all runoff along their western coasts. In the case of the Indian Ocean, the opposite is true (Baumgartner and Beichel 1975). A wealth of chemical and physical processes are dependent on, or controlled by, the global hydrological cycle. It assembles living systems with differing histories and often deviating genetic structures into both large and small habitat systems. The biota of the earth, when measured by its weight, is insignificant (only 0.1% of the earth, or 10^{14} to 10^{15} t); in the interplay with the abiotic factors, however, the highest-ranking ecosystem, the earth, is created.

The yearly productivity of the major ecosystems is decisively influenced by fluctuations in the abiotic and biotic factors. The largest part of the ultra-violet radiation transmitted by the sun to the earth is screened off by the ozone layer at the upper limit of the biosphere. Life without this shield would be impossible on our planet due to excessive ultra-violet radiation which has a photochemical effect on cells resulting in, among other things, mutations. The ozone content of the air shows considerable fluctuations and is dependent on the atmospheric circulation. In winter, the arctic air is ozone-rich, while the tropics show very low ozone concentrations throughout the year (Campbell 1977) (Table 1, Fig. 3, Fig. 4).

2.1.2 The Biosphere, Neighboring Planets, and the Moon

While the sun deeply affects all vital processes on our planet, our neighboring planets and the only satellite of the earth, the moon, at a distance of about 384,700 km, are of secondary importance. Although the moon played a special role in the early days of agriculture, particularly with respect to female deities, and although there was no lack of attempts to correlate many of the phenomena of human life with the lunar orbits, a direct influence could only be ascertained for aquatic organisms.

Thus, the migration cycle of the European eel (*Anguilla anguilla*) shows

TABLE 1 NET PRIMARY PRODUCTION IN VARIOUS BIOMES OF THE BIOSPHERE (according to Lieth and Whittaker 1975)

Biome Type	Area (10^6 km^2)	Net Primary Production of Dry Mass			Biomass (Dry Mass)		
		Range (g.m^{-2}.a^{-1})	Mean (g.m^{-2}.a^{-1})	Total (10^9 t.a^{-1})	Range (kg/m^2)	Mean (kg/m^2)	Total (10^9 t)
Tropical rain forest	17.0	1000–3500	2200	37.4	6 – 80	45	765
Tropical seasonal rain forest	7.5	1000–2500	1600	12.0	6 – 60	35	260
Temperate forest:							
— evergreen	5.0	600–2500	1300	6.5	6 –200	35	175
— deciduous	7.0	600–2500	1200	8.4	6 – 60	30	210
Taiga	12.0	400–2000	800	9.6	6 – 40	20	240
Timber- and bushland	8.5	250–1200	700	6.0	2 – 20	6	50
Savanna	15.0	200–2000	900	13.5	0.2 – 15	4	60
Steppe	9.0	200–1500	600	5.4	0.2 – 5	1.6	14
Arctic and alpine tundra	8.0	10– 400	140	1.1	0.1 – 3	0.6	5
Desert and semi-deserts	18.0	10– 250	90	1.6	0.1 – 4	0.7	13
Extreme deserts (rocks, sand, ice)	24.0	0– 10	3	0.07	0 – 0.2	0.02	0.5
Cultivated land	14.0	100–4000	650	9.1	0.4 – 12	1	14
Swamps	2.0	800–6000	3000	6.0	3 – 50	15	30
Lakes and rivers	2.0	100–1500	400	0.8	0 – 0.1	0.02	0.05
Continents	149		782	117.5		12.2	1837
Open ocean	332.0	2– 400	125	41.5	0 – 0.005	0.003	1.0
Littoral	0.4	400–1000	500	0.2	0.005– 0.1	0.02	0.008
Continental shelf	26.6	200– 600	360	9.6	0.001– 0.04	0.001	0.27
Intertidal zones and reefs	0.6	500–4000	2500	1.6	0.04 – 4	2	1.2
Estuaries	1.4	200–4000	1500	2.1	0.01 – 4	1	1.4
Oceans	361	–	155	55.0	–	0.01	3.9
Oceans and continents	510	–	336	172.5	–	3.6	1841

a clear lunar periodicity in the rivers as well as in the Baltic Sea. Their migratory instinct is strongest at the time of the waning half-moon and weakest at the time of the waxing half-moon. The larvae of the oyster (*Ostrea edulis*) along the Dutch coast only swarm at the time of the waning half-moon. The same applies to the North American chiton (*Chaetopleura*

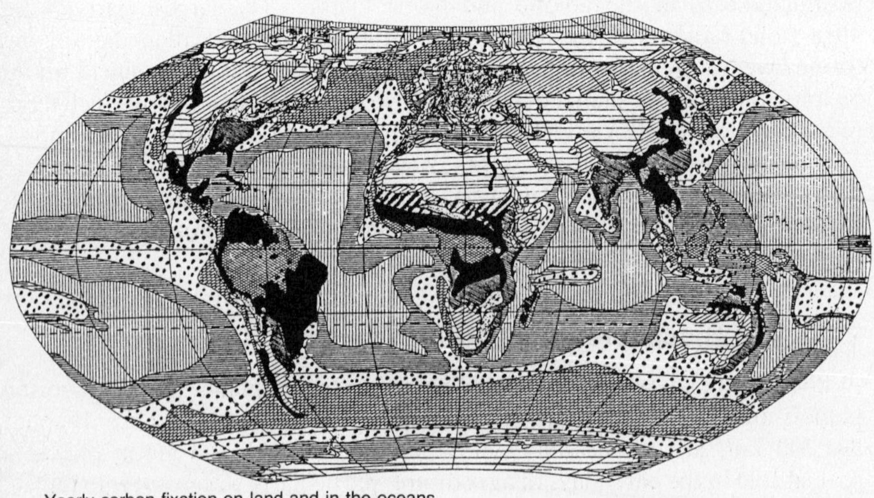

Figure 3 The distribution of annual carbon fixation on land and in oceans to which the carbon balance of the biosphere directly responds.

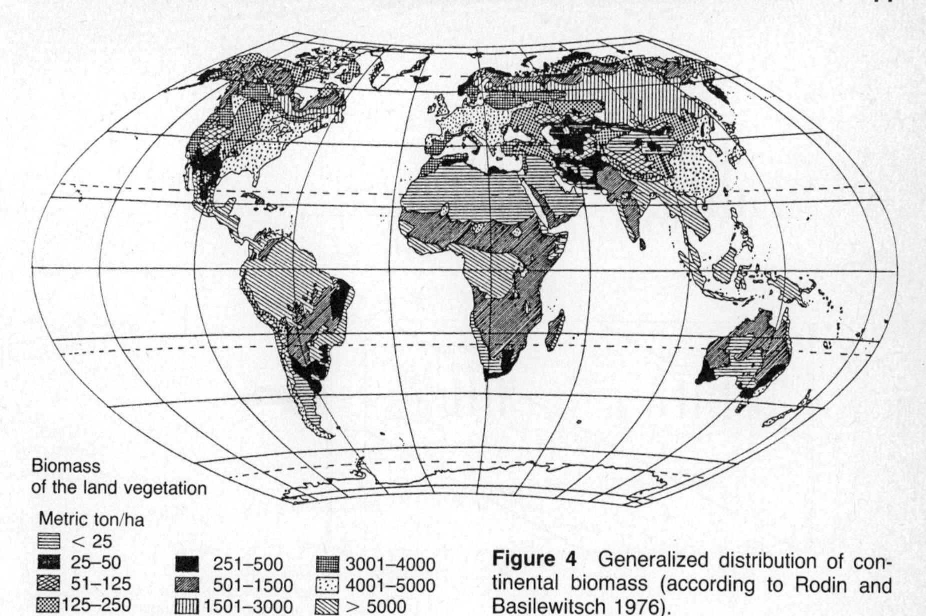

Figure 4 Generalized distribution of continental biomass (according to Rodin and Basilewitsch 1976).

apiculata), where the spawning period coincides with the time of the waning half-moon. The vertical movements of the pacific palolo worm (*Eunice virides*) — described by Collins in 1897 — are of economic importance; they, too, only take place at the time of the waning half-moon. In Bermuda, on the other hand, immediately prior to the new moon and to the full moon, crabs of the species *Anchistoides antiguensis* appear and form reproductive swarms, and the sea urchin *Centrechinus setosus* releases its gametes only during the time of the summer full moon. The larvae of the chironomid *Clunio marinus* (Fig. 5), inhabit the lower edge of the tidal zone and, in order to hatch, have to take advantage of the periods of the lowest water levels. This occurs about every fifteen days during the period of the spring tide. Here, too, the moon acts as a clock (Neumann 1965) (Fig. 5a).

2.1.3 Chemical Composition of the Atmosphere

The atmosphere can be broken down into:

1. Exosphere
2. Ionosphere
3. Mesosphere
4. Stratosphere
5. Troposphere (to an altitude of 20 km)

Examinations of the aeroplankton, however, show that the atmosphere does not contain life in all of the layers and that the composition of species differs considerably between layers (Müller 1977).

While, in the polar region, aeroplankton occurs only up to an altitude of 8 km, it is present up to an altitude of 18 km in the equatorial area. At an

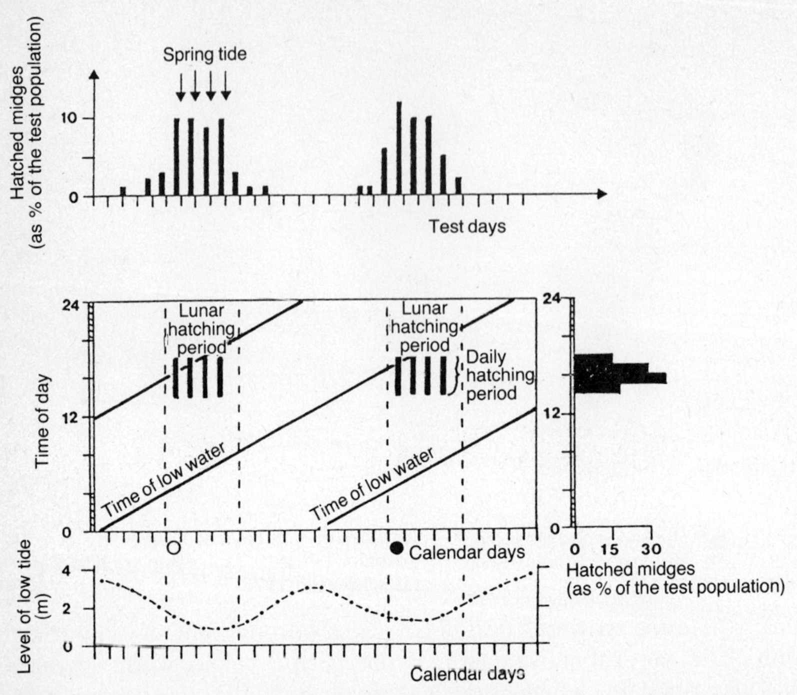

Figure 5 Dependency of the hatching cycle of *Clunio* on lunar phases (lunar periodicity; according to Neumann 1965). The tides controlled by the moon serve as a clock for the midges living in the littoral area.

altitude of 11 km, bacteria and spores of mold fungi are regularly found and at 7 km, condors have been observed. A hawk vulture collided with a plane at an altitude of 11,100 m (Brown 1979). In the Himalayas, geese were noticed from a plane cruising at 9,000 m, even though the normal flying height of birds is below 4,000 m (Jellmann 1979). The lack of oxygen is thereby much less of a limiting factor than the danger of icing which increases with higher altitudes.

The atmosphere represents a gas mixture (Table 2) which is held in place by the gravity of the earth (Fig. 5b). Atmospheric circulations and equilibria regulate the macrochemistry of our planet in many different ways.

2.1.3.1 Living Systems and Oxygen Most life is dependent on oxygen. Life without oxygen is possible for some animals and plants because they are able to liberate sufficient quantities of energy by fermentative decomposition of organic substances. Obligately anaerobic bacteria are responsible for butyric fermentation. The butyric acid causing perspiration odor is formed through the splitting of sugar. Nor are facultative anaerobic organisms a rarity. These include, for instance, the yeast fungus (*Saccharomyces cerevisiae*). With a normal supply of O_2, yeast respires aerobically and shows very strong growth (aerobic life mode). When the yeast culture is supplied with sugar, it ceases its aerobic mode of life due to the exclusion of oxygen. In a complicated

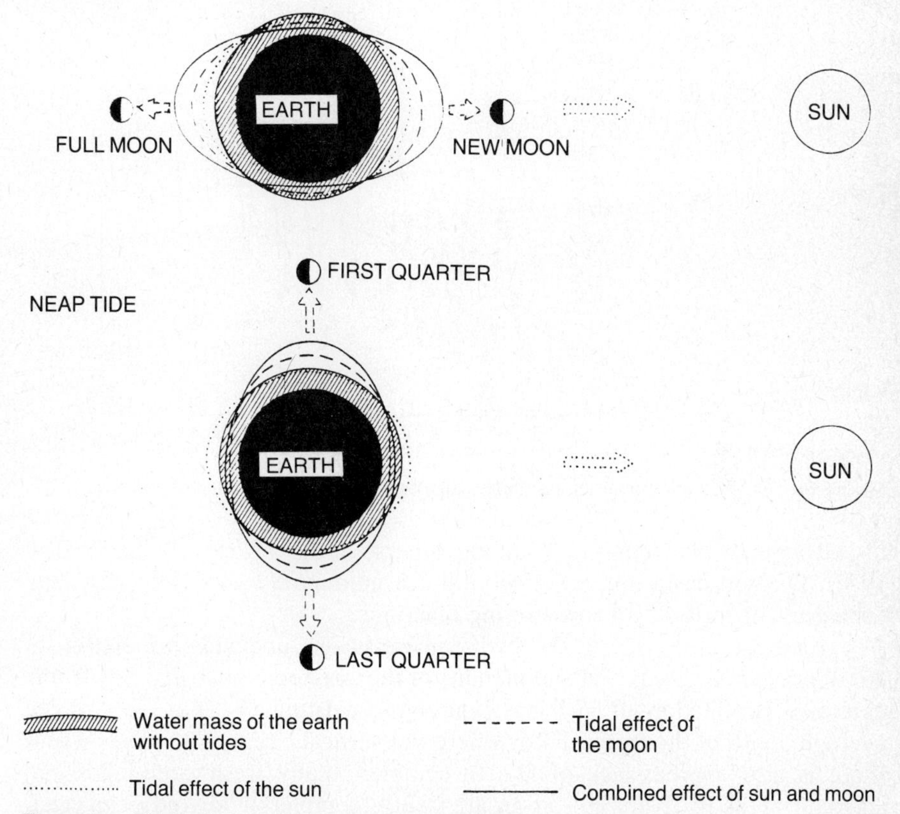

Figure 5a Effect of sun and moon on the tides to explain the spring tides at full and new moon (sun, moon, and earth are all facing in the same direction and their forces of attraction are multiplied) and of the neap tides (sun, moon, and earth are located at right angles, their forces acting antagonistically).

breakdown process (alcoholic fermentation), the sugar is converted into ethyl alcohol:

$$C_6H_{12}O_6 \rightarrow 2CO_2 + 2(CH_3\text{—}CH_2OH) + 21 \text{ kcal}^{1}$$

Lactic acid bacteria are also facultatively anaerobic and are responsible for lactic fermentation; they convert the sugar contained in milk into lactic acid:

$$C_6H_{12}O_6 \rightarrow 2(CH_3\text{—}CHOH\text{—}COOH) + 22 \text{ kcal}^{1}$$

In nature, *Lactobacillus* is present in abundance in green food. It also plays a significant role in the production of sauerkraut and in the ensiling of green fodder. The development of putrefactive bacteria is inhibited by the formation of lactic acid. A significantly more limited effect is produced by gases

[1] 1 kcal = 4.1868 kJ.

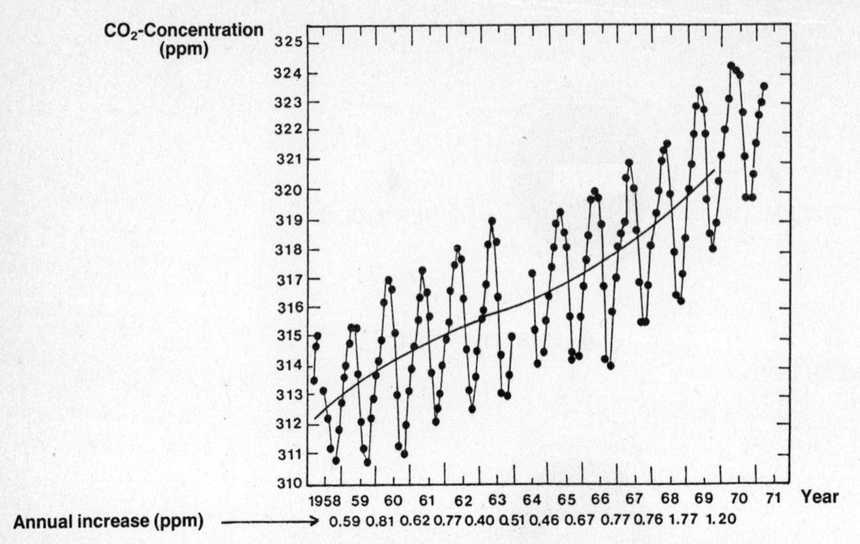

Figure 5b CO_2-Concentration of the earth's atmosphere since 1958 (various authors).

dissolved in water (compare, among others, Lange, Kappen and Schulze 1976). Oxygen has a limited effect and is a mimimum factor; the individual organisms nevertheless show varying reactions.

Like the atmosphere, the hydrosphere is also unevenly populated. In the depths of the Black Sea and in many of the isolated Norwegian fjords, the excessive production of H_2S has bound the existing O_2 (Fig. 5c); in the average depth of the tropical seas where the vertical circulation is very weak, there is also a great lack of O_2. In summer, many freshwater lakes and stagnant areas of rivers are, at greath depth, completely devoid of oxygen. However, heterotrophic organisms have been found on the floor of the deepest trenches of the Pacific at depths of more than 9,000 m. The oxygen necessary for vital processes can reach areas by such processes as diffusion or convection.

Disregarding caves and petroleum deposits, the lithosphere is inhabited by ground animals only to a depth of 5 m. There is no life in solid rock. Anaerobic bacteria enter petroleum deposits to a depth of 4,000 m. Earth-

TABLE 2 COMPOSITION OF THE ATMOSPHERE (as %)

Nitrogen	=	78.1
Oxygen	=	20.94
Argon	=	0.934
Carbon dioxide	=	0.03 (increasing by 0.7 ppm yearly)
Hydrogen	=	0.01
Neon	=	0.0018
Helium	=	0.0005
Krypton	=	0.0001
Xenon	=	0.000009
Water vapour	=	variable

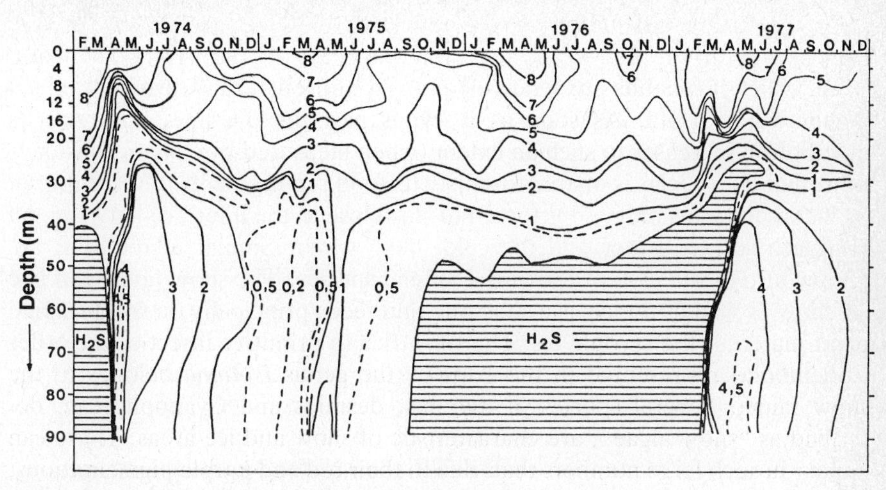

Figure 5c Changes in the oxygen concentrations (ml O_2/l) in the Frierfjord from February 1974 to December 1977 (according to Freeland *et al.*, 1980).

worms have been found in the southern Urals at depths of 8 m. South American leaf-cutting ants (Attini) build their nests at a depth of 6 m in the soil.

2.1.3.2 Living Systems and Temperature Limits In addition to oxygen, temperature is another significant limiting factor. Since Van't Hoff's rule applies to all physiological processes, ambient temperature is of great importance. On land, temperatures of between +59°C and −87°C occur regularly (+59°C in Death Valley, −87°C at Station Wostock in the Antarctic). The temperature range of the oceans, however, is much narrower. The summer and winter temperatures of the tropical seas differ by only 2.5°C. On 75% of the total ocean surface, the temperature fluctuations amount to less than 5°C. The standard reaction to changes in temperature is of varying importance for poikilothermic (cold-blooded) and homoiothermic (warm-blooded) organisms (Precht, Christophersen, Hensel and Larcher 1973). The temperatures measured in the deep sea of between 0 and −2.5°C can be considered as the lower temperature limit for poikilothermic organisms; however, the majority of organisms are not able to reproduce at temperatures which are permanently at such low levels. The deep waters of Hudson Bay which reach a maximum temperature of only 1.8°C in summer, contain no fish.

The freezing point of the blood plasma of Arctic and Antarctic fish which come into regular contact with sea ice at a temperature of −1.9°C, is at about −1.2°C. Fish from other climates, however, are unable to survive such a drop in their blood temperature. The "supercooled" polar fish have concentrations of plasma protein which are far higher than those of other polar animals. Glycoproteins (compare Kloft 1978) prevent the formation of ice down to a temperature of −2°C. The concentration of plasma protein increases with the decrease in the water temperature.

The bivalve *Pecten groenlandicus* is another good example of how finely the temperature adaptation can be regulated physiologically. This bivalve lives at a depth of 25 m in the sea along the coast of Greenland. The water strata which it inhabits are relatively poor in nutrients; thus, it tries to reach higher water strata. As soon as it swims past the 0°C-line, however, its metabolism increases to such an extent (when measured by oxygen consumption) that the species is unable to satisfy its food requirements. On account of its metabolic physiology, it is forced to stay close to the interface between the food-rich surface water and the cold, deep water. Similar adaptations are known also in the case of terrestrial animal species. The snow fauna can live on snow as well as on ice as a substrate and feeds principally on wind-carried plant material, the cryophytes. The snow flea (a primitive insect of the order Collembola) represented in the Alps by the genus *Isotoma* belongs to the snow fauna. Several species of diatoms, desmids and Cyanophyceae, described as "snow algae", are characteristic of snow and ice areas. These can appear in such large numbers that, due to their red and purple pigmentations, the snow appears to be red and they can be seen from afar in the form of "red snow". An extreme case is the alga *Haematococcus nivalis* which colors the snow of the Alps and of the polar regions red. It requires a temperature of around 0°C to develop and stops growing at temperatures above +4°C.

For most plants, the frost limit is of greatest vegetation-geographic importance. Various plants and animals, however, possess a greater resistance against cold and heat, which exceeds considerably the temperature fluctuations which at present exist in the terrestrial climate. As with the lower temperature limit, the upper limit is also of decisive importance. A rhizopod of the genus *Hyalodiscus* lives in hot springs at 54°C. The water snail *Bithynia therminalis* lives in waters at 53°C in the thermal springs of Rome. The entomostracan *Cypris balnearia* and the chironomid *Dasyhelea terna* endure a temperature of 51°C.

Synechococus lividus, a blue-green alga, lives in hot springs at temperatures of 74°C (Peary and Castenholz 1964), some of its relatives enduring even at 90°C (Njogu and Kinoti 1971). An endemic population of *Scardinius erythrophthalmus racovitzai* exists in springs at temperatures of 28°C to 34°C in Baile Epicopesti (formerly Pusztamerges), near Nagyvdrad, in Western Rumania. Temperatures below 20°C are lethal to animals of this population. In East African springs, the mosquito *Culex tenaguis* spawns at 42°C. Fish of the species *Tilapia grahami* have been found in Tanzania at water temperatures of 39°C (Beadle 1974). The amphibians *Hyla raniceps, Bufo paracnemis, Leptodactylus ocellatus, Leptodactylus pentadactylus* and *Pseudis bolbodactyla* spawn in warm springs and streams (38°C) in the vicinity of Pousada do Rio Quente (Brazil).

Under experimental conditions, the African chironomid *Polypedilum vanderplanki* was able to survive, after dessication, a heating up to +100°C and a cooling down to −196°C.

2.1.3.3 Living Systems and Hostile Substrates Substrates with extreme chemical composition are included in the biosphere as abiotic enclaves (Jeffreys

1976). It is noticeable, however, that, because of special adaptations, even such areas can be conquered by individual species. The "petroleum fly" *Psilopa petrolei* lives in petroleum pools and is dependent on animals and their remains falling into the pools (nutrition by allochthonous material).

Fumaroles are devoid of any form of life because CO_2 escapes in large quantities from the earth and, due to its greater density, it displaces the atmospheric air on the ground (for instance in the Grotta del Cane near Pozzuoli). Along the mofettes on the eastern banks of Lake Laach (Eifel), one quite often finds the carcasses of small birds and mammals (finches, mice) which, while in search of food, accidentally enter the CO_2-atmosphere and die there. The importance of the pH-value for the various organisms and their propagation was discussed in detail by Brock (1969), and Edwards and Garrod (1972).

2.1.4 History of the Evolution of the Biosphere

Except for abiotic enclaves (such as volcanoes, ice, abiotic areas at the bottom of lake and sea basins, extreme pH-conditions, poison deposits), the biosphere encompasses the hydrosphere consisting of individual water bodies, a relatively thin layer of the lithosphere (exceptions: caves, petroleum deposits), the pedosphere and the lowest stratum of the atmosphere. The crust of the earth is the birthplace of the biosphere. The saying by Humboldt that "the being in its size and its inner character can only be fully recognized as that which has become" applies to the biosphere just as it does to all other systems.

The evolution of the gas composition of our atmosphere is intrinsically linked to the historical development of the biosphere. A chemical evolution, however, also took place outside our planet, as can be concluded from the existence of organic molecules in interstellar dust clouds (Dose and Rauchfuss 1975).

About 3.5 billion years ago, algae with photosynthetic abilities already existed on our planet. They have been found in the most ancient sedimentary rock existing on earth. Accordingly, life and thus the biosphere must be older still (Miller and Orgel 1974).

Fox (1973), and Dose and Rauchfuss (1975) presented a synopsis of the problems of the origin of living systems. While Fox puts simple proteins at the beginning of evolution, Kuhn (1972) proposes nucleic acids, and Eigen (1971), on the basis of reaction-kinetics, regards nucleic acids and proteins in catalytic cycles as the basis for the first evolving systems.

2.2 THE GENETIC MACRO-STRUCTURE OF THE BIOSPHERE

Since the middle of the last century, the extensive (usually continental) arrangement of the distribution areas of supra-specific taxa (orders, families, genera) led to the definition of the floral and faunal realms.

Totally independent of whether this distribution is due more to strong phylogenetic or more to recent ecological causes, these realms can be considered as real units for each plant or animal taxon. Problems only appear when we attempt to make this phylogenetic macro-structure coincide with the ecological one. Certainly, numerous groups of organisms exist which can only be assigned to one major biome in a single region; however, there are many exceptions which warn against excessive generalization. Furthermore, totally different types of information with respect to the ecological abilities and the history of life on our planet are hidden behind the ecological and phylogenetic macro-structure of the biosphere. Their clarification is only possible as long as the differences between the two are borne in mind.

2.2.1 The Terrestrial Biorealms

Sclater (1858) and Wallace (1876) classified the biosphere into three large realms:

1. Arctogen, including North America, Eurasia, African, the Arabian Peninsula, India and Southeast Asia.
2. Notogea, including Australia, Oceania and New Zealand.
3. Neogea, including South and Central America and the Caribbean Islands.

The floristic classification of the biosphere was greatly influenced by

TABLE 3 THE BIOREALMS OF THE EARTH (according to Müller 1973, 1977)

Realm	Region	Geographical Areas
1. Holarctic	(a) Nearctic region	North America (in contrast to the floristic realm, including Florida and the California peninsula; Greenland and the Highlands of Mexico)
	(b) Palaearctic region	Eurasia (including Iceland, the Canary Islands, Korea, Japan) and North Africa
2. Palaeotropic	(a) Ethiopian region	Africa, south of the Sahara
	(b) Malagassy region	Madagascar and offshore islands
	(c) Oriental region	India — SE Asia to the Wallace Line
3. Australis	(a) Australian region	New Guinea and the islands east of the Wallace Line, Oceania, New Caledonia, the Solomon Islands
	(b) Oceanic region	Central and northern New Zealand and Hawaii are left here with Australis. These groups of islands are, however, marked by so much endemism and with such close relationships to the Palaeotropic that a classification into Australis is not appropriate for numerous animal and plant groups
4. Neotropic		South and Central America, including the Caribbean Islands
5. Archinotic		The Antarctic, SW South America and SW New Zealand

Engler (1879). He distinguishes a Boreal, a Palaeotropic, an Australian, and a Neotropic realm. When comparing more recent floristic classifications (Good 1964, Takhtajan 1969, among others), the basic structure remains distinct, in spite of additional differentiations.

It is not hard to recognize, however, that Good's classification also reflects the ecological structure. This also applies to the analysis by Takhtajan (1969), although, in some instances, he arrives at different classification proposals.

The more recent zoogeographical classifications also make a much stronger differentiation than that of Sclater and Wallace. A basic representation and reclassification was published by Udvardy (1975). He distinguishes eight biorealms which he subdivides into numerous regions:

1. Palaearctic Realm
2. Nearctic Realm
3. Afrotropic Realm
4. Indo-Malayan Realm
5. Oceanic Realm
6. Australian Realm
7. Antarctic Realm
8. Neotropic Realm

When comparing his classification with that by Müller (1973, 1974, 1977) (Table 3), it is easy to recognize the similarities and differences of the two schemes. Taking into consideration the available knowledge about the distribution of vertebrates as well as invertebrates and their phylogenetic relationship, we arrive at a classification of the terrestrial animal realms which, notwithstanding some striking special features and deviations, are in basic agreement with the plant realms of Good (1964) and Takhtajan (1969) (Fig.6).

Plant geographers group, with good reason, the southern part of Florida and the southern tip of the Californian peninsula with the Neotropics — which is also indicated by the distribution of a few animals. The majority of

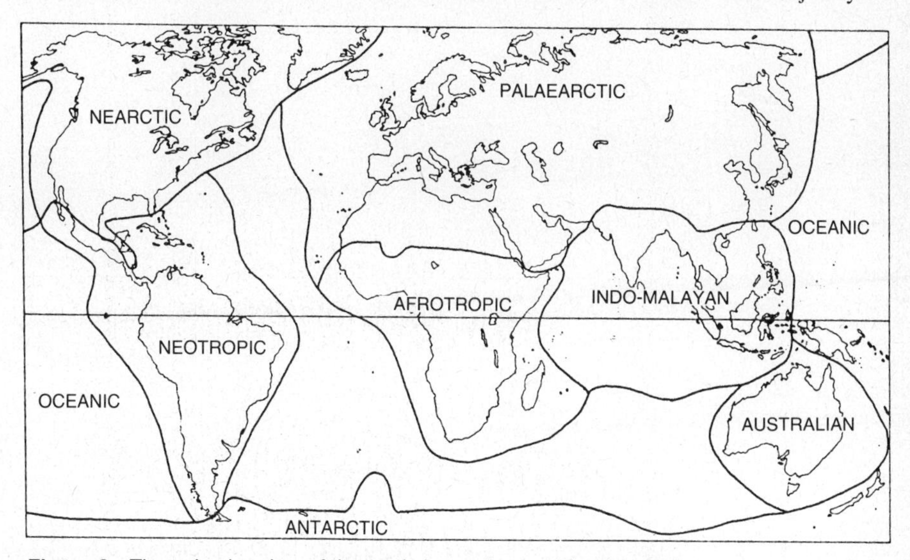

Figure 6 The animal realms of the earth (according to Udvardy 1975).

the animal species occurring in this area show, however, that Southern Baja California and Southern Florida belong to the Nearctic.

Plant and animal realms are only delineated by sharp borders in areas where there are high mountains, wide inlets in the sea, or ice deserts hostile to life. In general, there are wide transition and intermixing zones between the realms which, in many cases, have their own geological past and consist of faunas and floras of differing phylogenetic age and origin. Many transition areas, therefore, are regarded as "relic" areas by some biogeographers (Fig. 7).

2.2.1.1 Transition Areas

(a) *Oriental-Australian boundary (Indonesia).* The boundaries within the transition area of Wallacea are of only minor importance to the numerous groups of flying invertebrates and plants, although they are clearly adhered to by the vertebrates. Wallacea extends from the Lesser Sunda islands, Sulawesi (Celebes) and Lombok in the west to the Halmahera (Moluccas), Kei, and Aru islands in the east. It is separated by the Lydekker-line from the Australian and by the Wallace-line from the Palaeotropic regions (Oriental Region). Although Wallacea is marked by a mixed fauna of Oriental and Australian origin, it still contains some noteworthy endemics. These include, for instance, the black ape (*Cynepithecus niger*) which is related to the black apes of Sulawesi, and the babirusa (*Babyrousa babirussa*) of Sulawesi, a primitive representative of the pig-deer family. An analysis of the entire fauna of Sulawesi, however, shows that the zoogeographic conditions are considerably more complicated. Although the island contains 13 endemic mammal genera (of which 7 are rodents) which can probably be traced back to the Pliocene immigration of Malayan ancestors (Groves 1976), there exist, in addition, species which point toward considerably more recent connections with the Sunda islands. Moreover, not only the zoogeography of Sulawesi, but the

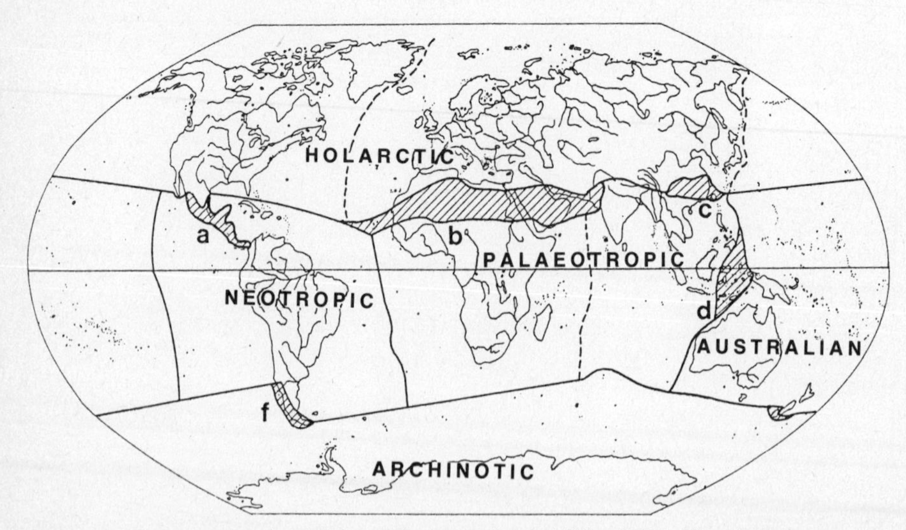

Figure 7 The animal realms of the earth (according to Müller 1973). The shaded areas are those of mixed affinity.

geology of this island, too, is differentiated. The eastern peninsulas are probably Mesozoic parts of the Australian continent, while the northern and the southern peninsula belong to Southeast Asia. The present configuration, which remained unchanged during the Pleistocene, most probably existed during the Pliocene. There is no evidence for a Pleistocene land bridge (caused by eustatic fluctuations in sea level) between Sulawesi and the Southeast Asian continent. The Weber-line running through Wallacea between the Halmahera islands and Sulawesi as well as between Tanimbar and the Lesser Sunda islands is a line inhabited with almost equal proportions of animals of Oriental and Australian origin. The fact that Salomon Müller had already recognized the Wallace-line in 1846 as an important zoogeographic boundary is worth mentioning. The Müller-line, however, still assigns Lombok and Sumbawa to the Oriental region. Furthermore, in contrast to Wallace, Müller was of the opinion that the borderline is more ecologically-dependent. The eastern limit for the dispersal of Australian marsupial groups generally corresponds to the Müller- and Wallace-lines. In the south, however, marsupials are absent west of Wetar and Timor.

(b) *Neotropic-Palaeotropic-Holarctic boundary (Central America).* The boundaries between Neotropic, Palaeotropic and Holarctic in the Central American transition area have recently been questioned.

Central America, considered by most biogeographers as part of the Neotropic, was regarded by others as a transition zone between the Neotropic and Nearctic and was finally made into a separate area adjacent to the Neotropic by Mertens (1952), Kraus (1964) and Savage (1966). The reason why individual researchers arrived at radically differing conclusions is at least partly due to the differing extent to which the animal groups being researched migrate. Animal groups which are capable of migrating some distance (e.g. mammals and birds), dealt with by Simpson (1966) Mayr (1964) and Howell (1969) obscure the charcteristics of the Central American region, whereas others (Chilopods, Diplopods, Amphibia, Reptilia, Gastropods) emphasize the characteristics.

Regardless of how individual researchers think Central America should be classified, they are all agreed that in the low-lying tropical rainforests the number of species of South American origin is astonishingly high. North American species penetrate far into South America along the Andes area. This fact, however, must not lead to the conclusion that the high mountain regions of Central America are part of the Nearctic, as suggested by many writers. If we take as our example the isolated Paramos of Sierra de Talamanca in Costa Rica, we can clearly see the way in which the different centres of origin of the populations living there are interrelated. Here North American taxa (among others, insectivores of the genus *Cryptotis* and the pigeon species *Zenaida macroura*, widespread in North America), related groups (among others, *Philydor rufus* [Funariidae]) and numerous endemics differentiated in these high mountains. Of the 758 bird species living in Costa Rica, 21 are represented by subspecific or even specific (3 species) differentiated populations in the paramos of Talamanca. In this connection, it is noteworthy that of the six bird species endemic to and living in Costa Rica, three (*Selasphorus simoni, Chlorospingus zeledone,* and *Acanthidos bairdi*) are nevertheless endemic to the Talamanca Paramo. When looking at the family relations in the 21 differentiated bird populations of the Sierra de Talamanca, Mayr (1964) considered only two subspecies of the families Funariidae and Contingidae (*Philydor rufus panerythrus* and *Pachyramphus versicolor costaricensis*) belonged to the bird families of South American origin.

For all other families, a Nearctic or Central American origin must be assumed. These findings are in complete opposition to our results obtained in the rainforests at the base of the mountains in Costa Rica, where bird populations belonging to Neotropic families are clearly dominant (Müller 1973).

(c) *Palaeotropic-Holarctic boundary (Africa-S.E. Asia).* Wide transition areas also exist between the Palaeotropic and Holarctic. This applies above all to the African and Southeast Asian regions. Continuing the boundary line between the Ethiopian region and the Western Palaearctic — as discussed by numerous biogeographers — through to North Africa and the Arabian peninsula, it becomes clear that the Sahara, for instance, is covered by a dense "network of boundary lines". These lines, on the one hand, reflect specific characteristics of animal groups, and, on the other, a host of ecological distribution anomalies. The Sahara is not a uniform desert area. Isolated mountain ranges (among others, Tibesti, Hoggar, Air) are dispersed among its dry plains and in the past, Holarctic species advanced southward while Ethiopian taxa wandered north in these mountain ranges. In oases in the western part of the Sahara, species are present (among others the laughing frog, *Rana ridibunda,* and the large-eyed lizard, *Lacerta lepida*) which are present in the European-Mediterranean area. These striking aberrations, however, are by no means limited to animals; similar examples have also been reported for plant species. Reasons for these highly interesting area connections of varying origin within the Ethiopian-Palaearctic transition area can be found, on the one hand, in the recent ecology of the Sahara and, on the other, in the recent history of the North African biomes. By means of analyses of the geobotanical tropical boundary (the southern limit of dominance of tropical plant species) in the Sahara, Lauer and Frankenberg (1977) clearly showed that the floristic composition of the Libyan desert and the western core of the Sahara was subject to only minor changes during the Holocene. "The oscillations of the floral regions must, therefore, have been of rather great proportions, in the north and south border regions of the desert, along the coastal areas and within the area of the humidity belts drifting through the Sahara. Such humidity bridges normally cross the Sahara in a N-S direction running along mountain ranges such as: Air-Hoggar-Tassili'n'Ajjer–Trademait-Atlas or through the Nile Valley. The Tibesti Mountains, too underwent a clear floristic change during the more recent climatic fluctuations" (Lauer and Frankenberg, 1977; also compare Schulz 1974, Court and Duzer 1976, Servant and Servant-Vildary 1973). The temperatures at the end of the last glacial period of North Africa are assumed to have been 6°C lower (Heine 1974). During this glacial stage extratropical, mediterranean species migrated to the southern edge of the present day Sahara. Sahara-Arab fauna and flora elements were pushed back to the hyperarid central areas of the Sahara. Tropical savanna species (*Bitis arietans,* among others) in the western part of the Sahara were able to advance as far as the Moroccan Sou Valley during the warmer post-glacial periods. After the expansion of the central arid plains about 3,000 years B.C., the Palaearctic elements were pushed northwards, i.e., to higher elevations within the Saharan mountains. The actual desert fauna of Africa (Fig. 8) consists of numerous species which, to a great extent, are also present in the arid areas of the Indian sub-continent. In the case of the flora, the percentage of Saharo-Sindhian species amounts to 70% (Good 1964). This close relationship with the Indian sub-continent also provides justification for the association of the Oriental and Ethiopian regions with the

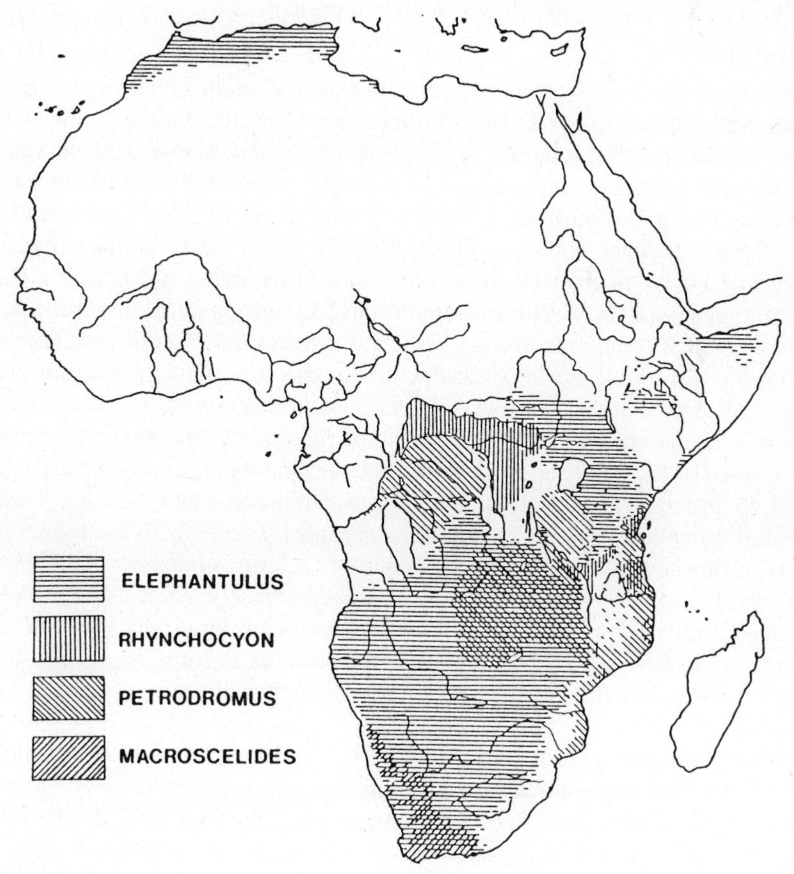

Figure 8 Distribution of the elephant shrew (Macroscelididae) in Africa (according to Corbet and Hanks, 1968).

Palaeotropic biorealm. The relationship between the Ethiopian and Oriental regions in the case of animals, is closer than between these two and the Holarctic. Thus, for instance, the mammal families Tragulidae, Rhinocerotidae, Elephantidae, Hyaenidae, Manidae, Pongidae, Cercopithecidae, and Lorisidae; the bird families Nectariniidae, Pycnonotidae, Indicatoridae, Bucerotidae, and Pterocidae; the reptile family Chamaelontidae; and the amphibian family Rhacophoridae are present in both regions, whereas the number of birds endemic to the Ethiopian (only 4 of 67 families) as well as to the Oriental regions (only Irenidae) is small.

267 bird genera are common to both Ethiopian and Oriental regions. 69 Ethiopian bird species reach India and 63 reach Europe.

A similar sized transition area to that in North Africa exists between the Palaeotropic and Holarctic in China. Through human influence, the original subtropical forests have to a large extent been destroyed. Species which have adapted to the open landscapes have, in part, replaced the original flora and fauna of the forest. The fauna of Taiwan, which during the ice ages was connected to the mainland, shows a pronounced mixture of Holarctic and Palaeotropic species.

(d) *Antarctic boundaries (South America and New Zealand).* On the southern continents, particularly at the southern tips of South America and New Zealand, there exist comparable transition areas also showing close relationships between each other as well as with the "old south", the Antarctic. These have been the subject of intensive examinations of a phylogenetic nature during the past few years (Brundin 1972, Illies 1965, among others). While most of the terrestrial vertebrates presently living here have close taxonomic ties with the northern populations, numerous older invertebrates and plant groups (compare Moore 1972, van Steenis 1972) indicate a closer relationship with those of the southern tips of the continents. Plecopterans, chironomids, and even crustacean families (Syncarida) already in existence in the early Tertiary justify the land bridging of these southern transition areas to the old south (= Antarctica). Some biogeographers in the last century had wanted to proclaim the Antarctic a separate biorealm and this has been confirmed by numerous well-founded examples established by biogeographers of our century. While the phylogenetic relations become clearer the further south in the Pacific we go, we are dependent to a great extent solely upon the confirmed results of geophysical examinations and upon coincidental conclusions when discussing the question as to how the recent distribution patterns of the above mentioned taxa materialized.

2.2.1.2 The Holarctic Since de Candolle (1855), Engler (1879) and Kobelt (1897), extensive material has been collected concerning the connections between the Holarctic taxa and the delineation of the Holarctic itself. This information was clearly arranged with some stimulating theoretical suggestions for the first time in Reinig's *Holarktis* (1936). Aspects which were further developed biogeographically after that, and which, in part, even had to be completely reassessed, show trends which were at least hinted at already in that study.

Numerous plant groups are of the Holarctic distribution type (among others Aceraceae, Betulaceae, Caryophyllaceae and Cruciferae) as well as quite a few individual species (among others, *Equisetum arvense, Cardamine pratensis, Zostera marina,* compare Böcher *et al.* 1968, Pimenov 1968, Hülton 1960, 1962). Animal groups typical of the Holarctic include, among others, the moles (Talpidae), the genus *Bison*, the beavers (Castoridae), the Ochotonidae, the Zapodidae, the pikes (Esoxidae), the crayfish (Astacidae), the fish of the family Coregonidae, the giant salamanders of East Asia (Cryptobranchidae), the North American salamanders (Salamandridae), the Proteidae and the bird family Alcidae (Fig. 9). Numerous carabid beetles live in the Holarctic areas (for instance, the subgenus *Cryobius*).

In the case of very many animal species, a number of Palaearctic-Nearctic sister groups are known; we give some avian examples:

Palearctic	Nearctic
Glaucidium passerinum	*Glaucidium gnoma*
Perisoreus infaustus	*Perisoreus canadensis*
Spinus spinus	*Spinus pinus*
Parus montanus	*Parus atricapillus*
Regulus regulus	*Regulus satrapa*
Nucifraga caryocatactes	*Nucifraga columbiana*

In the Nearctic and Palaearctic, species of the Eurasian tundra, the taiga and the temperate deciduous forests are represented by closely-related species or subspecies (= vicariants). The faunistic similarities in the northern continents are far greater than those in the southern continents and grow stronger with increasing northern latitude. In spite of these similarities, however, there is a remarkable number of endemics in both the Nearctic and Palaearctic which indicates that the fauna on the northern continents underwent a stronger and in some cases also a faster differentiation than did the flora. Although the faunistic differences between the Nearctic and Palaearctic are greater than the floristic ones, it is not appropriate to consider them as independent realms, but only as regions of the Holarctic, as already pointed out by Heilprin (1887), Blanford (1890), Lydekker (1896) and Reinig (1936).

The last land bridge connection between North America and Eurasia via the Bering Strait was during the last glacial and permitted an exchange of fauna, flora, and last but not least culture between the areas separated today. Animal groups limited to the Nearctic include, among the mammals, the rodents Aplodontidae, Geomyidae, and Heteromyidae (which, although similar to the Soricidae, are also present in northern South America), and the

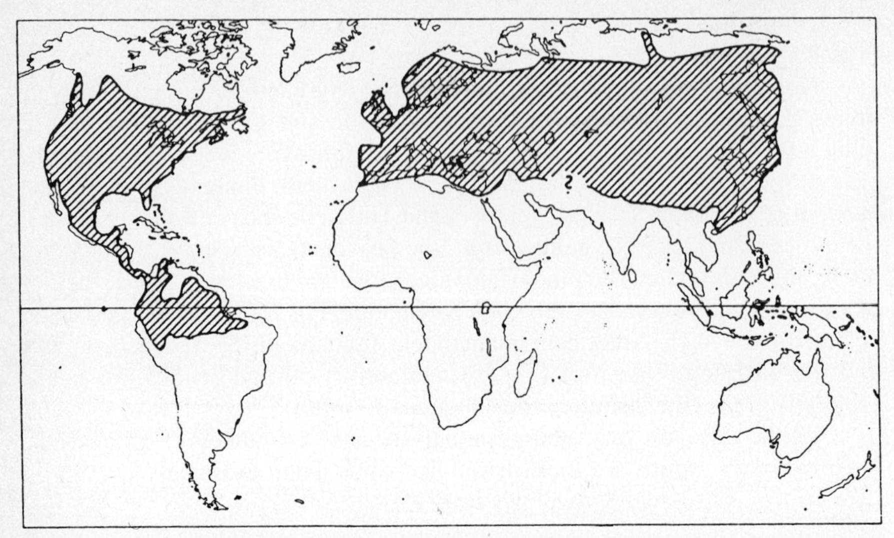

Figure 9 Global distribution of Urodela (Amphibia). The largest number of species of this order are found in the Holarctic. The genus *Bolitoglossa* penetrates South America through the Andes, and one species lives in the tropical lowlands (*Bolitoglossa altamazonica*).

pronghorns (Antilocapridae); among the reptiles, the Californian Anniellidae, the Helodermatidae which are adapted to the arid areas of southwestern U.S.A. and Mexico, the Gerrhonotinae which penetrated through the mountains of Central America far to the south; and, among the amphibians, the isolated Ascaphidae which were formerly regarded as in the Leiopelmatidae of New Zealand; and the Urodela families Ambystomidae, Amphiumidae, and Sirenidae.

The Palaearctic also has a number of characteristic animal groups.

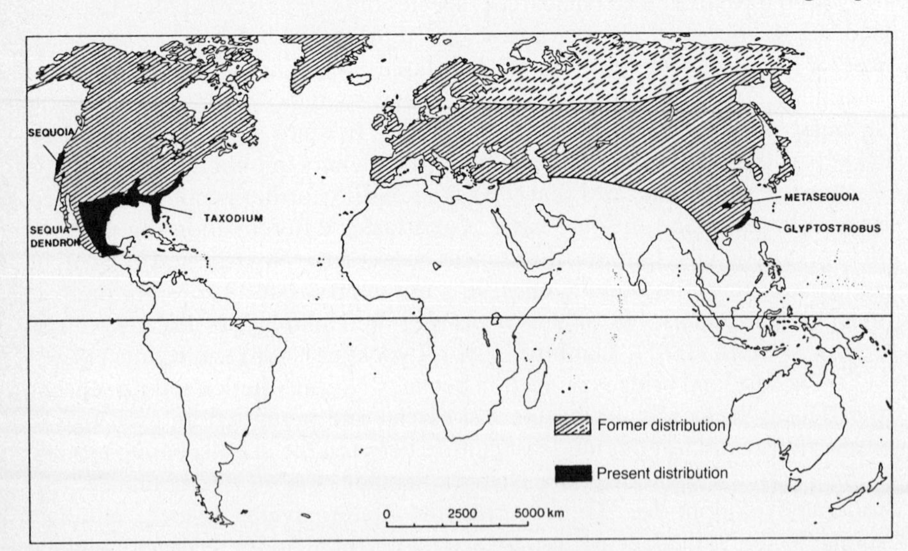

Figure 10 Present and former distribution of some Conifer genera.

These include the Siberian Hynobiidae, the disk-tongued frogs (Discoglossidae) present as far east as the Philippines, the real blindworms (Anguinae), hedge sparrows (Prunellidae), Glirinae and Spalacidae, the rodent family Seleviniidae which is represented by only one species, the Pandas (Ailuriunae) and the chamois (Rupicaprinae).

2.2.1.3 The Neotropic The Neotropic encompasses South America, the Caribbean Islands, and large parts of tropical Central America. Neotropic plant families are also found extensively outside the boundaries in Florida and the southern tip of the Californian peninsula. The latter is characterized by a very rich fauna and flora both in terms of numbers of species and endemics. The Neotropic plant families include the Cannaceae, Caryocarraceae, Columbelliaceae, Julianiaceae, Marcgraviaceae and Tropaeolaceae (Table 4).

Amongst the animals the marsupial families Didelphidae and Caenolestidae are endemic (remembering that the opossum in the past was able to conquer large areas of North America), as well as anteaters (*Myrmecophagidae*), sloths (*Bradypodidae*), armadillos (*Dasypodidae*), wide-nosed apes (*Ceboidae*) the rodent families, *Caviidae, Hydrochoeridae, Dinomyidae, Dasyproctidae, Chinchillidae, Capromyidae, Octodontidae, Ctenomydae, Abrocomidae* and *Echimyidae,* white lipped pigs (*Tayassuidae*) and the bat families *Desmodontidae, Natalidae, Furipteridae, Thyropteridae* and *Phyllostomatidae.* The agouta (*Solenodontidae*) are confined to the Caribbean Islands. 2926 bird species alone are found in South America in 2 endemic orders (the *Nandus=Rheiformes* and *Tinamus = Tinamidae*) and 31 endemic families. Colibriates, endemic to the New World, have 242 species in South America, where they populate all realms from the highest points of the Andes to the Amazonian lowlands; they are, however, absent in the Galapagos. 15 colibriates reached North America, and the extensively allopatrically distributed *Selasphorus Rufus* (West) and *Archilochus collibris* (East) penetrated as far as the mouth of the Lawrence or to south west Alaska. Other animals worth noting are the marsupials which otherwise are only found in Australia (87 species, 2 families), the endemic edentates with the sloth (*Bradypodidae*), ant-eaters of the genus *Myrmecophaga, Tamandua* and *Cyclopes* and armadillos (20 species), wide-nosed apes, the tree prickers, chinchillas, white-lipped pigs and the high proportion of endemic bat families. Deer (*Cerviden*) are represented by 11 species in the paramo of Ecuador and Colombia (*Pudu mephistophiles*), the *Nothorfagus* woods of Argentina and Chile (*Pudu pudu*), the puna of the Central Andes (*Hippocamelus antisensis*) to the flood plains on the Mexiana island in the Amazon delta (*Mazama gouazoubira mexianae*).

The Jaguar, puma, rattlesnake, and Virginian deer are species whose ancestors immigrated only during the ice ages from the north when, during the North-Andean rise, the Strait of Panama was dry land. For long periods during the Tertiary, South America was an island continent.

The evolution of most cricetid rodents of the Neotropic is similar to that of the Neotropic iguana in that it is characterized by a reduction in the

TABLE 4 PRODUCTIVE PLANTS OF THE NEOTROPIC

Family	Species	Family	Species
Juglandaceae	*Juglans nigra*	Umbelliferae	*Arracacia xanthorrhiza*
Moraceae	*Brosimum galactodendron*	Sapotaceae	*Calocarpum sapota*
Proteaceae	*Guevina avellana*	Rubiaceae	*Cinchona officinalis*
Oleaceae	*Ximenia*	Convolvulaceae	*Ipomoea batatas*
Basellaceae	*Ullucus tuberosus*		*Ipomoea trifida*
Chenopodiaceae	*Chenopodium quinoa*	Solanaceae	*Lycopersicon esculentum*
Amaranthaceae	*Amaranthus edulis*		*Physalis peruviana*
Annonaceae	*Annona cheromola*		*Nicotiana tabacum*
	Annona muricata		*Solanum tuberosum*
Mecitidaceae	*Bertolletia excelsa*		*Solanum muricatum*
Lauraceae	*Persea lingua*		*Solanum quitoense*
	Persea americana		*Solanum topiro*
Rosaceae	*Prunus* spec.		*Capsicum annuum*
	Rubus glaucus	Bignoniaceae	*Crescentia cigete*
	Fragaria chiloense	Erythroxylaceae	*Erythroxylum coca*
Leguminosae	*Arachis hypogaea*	Euphorbiaceae	*Hevea brasiliensis*
	Lupinus mutabilis		*Manihot esculenta*
	Lupinus perennis		*Manihot glazovii*
	Phaseolus vulgaris	Anacardiaceae	*Anacardium occidentale*
	Phaeolus lunatus	Aquifoliaceae	*Ilex paraguayiensis*
Oxalidaceae	*Oxalis tuberosa*	Compositae	*Madia sativa*
Malpighiaceae	*Malpighia punicifolia*	Araceae	*Colocasia esculenta*
Tropaeolaceae	*Tropaeolum tuberosum*	Bromeliaceae	*Ananas comosus*
Malvaceae	*Gossypium hirsutum*	Gramineae	*Bromus mango*
Serculiaceae	*Theobroma cacao*		*Sorghum almum*
Passifloraceae	*Passiflora edulis*		(*Sorghum almum* originated in Argentina by hybridization between *S. halepensis* and Sudan grass *S. sudanense*)
Bixaceae	*Bixa orellana*		
Cactaceae	*Opuntia ficus-indica*		
Caricaceae	*Carica candamarcensis*		
	Carica microcarpa		
	Carica papaya		*Zea Mays*
Cucurbitaceae	*Cucurbita andreana*	Palmae	*Cocos nucifera*
	Cucurbita pepo	Dioscoreaceae	*Dioscorea trifida*
	Cucurbita maxima	Cannaceae	*Canna edulis*
	Cucurbita texana	Marantaceae	*Maranta arundinacea*
	Cucurbita mixta	Amaryllidaceae	*Agave atrovirens*
	Cucurbita ficifolia		*Agave sisalana*
	Cucurbita moschata	Orchidaceae	*Vanilla planifolia*
Myrtaceae	*Ugni molinae*		

numbers of chromosome pairs. Phylogenetically older forms have higher chromosome numbers than do the younger ones. *Phyllotis osilae* occurring in isolation in northern Argentina and southern Peru, for instance, has 68 chromosome pairs, whereas the desert species derived from it, *Phyllotis gerbillus,* has only 38 pairs (Pearson and Patton 1976, Gardner and Patton 1976).

Among the reptiles, predominant species are the snake-necked turtles, the iguana which elsewhere only occurs in Madagascar and on the Fiji and Tonga islands, the coral snakes (*Micrurus, Leptomicrurus, Micruroides*), the rat-tailed serpents (*Bothrops, Lachesis*), the giant boa (*Boinae*) and caiman. The poisonous coral snakes reaches the U.S.A. in the form of two species

(*Micrurus fulvius, Micruroides euryxanthus*), while *Bothrops* and *Lachesis* are limited to the Neotropic.

Amongst the amphibians, the leaf frogs (*Hylidae*), bronze frogs (*Leptodactylidae*) and Atelopodidae dominate. *Pipa* species which live in water and which, together with the African platannas, belong to the Pipidae family, are of particular historical-biogeographical interest. The brood care behaviour of many South American frogs is remarkable. The building of mud nests (*Hyla faber,* among others) and leaf nests (*Phyllomedusa,* among others) and the carrying of the eggs of their young ones on their backs (*Hyla goeldii, Pipa pipa, Dendrobates*) or in special pouches on their chests (marsupial frogs of the genus *Gastrotheca*) are only a few of the characteristic features of caring for their young. Only one species of the true frogs (Ranidae) reached South America, whereas 12 species of lungless salamanders arrived (one in Amazonia, the rest in Colombia).

The siluroids are represented by the endemic shell siluroids (Callichthyidae) and the catfish (*Loricaridae*). Callichthyides are fossils known from the Tertiary in Argentina (*Cirydoras revelatus* from the Jujuy province). At present, we have knowledge of 94 species.

In contrast to the Callichthyidae, the cichlids (Cichlides, 260 species, 16 genera) common in South and Central America, are also found in the Caribbean Islands, Madagascar, and India. This type of dispersal stimulated Eigenmann (1909) and Ihering (1907) to reconstruct hypothetical continents and land bridges. Fossil findings and the fact that many cichlids are able to tolerate brackish or even salt water (a species of the genus *Tilapia, Cichlasoma haitiensis* was caught in one of the salt lakes of Haiti today), however, forces us to have many reservations with respect to these hypotheses. Similar to the cichlids, however, others of the 2,700 South American fish species and their parasites show remarkable ties to Africa: Characidae; *Arapaima gigas* (Osteoglossidae) which grows to a length of over 3 m; the South American lungfish (*Lepidosiren paradoxa*). The distribution of sub-groups of the Glossoscolecinae is shown in Fig. 11.

In the case of invertebrate groups (such as millipedes, *Spirostreptidae, Ostracods, Bathynellacea, Arachnids, Chironomids, Plecoptera, Onychophora,* molluscs), close relationships exist with Africa and the other southern continents.

2.2.1.4 The Australian Realm As in the Neotropic, the Australian realm is also characterized by an animal and plant world rich in endemics.

Of all the floral assemblages the Australian is the most peculiar (Schmithüsen 1968). The tree-shaped Liliaceae (*Xanthorrhoea preisii*), the Casuarinaceae and the genus *Eucalyptus* with its many species are some of the characteristic plant groups of this realm. The following plant families are endemic to the region: Akaniaceae, Austrobaileyaceae, Baueraceae, Byblidaceae, Cephalotaceae, Chloanthaceae, Davidsoniaceae, Dyspaniaceae, Eremosynaceae, Gyrostemonaceae, Petermanniaceae, and Tremandraceae.

The Casuarinaceae occur in Australia as well as in New Caledonia, the

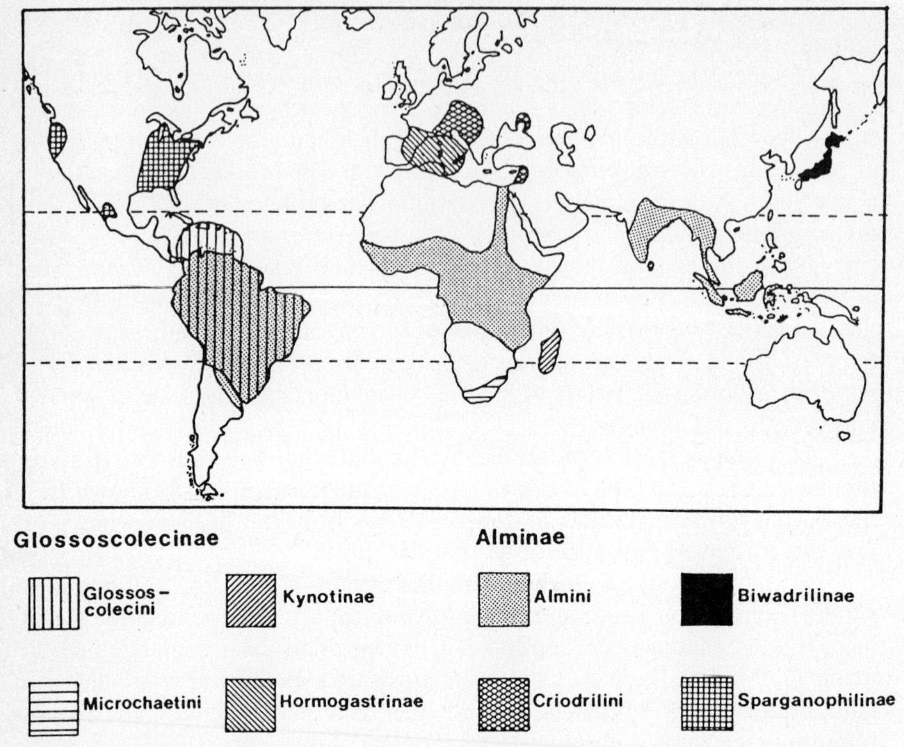

Figure 11 Disjunct distribution of systematic sub-groups of the Glossoscolecinae (according to Janieson, 1971).

Fiji islands and in Burma. Australia shares with New Caledonia the Strasburgeriaceae, Xanthorrhoeaceae, Oceanopapaveraceae, and Amborellaceae.

Among the higher flowering plants there are only a few genera which are also found in the northern hemisphere. Many moss species, however, have relatives known to us from Central European or South American habitats (Scott and Stone 1976). Apart from the Australian mainland, which has been separated from Asia since the early Mesozoic, and Tasmania, the regions of New Guinea, New Zealand, New Caledonia, East Melanesia, Micronesia, French Polynesia, and Hawaii are also considered part of the Australian biorealm. On the basis of the vertebrates, the assignment of New Guinea to Australia — as has been suggested by earlier authors — has been confirmed. This, however, does not apply to the same extent to the Fiji islands, the Solomon islands, Micronesia, French Polynesia (including Easter Island), and Hawaii, which, due to their isolation developed strong evolutionary centers of their own, and which, consequently, have equally strong Palaeotropic as well as Australian characteristics (cf. Holloway and Jardine 1968).

The close relationship between New Guinea and Australia is generally regarded as due to the repeated eustatic draining of the Torres Strait. The last firm land connection between Australia and New Guinea was not severed until about 8,000 to 6,500 years ago (Walker 1972, among others), and during the Pleistocene, an exchange of fauna took place several times between these two land masses.

Homo sapiens reached the Australian realm via the sea at least 30,000 years ago (Bowler, Jones, Allen and Thorne 1970). The high mountains of New Guinea were, of course, glaciated during the Pleistocene (Löffler 1970, Paijmans and Löffler 1972). Nevertheless, many of the plant families occurring in Australia never reached New Guinea (Van Balgooy 1976).

Examining the mammal fauna of Tasmania — which was discovered in 1642 and measures 55,000 km^2 — the placentals play a very insignificant role. Apart from some introduced species (*Oryctolagus cuniculus, Lepus europaeus, Mustela putorius, Rattus norvegicus, Mus musculus, Cervus dama*), the marsupials are dominant (the Tasmanian wolf, *Thylacinus cynocephalus* was extirpated) (Table 5).

In Tasmania, the systematic and ecological composition of the avifauna (104 species) is similar in character to that of Australia.

From a biogeographical point of view, the amphibians and reptiles are of particular interest since, being poikilothermic organisms, not only do they accurately reflect the present climatic conditions, but (at least certain individual species) also endured a glacial phase in central Tasmania during the last glacial period (the last firm land connection between Tasmania and Australia being around 12,750 B.C.) (Littlejohn and Martin 1974, Rawlinson 1974). Although the fish fauna of the Australian realm does not include any primary freshwater fish, it does contain a remarkable number of endemics which are compulsory freshwater spawners (Whitley 1959). On Tasmania, the Petromyzoniformes, for instance, are represented by the two species *Geotria australis* and *Mordacia mordax*. The order Salmoniformes consists of a great many species. All of the species, however, belong to the group Galaxioidei which are also present in New Zealand. The Perciformes are represented, among others, by the families Percichthyidae, Kuhliidae, Gadopsidae, and Bocichthydae. *Salmo trutta, Perca fluviatilis,* and *Tinca tinca,* among others were, however, introduced. Trout fishing is widespread in Tasmania today.

As a result of its rich avifauna, the Australian realm (= Notogea) was also referred to as Ornithogea. However, due to the numerous species of parakeets and parrots, it was also called *Terra psittacorum* during the 17th century. The Australian continent, with its 464 bird species, however, is

TABLE 5 ORDERS OF NATIVE MAMMALIA FOUND IN TASMANIA (according to Williams 1974)

Order	Families	Genera	Species
Monotremata	2	2	2
Marsupialia	5	16	20
Rodentia	1	4	4
Chiroptera	1	4	5
Carnivora	2	3	3

considerably poorer in species than tropical Africa (1,481 species) or even South America. Bird families endemic to the Australian realm are: the bowerbirds (Ptilonorhynchidae), the emus which have adapted to living in the open landscapes (Dromicidae), the forest-dwelling cassowaries (Casuaridae), the lyrebirds (Menuridae), the scrub fowls (Megapodiidae) characterized by their exceptional brood care, the sugarbirds (Meliphaginae), the flower-peckers (Dicaeidae), the Aegothelidae, the flightless Atrichornithidae, the flutebirds (Cracticidae), the Grallinidae, the Pedionominae, the Kakatoeinae, and the Paradisaeidae. 35% of the total Australian bird fauna consists of these endemic groups (Keast 1961). The various distribution areas of the avifauna are well documented and have served as a basis to many authors in their proposals for regional classifications of the Australian realm (Kikkawa and Pearse 1969, Horton 1973, and others). In the case of the zoogeographically important amphibians, total absence of entire orders (e.g. Urodela), and prevalence of such orders as Hylidae and Leptodacylidae is most noteworthy (similar to South America). Endemic Australian reptiles are the Carettochelyidae, mainly living in New Guinea, and the Pygopodidae. The lack of Viperidae and the predominance of poisonous Elapidae is also striking.

The snake-necked turtle (Chelidae) lives only in Australia and South America and has a fossil parallel in the Meiolaniidae, the extinct land tortoise of Australia and South America. Some invertebrate groups of the Australian realm have old southern hemisphere relatives (the crustacean family Anaspidaceae, for instance, which is known to have existed since the Carboniferous period, the cicada family, Peloriidae, the stone fly family, Gripopterygidae, and the chironomid family, Aphrotaeniidae; cf. Schminke 1974, among others) (Fig. 12).

New Caledonia The flora of New Caledonia is characterized by a wealth of endemics (more than 100 endemic plant genera) and by relations to Australia and New Zealand.

Among its 68 bird species, there is only one family, the kagu family (Rhynochetidae), which is endemic to New Caledonia. It has been confirmed that 18 of the bird species probably came from Australia to colonize New Caledonia. These include, among others, the swifts of the Aegothelidae family. The six endemic giant geckos of the genus *Rhacodactylus* and the endemic wealth among the invertebrates is also noteworthy. On the other hand, amphibians, native mammals (with the exception of Chiroptera), the land tortoises (with the exception of the Meiolaniidae which are known as fossils from the Pleistocene in New Zealand), and snakes are apparently absent from the original fauna of these strongly ecologically differentiated islands.

New Zealand (cf. Gressit 1961, Hennig 1960, Munroe 1965, Bull and Whitaker 1975, and others) — 267,800 km^2 in area — occupies a special biogeographical position. The number of endemics among the plants and the

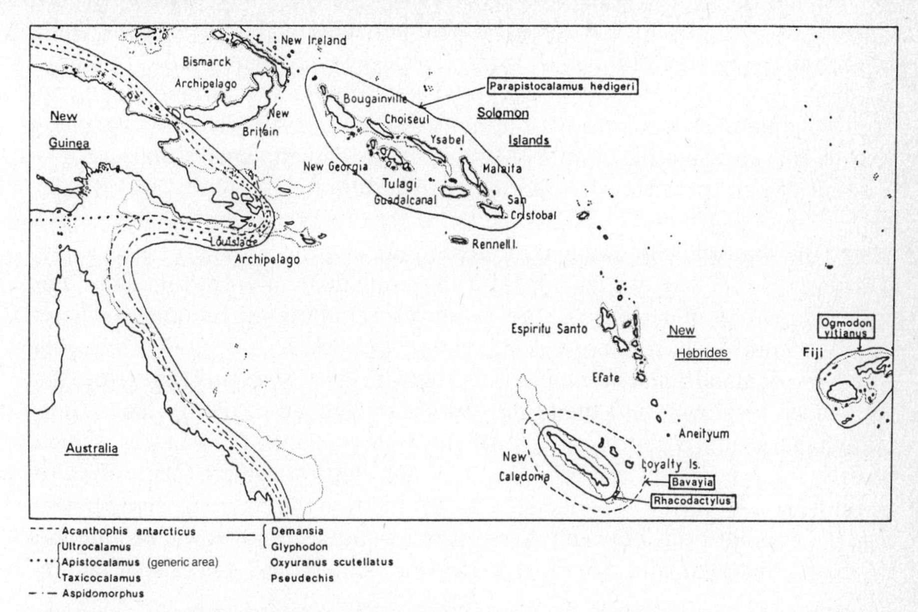

Figure 12 Distribution areas of endemic reptiles between the Fiji archipelago and Australia.

animals is exceptionally great (Table 6).

The amphibians are represented by the primitive frogs (Leiopelmatidae) of which three species have been preserved. *Feiopelma* was recently discovered on North Island and on the offshore islands to the north of South Island. Fossil findings show that these species lived in many areas of the northern island 1,000 years ago. The Australian greenbacks, *Litoria aurea* and *Litoria ewingi* were, however, introduced.

Among the reptiles, another relict, the tuatara (*Sphenodon punctatus*) is noteworthy. Three gecko genera are endemic, two Scincid genera (*Leiolopisma, Sphenomorphus*) belong to the native fauna. Primary freshwater fish are absent.

Apart from the extinct Moas (Dinornithidae, Anomalopterygidae) and their more recent relatives, the kiwis (three species of Apterygidae: *Apteryx*

TABLE 6 GENERA, SPECIES, AND ENDEMICS OF HIGHER PLANTS IN NEW ZEALAND (according to Godley 1975)

	Genera			Species		
	Total	Endemic	%	Total	Endemic	%
Ferns	47	1	2	163	67	41
Gymnosperms	5	0	0	20	20	100
Dicotyledons	235	32	14	1268	1131	89
Monocotyledons (excluding grasses)	75	3	4	339	214	63
Gramineae	31	3	10	206	186	90
Total	393	39	10	1996	1618	81

australis, A. haasti, A. oweni), two additional endemic families, the Xenicidae (3 species) and the Callaeatidae (2 species) are part of the endemic avifauna. The kakapos (*Nestor, Strigops*) are striking. The avifauna confined to the mainland shows links with both Tasmania and Australia. This applies also to the endemic bat family (Mystacinidae) and the snail family (Atoracopharidae). Apart from cosmopolitan invertebrates (cf. Jolly and Brown 1975, Winterbourn and Lewis 1975), there are others present that have close ties to the South Pacific islands, the Antarctic, and the southern tip of South America. This is true, for instance, for the stone fly family, Eustheniidae, the crustaceans, Stygocaridacea, the freshwater pulmonate molluscs of the Latiidae family whose closest relatives are the South American Chilinides (the New Zealand *Latia neritoides* is the only freshwater animal that possesses the ability to glow), and many mosses (Scott and Stone 1976). The Chilibathynella group is representative of the Bathynellacea (Crustacea), which mostly live in freshwater and have a distribution which, according to Schminke, can only be explained on the assumption that there existed "mesozoic land connections between Australia/New Guinea and New Zealand and between Australia and South America via Antarctica" (cf. also Fleming 1975).

2.2.1.5 The Palaeotropics The Palaeotropics, which can be categorized into the Ethiopian (= Afrotropical), Madagascan, and Oriental regions, is inserted between the Australian and Neotropic realms, covering the area of India and Africa. Numerous animal and plant families and genera characterize this area (Ancistrocladaceae, Pandanaceae, Nepenthaceae, Melianthaceae, *Corypha, Nypa, Areca, Borassus, Hyphaene, Phoenix,* and *Raphia* palmtrees). The individual regions are, however, characterized by an abundance of their own taxa which justify their special position.

Among the plants, the following families are limited to Africa: Barbeyaceae (also found in Arabia), Cyanastraceae, Dioncophyllaceae, Dirachmaceae, Medusandraceae, Melianthaceae, Nectaropetalaceae, Octoknemataceae, Oliniaceae, Pandaceae, Pentadiplandraceae, and Syctopetalaceae. The following families are common to both the Neotropics and Ethiopian regions: Bromeliaceae, Cariaceae, Humiriaceae, Loasaceae, Mayacaceae, Rapateaceae, and Vochysiaceae. The Ethiopian families Hydrostachyaceae, Montiniaceae. Myrothamnaceae, Selaginaceae, and Sphenocoeaceae are also present in Madagascar. Numerous families found in Africa, however, are characteristic of the Paleotropics. Some of the characteristic animal groups in the Ethiopian region include: the platannas of the genus *Xenopus* (family Pipidae, related to the South American genus *Pipa*), the frog family Phrynomeridae, the Mambas (Dendroaspinae), the bird families Struthionidae, Sagittaridae, Scopidae, Musophagidae, Coliidae, and Balaenicipitidae, as well as the mammal families Potamogalidae, Chrysochloirdae, Macroscelididae, Pedetidae; the Graphiurinae belonging to the Gliridae, the Thryonomyidae, Petromyidae, Hyanidae, Orycteropodidae, Procaviidae, Hippopotamidae, and Giraffidae.

Among the fish, the presence of the ancient lobe-finned fish (Polypterini), the Gymnarchus (Mormyridae), the electric catfish (Malapteruridae), and the lung-fish (*Protopterus*) are especially striking (Fig. 12a). The Ailyidae are an endemic Ethiopian snail family. The vast savanna areas are inhabited by a unique fauna of large animals living in huge herds, characterized by the absence of deer and a preponderance of antelopes, gazelles, bovines, and giraffes.

South Africa, containing about 7,000 flowering plants, represents an independent vegetable kingdom (the Cape). The heathlands of the Cape with their abundance of species, include numerous *Erica* species and small-leafed Rutaceae. The landmarks of the Cape countryside, the silver tree (*Leucadendron argenteum*) and the Cape cedar (*Widdringtonia juniperoides;* Proteaceae) make up this landscape just as much as the striking Proteaceae and the vigorously blooming *Mesembryanthemum* species. In addition, there are those families which have a disjunct distribution and live on the southern tip of the southern continents: Cunoniaceae, Escalloniaceae, Gunneraceae, Philesiaceae, Proteaceae, and Restionaceae. *Amaryllis,* of the Amaryllidaceae, and *Clivia* belong to the Cape as do the genera *Stapelia* (Asclepiadaceae), *Freesia* (Iridaceae) and *Aloë* (Liliaceae).

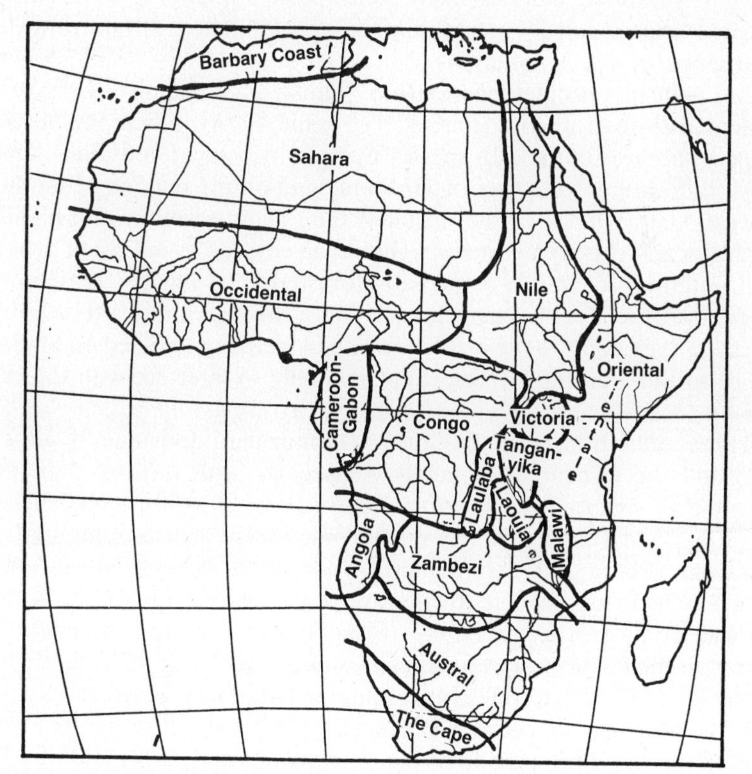

Figure 12a Regional classification of the Ethiopian region according to the dispersal areas of freshwater fish (according to Poll 1974).

In contrast to the vegetable kingdom, no independent 'Cape' animal kingdom can be delineated.

The Malagasy region consisting of Madagascar (about 592,000 km^2; highest elevation: 2,876 m), the Seychelles (405 km^2), the Comores and Mascarene islands, distinguishes itself from the Ethiopian region by the absence of true monkeys, Artiads (an extinct species of hippopotamus lived on Madagascar during the Pleistocene; *Potamochoerus porcus* was introduced by man), the Perissodactyls, elephants, aardvarks, pangolins, poisonous snakes (Elapidae and Viperidae), Leptotyphlopids, Agamids, Varanids, soft-shelled turtles, toads (Bufonidae) and Gymnophions (with the exception of the Seychelles) and the presence of an animal world exceptionally rich in endemics.

The flora of Madagascar (Malagasy) is also rich in species (about 10,000 different Angiosperm species) and endemics (nearly 80%). The wealth of endemics, however, shows a varying pattern of distribution. Only about 21% are found in the littoral zones, whereas in the forests, they amount to 89% (Koechlin 1972). Here, 94% of the trees and 85% of the perennial grasses are endemic. A particularly interesting endemic family is the Didieraceae found in the arid southern region of the island. More than 40% of the phylogenetically more primitive families have pantropical affiliations; 27% show an obviously African, and 7% an Indo-Malaysian influence. Affinities with the southern tips of the southern continents (South Africa, South America, Oceania) indicate an older relationship. Similar area disjunctions have also been demonstrated for generic areas. The genus *Pachypodium,* belonging to the Apocynaceae, contains 18 species living in two separated areas. One of the distribution areas consists of arid and semi-desert regions in south and southwest Africa, and the other includes the savannas and central highlands of Madagascar. Although vast areas should be wooded as indicated by recent growth potential, only 28% is actually forested. The Lavaka erosion types (Bourdie 1972) reflect human influence (since the 4th century). However, in some cases, they point to the fact that an open landscape existed on Madagascar prior to the time man started fundamentally to interfere with the ecosystems.

Three autochthonous lemuroids (Lemuridae, Indridae, Daubentoniidae) and the insectivorous family Tenrecidae with the two subfamilies Tenrecinae (6 species) and Oryzoryctinae (21 species) are included in the endemic fauna of Madagascar. Seven endemic rodent genera of the subfamily Nesomyinae (Petter 1972) show marked adaptation to the damp woods of eastern Madagascar, but also to the arid areas of the south. The endemic fauna also comprises a bat family (Myzopodidae), eleven endemic ferret species with the Fossa (*Cryptoprocta ferox*) probably being the fore-runner of both the cats and the ferrets, Mesoenatidae which are about the size of a thrush, the Madagascan cuckoos (Couinae), the Leptosomatinae, the Brockypteracinae, the Philepittidae, the Vangidae, and the giant ostriches (Aepyornithidae) which did not become extinct until about 800 A.D. With the Aepyornithidae, Madagascar was also inhabited by the giant lemurs

(*Megaladapis edwardsi*), confirmed by fossils; by the giant tortoises (*Testudo grandidieri, Testudo abrupta*); and by the hippopotamus *Hippopotamus lemerlei*, approximately 2 m in length.

The Malagasy herpetofauna is also noted for its large number of endemics (Blanc 1972, Bourgat 1973), although the amphibians have not as yet been sufficiently studied.

The Malagasy arachnids are predominantly related to those of Ethiopia, the other invertebrate groups in Madagascar apparently having their own secondary differentiation center.

The more primitive animal and plant groups have connections with the Neotropic realm (for example, the fan palm *Ravenala* belonging to the Musaceae; the Malagasy boas; the Iguanidae including the two Malagasy genera *Chalarodon* and *Oplurus;* as well as the tortoise genus *Podocnemis,* classified among the Pelomedusidae). The relationship between the Malagasy and Oriental regions, however, has often been over-emphasized (compare Günther 1970, Mertens 1972).

The Seychelles, belonging to the Malagasy region, have 11 endemic plant genera (among others: *Medusagyne, Deckenia, Lodoicea*). *Medusagyne* constitutes a monotypic family which, in the form of Medusagynaceae, is endemic to the Seychelles. Among the animal species, 14 endemic inland bird species, one giant tortoise (*Testudo gigantea*), the endemic Gymnophionae genus *Hypogeophis* (6 species) and the ancient, endemic Sooglossinae frogs should be noted. The *Sooglossus* frogs pass through their larval development outside water, attached to the back of the male. They are the basic form of ranoid frogs which are phylogenetically related to the South African Arthroleptinae.

The closest relatives of the six *Hypogeophis* species live in Africa. The zoogeographical relationship between the Oriental and Ethiopian regions is far greater than the differences, which have become distinct through the special, individual development in the separate regions.

The plant families Actinidiaceae, Crypteroniaceae, Daphniphyllaceae, Erythropalaceae, Pentaphragmataceae, Peripterygiaceae, Petrosaviaceae, Sarcospermataceae, and Siphonodontaceae are endemic to Southeast Asia and Malaysia. According to Good (1964), the Gonystylaceae (as far as the Solomon and Fiji Islands), Lowiaceae, Scyphostegiaceae, Stenomeridaceae, and Tetrameristaceae are endemic to the Malayan region. The semi-arid to arid belt extending through the Thar Desert delimited by Mysore, Ahmadanagar, Jodpur and Ganganagar forms a distinct barrier against forest species. While within this arid belt, taxa with Ethiopian and Palaearctic connections are dominant among the verterbrates, species often occurring in the forests of southwest India and Sri Lanka have their closest relatives in Southeast Asia.

Comparing the African savannas with those of India and, in particular, those of Sri Lanka, distinct characteristics can be noted. Although elephants, leopards, and the Palaeotropical rhinoceros birds can be observed in the Wilpattu National Park (Northwestern Sri Lanka), just as in the Tsavo National Park (Kenya), huge herds of axis deer, intermingled with individual royal sambars, take the place of the antelopes and gazelles absent from Sri Lanka. This noticeable difference is intensified at the species level by a number of Sri Lankan endemics. The special position Sri Lanka occupies in South Asia from a zoogeographical point of view is closely tied not only to the historical development of the area, but also to the recent ecological features which offer a habitat not only to the fauna but simultaneously to that of the savannas.

Reptile families endemic to the Oriental region include the beaded lizards (Lanthanotidae) and the Gavialidae. Among the avifauna, only the family Irenidae and the Hemiprocninae (swifts) are endemic, and, among the native mammals, only the squirrel shrews (Tupaiidae), the gibbons (Hylobatinae); and the Platacanthomyidae (Rodentia) are endemic. The Trictenotomidae also belong to the Oriental fauna. The total proportion of carnivores in the Oriental region is remarkably high.

2.2.1.6 The Antarctic At present, Antarctica is the poorest land mass for plants and animals. Fossil finds show that Antarctica (12.4 million km^2 of continental mass, 1.5 million km^2 of shelf ice, 70,400 km^2 of islands) has not always been as hostile to life as it is today. Coal deposit sites yielding Labyrinthodontia from the lower Triassic, Gondwanaland flora (including *Glossopteris*) and large saurians, such as *Lystrosaurus* confirm this assumption.

The volume of mainland ice is estimated at about 24 million km^3. Dry valleys containing moraines and glacier striations point to an even greater glaciation during the glacial epoch. About 200,000 km^2 of Antarctica is, today, free of ice. This includes, in part, the Markham Mountains in eastern Antarctica at an altitude of 4,000 m and 3,000 km in length. In 1965, 42 mite and 12 insect species were recorded in Antarctica.

According to Janetschek (1967), the algae flora of Victoria Land consists of 80 Cyanophyceae, 50 Diatomeae and 20 Chlorophyceae (also cf. Hirano 1965). 32 endemic lichen species live around the coastline. The genus

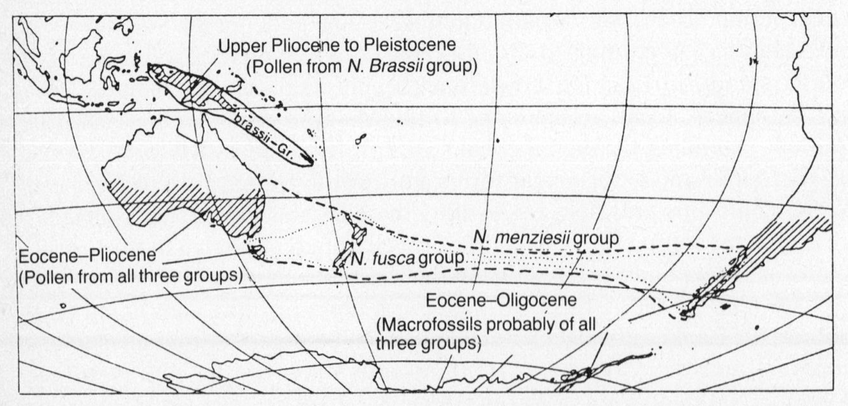

Figure 13 Recent and fossil distribution of the southern beech genus *Nothofagus* (according to Müller and Schmithüsen, 1970).

Usnea grows only in protected locations (Stonehouse 1972). The most southern occurrences of more advanced plants (*Colobanthus crassifolius* and *Deschampsia antarctica*) are found at 68° 12′S. Among the microfauna Tardigrada, Acarina, and Collembola should be particularly mentioned.

The presence of vertebrates is made possible by the food chains having their origins in the ocean. This is particularly true for penguins (Spheniscidae), an old group of birds, known since the Eocene, whose relatives are still unknown to a large extent and whose 17 species are exceptionally well adapted to life along the Antarctic coasts, thanks to their waterproof plumage, a 2 to 3 cm thick fat layer and various behavioural patterns resulting in an increase in basic metabolism. Most Antarctic species cannot endure sudden fluctuations in temperature. The emperor penguin (*Aptenodytes forsteri*) will only breed on the coast of the Antarctic mainland (to a distance of 1,400 km from the South Pole). Further common Antarctic penguins include *Eudyptes chrysolophus, Spheniscus magellanicus* (which, however, does occur as a stray visitor as far as Bahia Laura along the Brazilian coast), *Pygoscelis antarctica, Aptenodytes patagonica, Megadyptes antipodes* and *Pygoscelis adebiae.* The more advanced nonpasserines are represented by *Diomedea exulans, Stercorarius skua, Pelecanoides magellani, Oceanites oceanus,* and *Chionis alba;* whereas the passerines are represented only by a pipit (*Anthus-antarcticus*) which evidently is closely related to the South American *Anthus correndera.*

The Antarctic mammals also owe their existence to food chains derived from the ocean. This particularly applies to the seals (*Ommatophoca rossi, Hydrurga leptonyx, Lobodon carcinophagus, Leptonychotes weddelli*), and the whales (*Balaenoptera physalus, Balaenoptera musculus, Orcinus orca*). An important food source for the *Balaenoptera* species is *Euphausia superba*

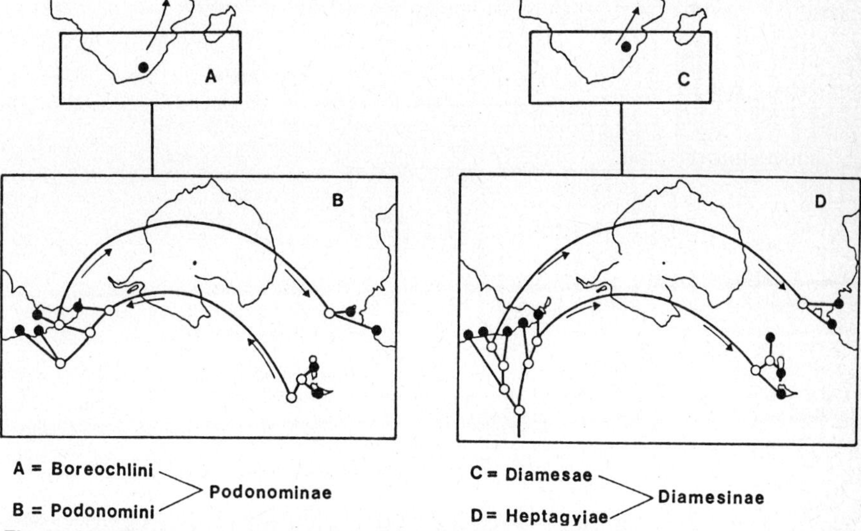

Figure 14 Phylogenetic affinities and dispersal routes of southern hemisphere Diptera of the Chironomidae family (according to Brundin 1972).

which grows to a length of 6 to 9 cm. The fish fauna is a useful basis for classifying the Antarctic Ocean into individual, well-defined regions (Andriashev 1965).

The South Pacific southern beech forests (*Nothofagus*) (Fig. 13) of Chile and New Zealand today contain a fauna which probably already existed during the Tertiary and dispersed in an arc shape with the forests over the Antarctic (Harrington 1965, Brundin 1972, Müller and Schmithüsen 1970). The geological structures of the eastern and western halves of the Antarctic are notably different. The Precambrian rocks of Antarctica point toward an old complex of land, which is in contrast to the western side. Younger fold

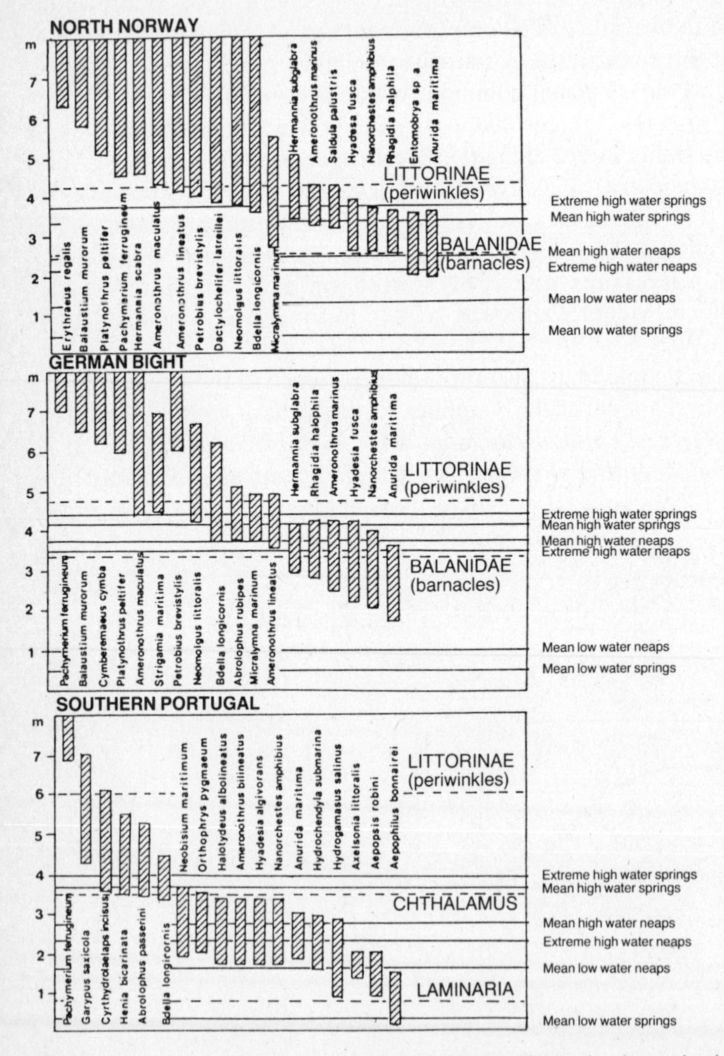

Figure 15 (above and opposite) Vertical distribution of littoral arthropods along various coastlines of the Northern Hemisphere (according to Schulte 1977). The littoral zone characterized by the Littorinae (periwinkles) is recorded as well as the barnacle zone flooded during high tide.

mountains connecting to the Andean sytem (South Antilles arc) emerged during the Jurassic and the early Tertiary.

2.2.2 The Biorealms of the Oceans

The plant and animal realms of the mainland are in complete contrast to those of the oceans. The genetic structure of this habitat, the largest within the biosphere, shows a much stronger adaptation to its ecological environment than is known on the mainland. Thus, most authors undertake a classification based on the major habitats (littoral, abyssal, pelagic). Analysis of the relationships among the abyssal fauna shows, however, that closer connections may exist between the littoral and the abyssal biota of some seas than between the littoral biota of various oceans. In spite of this, however, more

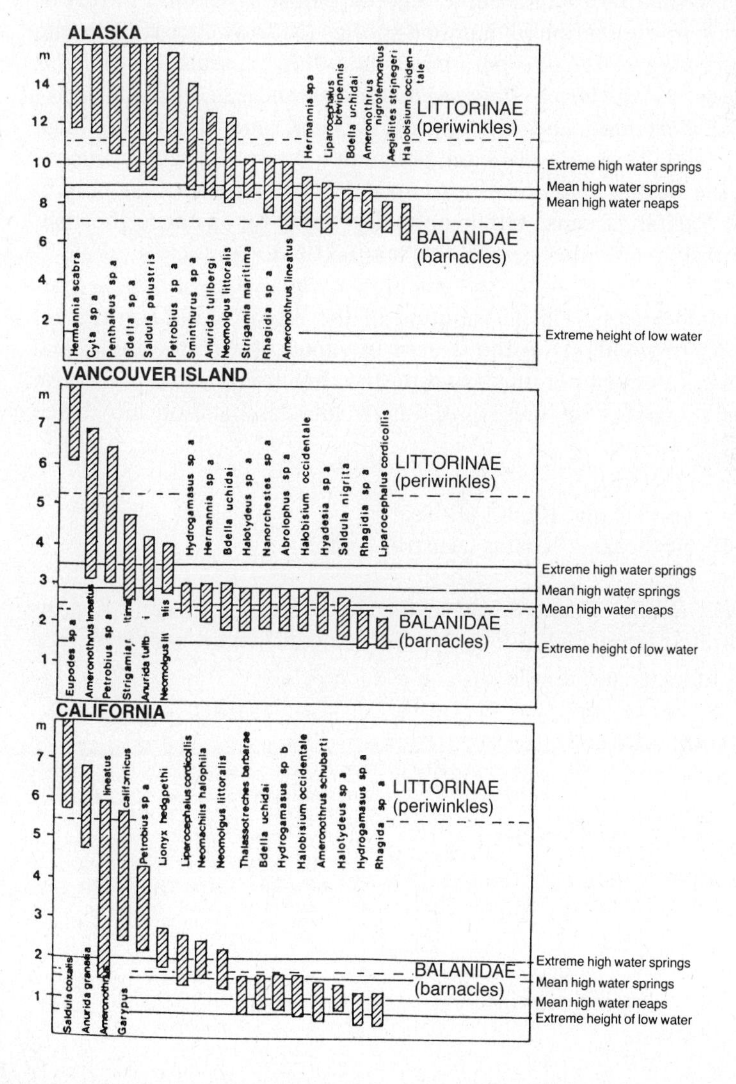

recent biogeographical works (Briggs 1974, for instance) continue to retain the major classification proposed, mainly by Ekman in 1935. On the basis of this concept, the three realms considered are:

(1) Littoral Realm (= continental shelf)
(2) Pelagic Realm (= open seas)
(3) Abyssal Realm (= deep sea)

Although this classification is by no means satisfactory, we shall continue to abide by it here, too. The genetic composition of both the flora and fauna differs entirely from that of the land (compare Marine Biomes, Chapter 5.2). Many animal species, for example the insects, are to a large extent absent from the open seas (exceptions: the waterskipper, genus *Halobates*), only marine groups having developed within the littoral area (Cheng 1976, Schulte 1977). On the other hand, numerous ectoparasites also live purely or predominantly on marine bird and mammal species (Murray 1976). Thus, the anoplura *Echinophthirus horridus* parasitizes the following seals: *Phoca Vitulina, Pusa hispida, Pusa sibirica, Pagophilus groenlandicus, Halichoerus grypus, Erignathus barbatus,* and *Cystophura cristata,* and the mallophaga *Austrogoniodes* and *Nesiotinus* live on penguins. While *Halobates micans* inhabits all of the oceans, *Halobates ericeus* and *H. germanus* occur only in the Indian and Pacific Oceans. Other animal groups, for instance the fish, have their greatest species diversity in the sea (63.6%).

2.2.2.1 The Littoral Realm The distribution of the biota on the continental shelves is directly dependent on the degree of separation of the individual ocean basins, on water temperatures and on the direction of cold and warm ocean currents (Fig. 15). On these grounds, a sub-classification into three realms seems appropriate:

(1) The Tropical Realm
(2) The Northern Oceanic Realm (Boreal)
(3) The Southern Oceanic Realm (Austral Realm)

The Tropical Realm The tropical realm can be divided into four regions which are separated from each other by ocean or continental barriers:

(a) Western Indo-Pacific region
(b) Eastern Pacific region
(c) West Atlantic region
(d) East Atlantic region

Western Indo-Pacific Region

(i) Malayan province
(ii) South-Central Pacific province
(iii) Hawaiian province
(iv) South Japanese province
(v) Northwest Australian province
(vi) Indian province

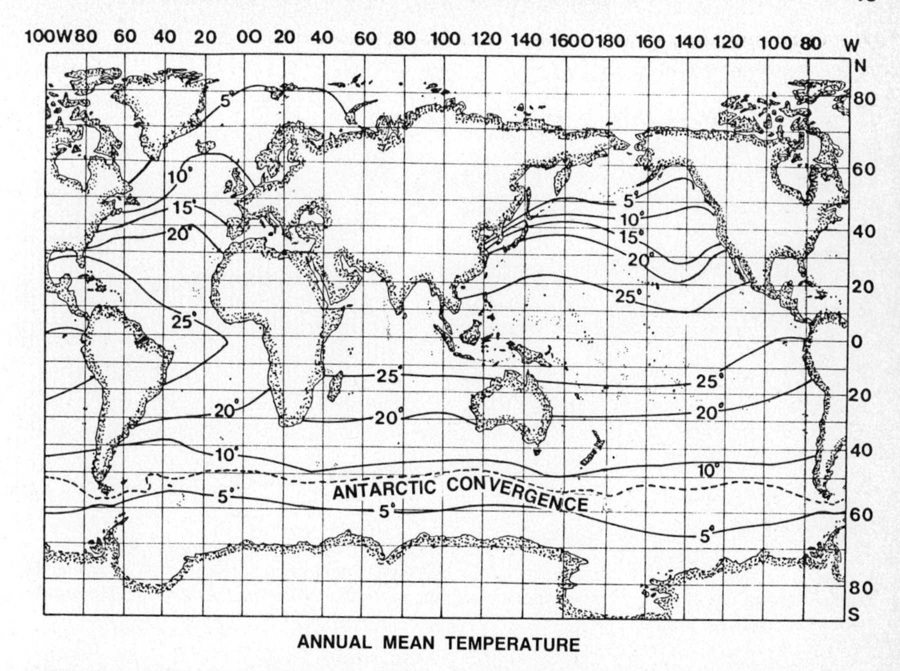

ANNUAL MEAN TEMPERATURE

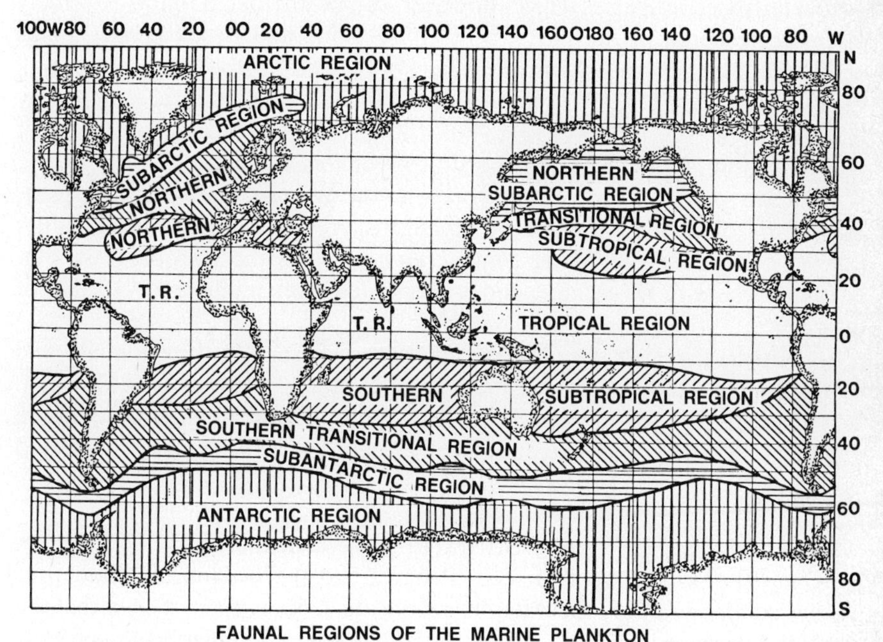

FAUNAL REGIONS OF THE MARINE PLANKTON

Figure 15a Mean annual temperatures and faunal regions of marine plankton.

The tropical realm is the habitat of the coral reefs (cf. Marine Biomes, Chapter 5.2) and the mangrove swamps along its coasts (cf. Mangrove Biomes, Chapter 5.1.11). The marine fauna in the Hawaiian province is

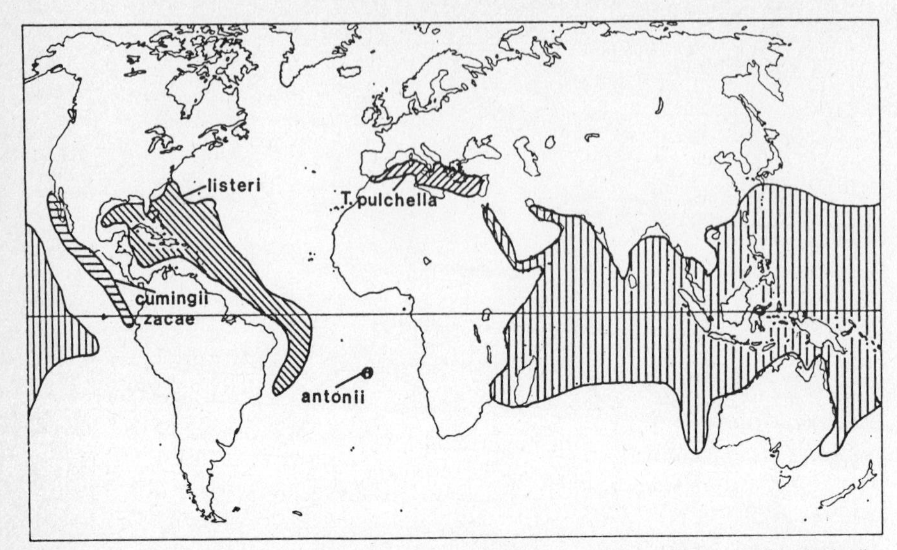

Figure 16 Distribution of the bivalve genus *Tellinella* (according to Boss, 1969). Vertically hatched areas are inhabited by the species *virgata, staurella, crucigera, rastellum, pulcherrima, asperrima,* and *verrucosa.*

particularly rich in species which, however, can be attributed to the extremely intensive research of the past few years. Gosline and Brock (1960) list as endemics 34% of the shellfish, Ely (1942) 30% of the Asteroidea and Ophiuroidea, Banner (1953) 45% of the Crangonidae (Crustaceae), and Kay (1967) 20% of the molluscs (Fig. 16). The Marquesas, located southeast of Hawaii, show a similarly high portion of endemic molluscs. Most of the Marquesan species, however, also occur in the Western Indo-Pacific region. Although some other groups of organisms are represented by endemics (among others, the Blenniidae genus *Entomacrodus*), the Marquesas certainly cannot claim to be an independent region of comparable status to that of Hawaii.

During the past few years, numerous Australian taxonomists and biogeographers have examined the littoral biota of Northwestern Australia. About 40% of the echinoderms occurring in this region are considered to be endemics. Similar high percentages are recorded for the gastropods and sponges.

Eastern Pacific Region Briggs (1974) divided the Eastern Pacific region into the provinces of Mexico, Panama and the Galapagos. Numerous species common to both the Atlantic and Pacific regions make it clear that the Central American landbridge separating the two regions must be geologically young (Pliocene), at least in some areas (Table 7).

The Mexican province extends from the southern tip of the Baja Californian peninsula to Tangola-Tangola Bay, the northern border of Panama province in the plains of Tehuantepec.

The littoral fauna of the Galapagos is particularly interesting since it

TABLE 7 ATLANTIC-PACIFIC SPONGE SPECIES (according to Wiedenmayer 1977)

Atlantic West Indian Region	Western Indo-Pacific Region
Aplyssina lacunosa	*A. spongelii, A. calyx*
Cribrochalina vasculum	*C. bilamellata*
Xestospongia muta	*X. testudinaria*
Callyspongia fallax	*C. peroni*
Niphares digitalis	*N. olemda*
Spinosella vaginalis	*S. muricina, S. aspericornis*
Thalysias juniperina	*T. juniperina*
Ectyoplasia ferox	*E. tabula*
Epipolasis lithophaga	*E. salomonensis*

appears to have been isolated for a longer period of time than the terrestrial biota. The molluscan fauna, however, shows a close relationship to that of the Panama area (Emerlon and Old 1965). 138 species of the Galapagos archipelago are also members of the Panama littoral fauna species, 8 having Western Indo-Pacific relationships and 29 are endemic. Of the 120 *Brachyura* species, 18 have been proved to be endemic. The reef corals of the Galapagos Islands (32 species) analyzed by Durham (1966) were found to include 13 endemics. The rest again show either a relationship to Panama or to the Western Indo-Pacific region. Of the 311 algal species confirmed so far (Silva 1966), 36% are endemic, whereas of the 223 littoral fish species (Walker 1966), only 23% are native. Both North and South American relationships among the littoral Galapagos groups are noteworthy. The endemic fish, *Archosargus pourtalesi* (Sparidae), the knife fish, *Arcos poecilophthalmus* (Gobiesocidae), and the fish *Labrisomus dentritus* (Clinidae) have their closest relatives in the Atlantic (*Arcos macrophthalmus, Labrisomus filamentosus*), as do numerous littoral crustaceans.

West Atlantic Region Briggs (1974) divides the West Atlantic region into the Caribbean (along the coast of Mexico to the Orinoco River), the Brazilian, and West Indies provinces. The Brazilian province extends from the Orinoco Delta, along the coral reefs to Rio de Janeiro (Cabo Frio). Isolated islands such as Fernando de Noronha or Trinidad, have endemic species. This applies in particular to the West Indies province, which generally follows the arc of the Caribbean islands from the Bermudas in the north to Grenada in the south. The north–south diversity gradient is very apparent. Work (1969) showed that the molluscan fauna of Bermuda represents an impoverished fauna of the tropical West Indian islands. Only 13 of the 265 littoral fish species around Bermuda are endemic.

East Atlantic Region The West African province is regarded as the center of the East Atlantic region (from 12°S to 15°N, around Senegal). As with the West Atlantic region, isolated groups of islands (Cape Verde, St. Helena, Ascension) have their endemic species which, among other things, have led to the development of distinctive provinces (e.g. Ascension – St. Helena – Province). Within the littoral zone of St. Helena, an island dating from Miocene times situated about 560 km east of the Atlantic ridge, 55 littoral fish have ben identified, 12 of which are endemic. 4 out of 23 Decapoda (Chace 1966) and 13 of the 26 echinoderms have been proved to be endemic. Ascension Island, on the other hand — being volcanic, and probably only of Pleistocene origin — has only two endemics among its 54 littoral fish species.

The Northern Oceanic Realm At least six regions can be distinguished in the Northern Oceanic (Boreal) Realm for the littoral biota:

(1) Mediterranean–Atlantic region
(2) Sarmation region
(3) Atlantic–Boreal region
(4) Baltic region
(5) North Pacific region
(6) Arctic region

Depending on the location and history of partially isolated, subsidiary oceans, these regions are subjected to greater subdivision than the Southern Oceanic realm. Isolated groups of islands (such as the Azores and Madeira within the Mediterranean-Atlantic region) are characterized by a smaller number of local endemics.

Endemics are apparently absent from the littoral fish of the Miocene islands. Madeira is also characterized by a smaller number of endemics.

The Arctic region is economically important, because of its fish fauna which is poor in species but rich in numbers. A typical representative of the Arctic region is the bottom dwelling fish *Iceleus spatula.*

The Southern Oceanic Realm The Southern Oceanic realm limits the inhabitants of the tropical littoral zone towards the south. It can be subdivided into six regions:

(1) South Australian region
(2) North New Zealand region
(3) Peruvian region
(4) Uruguayan region
(5) South African region
(6) Antarctic region

In the South Australian region, the Tasman herring (*Clupea bassensis*) is found along with various endemic fish species. In southwest Australia alone, 12.6% of the echinoderms (Clark 1946) and 28.4% of the 253 littoral fish have been reported as endemic. The littoral fauna of New Zealand, due to its long isolation, has an even greater wealth of endemics. From the Auckland Province, 80% of the echinoderms, 46% of the thecate hydroids, 31% of the polychaetes, and 40% of the 649 mollusc species (Powell 1940) have been described as endemics.

The distribution within the Peruvian region is directly influenced by the cold Humboldt Current. Because of their endemics, Briggs (1947) established the Juan Fernandez Islands as an independent province. The Uruguayan region (from the mouth of La Plata to the State of Saõ Paulo) is the Atlantic equivalent of the Peruvian region. The South African region can also be subdivided into individual provinces (Southwest African province, Agulha province, Amsterdam–St. Paul province). Their boundaries make clear — as does the isolated Kerguelen region — the great influence ocean currents of varying temperatures and also the open seas have on the littoral biota and their development. With the Kerguelen province, however, we have already entered the cold waters of the Subantarctic and Antarctic whose genetic structure has its own particular features. The southern tip of South America (Magellan province, Tristan-Gough Islands), of Tasmania, of southern New Zealand, and the Antarctic region are included herein. They are genetically connected to the Northern Oceanic realm and to the Arctic via those taxa with a bipolar distribution.

2.2.2.2 The Abyssal Realm The biota of the lightless zone of the Archibenthos can be much more easily subdivided in accordance with the classification of the ocean basin than the organisms of the bathypelagic habitat which are capable of significant vertical movements. According to de Lattin (1967), four regions can be delineated:

(1) Arctic abyssal region
(2) Atlantic abyssal region
(3) Indo-Pacific abyssal region
(4) Antarctic abyssal region.

Deep-sea trenches often possess their own distinctive fauna, thus delineating some very clear habitats.

2.2.2.3 The Pelagic Realm A subdivision between the epipelagic and the bathypelagic habitat is necessary to correspond to the varying light conditions between the surface and the depths. The classification of the epipelagic biota into tropical, boreal, and austral pelagic regions is useful. The pelagic realm, however, is by no means a uniform habitat. The production of phytoplankton varies considerably and the zooplankton also presents spatially differing patterns of distribution. The terminal links in the pelagic food chain are the toothed whale and large predatory fish which are as important to the deep-sea fishing industry as cod and tunafish. The Sargasso Sea represents a spatially definable region of the pelagic realm; its name stems from the seaweed (*Sargassum*) prevalent in that area. 67 freely-swimming animal species have been described as being inhabitants of the seaweed, such as the sargasso fish (*Histrio histrio*), and the pipefish *Sygnathus pelagicus* (Fig. 15a).

Apart from the great economic importance of marine biota, more and more emphasis is today being placed on its ecological function as the garbage disposer of the biosphere. Increasingly, marine organisms are being raised on artificial marine farms; this may be the reason why the economic significance of the lack of fish numbers within the open seas may have a less serious consequence in the future. Modern water analyses of large marine habitats, however, point more and more to the significance of individual marine organisms even in connection with the atmospheric composition of our planet.

Chapter 3

Propagation Areas

Propagation areas are the central research subject of biogeography. Under propagation areas, we understand an adaptive subsystem of the biosphere determined by the ecological potential, genetic variability and phylogeny of populations, as well as by a temporally and spatially changing mode of action; this subsystem possesses ecological as well as phylogenetic functions and its spatial extension can be described by a three-dimensional disperal area of differing size and structure (Müller 1976) (Fig. 17).

The species is the smallest, real, fundamental unit in evolutionary terms. In biogeography, the propagation area is its equivalent. It encompasses only that part of the distribution area of a species in which it is able to propagate itself permanently without constant external influx (cf. area of propagation; de Lattin 1976, Müller 1977). Although areas outside the propagation area may be visited regularly (cf. Animal Migrations, Chapter 3.1.5.2) or irregularly (cf. invasions) by organisms, these displacements take place in an ecological, rather than phylogenetic, context. Due to the spatial limitation of the propagation area concept a clear distinction between habitat, propagation, and migration areas of one species is made; on the other hand, that area in which the flow of characteristics of a species occurs by means of propagation will receive the due attention it deserves during the discussion of the developmental history of a population.

The behaviour of the propagation area is characterized by its biotic elements and the latter's genetically determined capabilities. The following three parameters are of importance to the structure of the propagation area:

(a) Type of linkage between organisms (taking into account functional correlations);

(b) Linkage density (a measure for the cross linkage, taking into account the number of inputs and outputs of the system's elements);

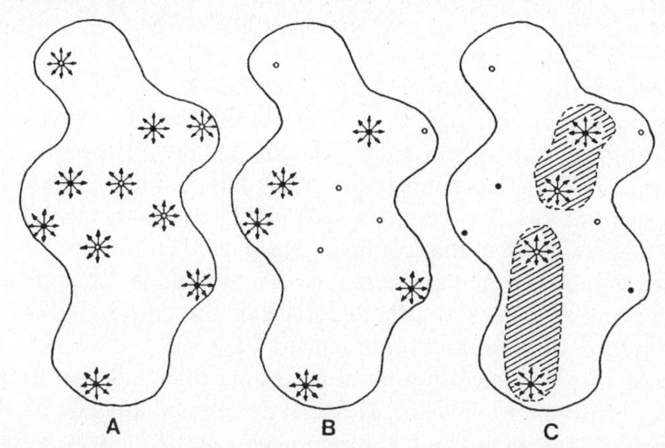

Figure 17 Theoretical propagation areas, constructed of elements of varying genotypes (two) and capabilities. Area A is based on the assumption that, due to intrinsic factors, closed as well as open circles may behave in an expansive mode; in area B, closed circles only have this capability; and in area C, closed and open circles only become expansive if corresponding external factors (habitat conditions = hatched) are present. When discussing area expansions or contractions a distinction must be made between these three types.

(c) Cross linkage type (represents the relationship between the inputs and outputs of inherent elements and systems, and alien systems).

The importance of an organism or population to the overall structure of its propagation area can only be adequately analyzed via these three structural parameters. The prerequisite for an analysis of these parameters is the determination of the spatial distribution and the functions of the elements.

3.1 SPATIAL DISTRIBUTION OF THE ORGANISMS

When looking primarily at the area boundaries and less at the frequency distribution of single individuals within a given area, the shape and the size of the area are of importance. The shape of the area is three-dimensional as far as the adaptation of the elements to a given space is concerned. It shows, therefore, a horizontal and a vertical structure. It can join either homotopographically or heterotopographically to one or several other ecosystems (cf. compare Biomes, Chapter 5).

3.1.1 Area Size

As a structural characteristic, the size of the propagation areas of individual plant and animal groups may be totally different from each other. Area size is by nature dependent, among other things, on ecological potential, competition between species, dispersal opportunities, history of propagation success and phylogeny, as well as on the geographic location of the origin of the elements constituting a given area. Only in rare cases is there a general correlation between area size and the evolutionary history of the taxa (cf. Anderson 1974).

Willis (1922), in his famous book *Age and Area — A Study of Geographical Distribution and Origin of Species*, an impressive work thanks to his examination of the flora of Sri Lanka, recognized that: "The area occupied at any given time, in any given country, by any group of allied species... depends chiefly, so long as conditions remain reasonably constant, upon the ages of the species of that group in that country, but may be enormously modified by the presence of barriers such as the sea, rivers, mountains, changes of climate from one region to the next, or other ecological boundaries, and the like, and also by the action of man, and other causes." (p. 63).

Nor can the ecological potential (= range of living conditions within which a species is able to develop; Hesse 1924) immediately be derived from the area size. Many plant and animal species settle only in very specific habitat areas within their cosmopolitan distribution area, whereas others may appear anywhere due to their ecological fitness (ubiquists).

3.1.1.1 Cosmopolitan Groups Numerous animal and plant families, genera, and species are cosmopolitan; i.e., they are represented by populations in the entire biosphere of the world. On the basis of their ecological potential, dispersal opportunity, or their close ties to man (among others domestic animals, parasites, or species in transit) they are not necessarily restricted to a biogeographic area. However, there are no species or genera common to the oceans, inland waters and the land. The following are examples of cosmopolitan plant families: Grasses (Gramineae), the composites (Compositae), the Cyperaceae, Orchidaceae, Papilionaceae, Labiatae, Scrophulariaceae, Liliaceae, Boraginaceae and Gentianaceae. Among the animals, dogs (Canidae), mice (Muridae), pipits (Motacilladae) and falcons (Falconidae) should be mentioned as examples.

Naturally cosmopolitan genera and species areas are considerably more interesting. Cosmopolitan plant genera include for example: *Andropogon, Anemone, Bromus,* and *Drosera.*

Sub-cosmopolitan genera (Fig. 18) tend to inhabit areas which exclude certain realms totally. In the case of plants, examples of these genera include, *Impatiens, Salix, Typha,* and *Vaccinium.*

Among the animals, too, there are numerous cosmopolitan or sub-cosmopolitan species. The following can be mentioned as examples: barn owl (*Tyto alba*), sea eagle (*Pandion haliaetus*), peregrine falcon (*Falco peregrinus*), painted lady (*Vanessa cardui;* inhabits all of the continents with the exception of South America), the microlepidopteran *Plutella maculipennis* and *Nonnophila noctuella,* and numerous tardigrades (e.g. *Macrobiotus hufelandi*). In addition, the Central European common swallow (*Hirundo rustica*) forms a superspecies complex with a cosmopolitan distribution (with the exception of South America which it only visits during its migrations).

The high number of cosmopolitan species among the tardigrades is remarkable. Morgan and King (1976) found 15 species in the British Isles

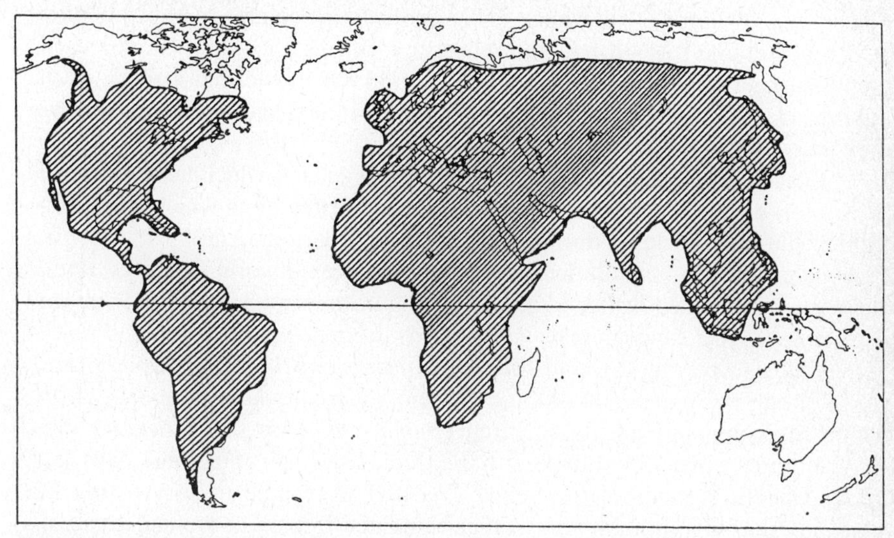

Figure 18 Sub-cosmopolitan area of the toads (Bufonidae). *Bufo marinus* was introduced into the Australian realm.

alone. The windborne drifting of their resistant shells is primarily responsible for their worldwide distribution. Numerous pests destructive to useful plants also have a cosmopolitan distribution (cf. Hill 1975, among others).

3.1.1.2 Species Occupying Restricted Areas In contrast to the cosmopolitan groups, there are species which at present are known to inhabit only very restricted areas. For many of these populations, the habitat is at the same time their center of origin; for others, it is a place of refuge, a relict area of a formerly wider distribution.

The snail species *Laminifera* (Clausiliidae) is an endemic of the mountain peak La Rhune (Western Pyrenees). About 20 Nearctic Urodela are known in a single location, among them, numerous speleological forms. The Kabyle nuthatch (*Sitta ledanti*), together with the Numidian fir *(Abies numidica)*, occurs only on the Djebel Babor in the northern Algerian Atlas Mountains above 1,500 m (Vielliard 1978).

Relict organisms are those with a former wider distribution which, in the course of a reduction, a fragmentation, or a displacement of an area, caused by a change in the environmental conditions, were able to survive only in particularly favorable locations. Relicts can date from various periods (Tertiary, Pleistocene, Holocene, etc.) and, on the basis of their relationships to the habitat, may be indicators of formerly different ecological conditions (e.g. xerothermal relicts, steppe relicts). Taxa of Tertiary relicts, for example, have generally survived essentially unchanged until today, at least since the Pliocene, at a favorable place of refuge.

The term Tertiary relict can only really be applied in those cases where the pre-glacial relict characteristics of a population have been proven with respect to its phylogenetic (constancy of characteristics) and biogeography (constancy of habitat) relevance. This applies particularly to those habitats which have been less strongly influenced by, for instance, the climatic variations during the Pleistocene (old lakes, ground water, thermal springs, caves, islands: cf. Canary Islands, Section 6.4.2), and animal species having a slower rate of evolution.

Glacial relicts consist of plant and animal species which have existed at their most recent location as remains of stenothermic biota which adapted to colder climates, at least during the last glacial period. Among the Central European animals, the following are examples: the flatworm *Crenobia alpina,* belonging to the headwater fauna and living in the rhithral of the low mountain ranges, the cave animals *Onychiurus sibiricus* (Apterygota), *Pseudosinella alba* (Apterygota), and *Choleva septemtrionis holsatica* (Coleoptera), and, among the plants, the Alpine clubmoss (*Lycopodium alpinum*) on the Hohe Meissner and the dwarf birch (*Betula nana*) on the Brocken. Some invertebrates which have adapted to the treeless regions of the high Alps (e.g. ground beetle species of the genus *Trechus*) were able to survive the last glacial period in mountain areas which remained free of ice (so-called *massifs de refuge*), or on some mountain peaks jutting out above the glacial ice (nunataks) (Holdhaus 1954, Besuchet 1968, Nadig 1968; cf. High Mountain Biomes Chapter 5.1.9). More recent results from glacial research in Scandinavia concerning the spatial and temporal duration of the last glacialperiod cast some doubt on the existence of real glacial relicts within this area, and the question of the smaller refugia from the last glacial period in Scandinavia (Lindroth 1939, 1969, Stop-Bowitz 1969) has, in particular, to be studied further. Lindroth (1969) was able to show that some carabids which had previously been assumed to be glacial relicts only survived the earlier Dryas period or, at best, Würm II (which is separated from Würm I by an interstadial) and thus, in the strictest sense, are not glacial relicts at all. Other species, however, were able to continue to exist as glacial relicts up to the present in isolated lakes of the northern Holarctic (Segerstrale 1966). Among the crustaceans, these include the following species: *Mysis relicta, Mesidotea entomon, Gammaracanthus lacustris, Pontoporeia affinis, Pallasea quadrispinosa* and *Limnocalanus macrurus-grimaldii.* These Baltic animals represent a relict inland water fauna from the northern Holarctic whose most closely related populations live in the ocean or which are derived from marine groups (although the genus *Pallasea* is limited to freshwater; *Pallasea kessleri,* for instance, can be found in Lake Baikal). Populations of Baltic fauna inhabit the Baltic Sea as well as numerous Scandinavian, English, Finnish, North Russian, Siberian, and North American continental lakes. During the postglacial period, the Scandinavian limnic populations were directly connected with the Baltic Sea and were isolated only when the Baltic sea level receded. This opinion was first expressed by Ekmann (1940) and was, therefore, termed the Ekmann Theory.

According to this theory, the limnic and Baltic occurrences of Baltic fauna represent relict populations which originally occupied an arctic-marine area and which migrated, either during or directly preceding the cool Yoldia period, through an ocean link from the White Sea via Lakes Onega and Ladoga to the Yoldia Sea (cf. Baltic Sea Stages, Chapter 4.1.4, Figs. 47 and 48) where, with the changing water level, they entered the various Scandinavian inland lakes, and, after the water level there fell, remained as relicts.

In contrast with the glacial relicts, the Central European xerothermic relicts are considerably younger. These are thermophilic populations which, at one time, had a wide distribution (*Littorina* period) and which, during the course of a reduction, a fragmentation, or a displacement of the area, caused by a climatic deterioration, were able to survive on particularly favorable "warm islands". The xerothermic relicts in the Mosel and Middle Rhine area have been particularly well examined (Warnecke 1927, de Lattin 1967, Müller 1971). The xerothermic relicts of the Mosel area reached their recent habitat during the climatic optimum (5,000–1,000 B.C.), presumably via the Mosel and Rhine rift valley (Müller 1971), which served as an ecological corridor. During the subsequent beech period, initiated by a reduction in temperature and, therefore, an expansion of beech, the post-glacial migration path was broken and the populations of the Mosel and Middle Rhine area were separated from their original populations in the Mediterranean area. The short isolation phase (about 3,000 years) in many areas led to the development of subspecies within the isolated populations which are easily recognized today. Thus, the apollo *Parnassius apollo*, for instance, is present at Winningen on the Mosel, in the form of a subspecies (Fig. 19) (*Parnassius apollo vinnengensis*) endemic to the area. These findings, however, cannot be confirmed for all animal groups. Thus, Nagel (1975) was able to show that the Coleoptera with their xerothermic habitat in the Saar-Mosel area show no sub- or semi-specific differentiations and in many cases are elements of a warmth-loving forest fauna which occurs at the northern boundaries of their areas and, for thermal reasons, in the open countryside (cf. also: Becker 1975).

3.1.2 Disjunct Distributions

Species, genera, or family areas which are classified into isolated sub-areas are called disjunct areas (discontinuous distribution). Due to ecological and/or historical reasons, these populations are absent in intermediate areas separating the individual disjunct areas. Since a separation of populations originally in the process of exchanging genes is an important prerequisite for geographic speciation, disjunct species has always been a rather hotly debated biogeographical topic. Depending on the spatial size of the disjunction, we speak of continental, regional, or local disjunctions. Numerous plant and animal families have continental disjunctions. One of the most prominent disjunct dispersal types is the bipolar one.

Taxa with bipolar distributions are absent in the tropics and only exist in higher latitudes of the northern and southern hemispheres (Berg 1933). Bipolar distribitions are found in the following Angiosperm families: Fagaceae, Juncaginaceae, Orobanchaceae, Papaveraceae, Posidoniaceae,

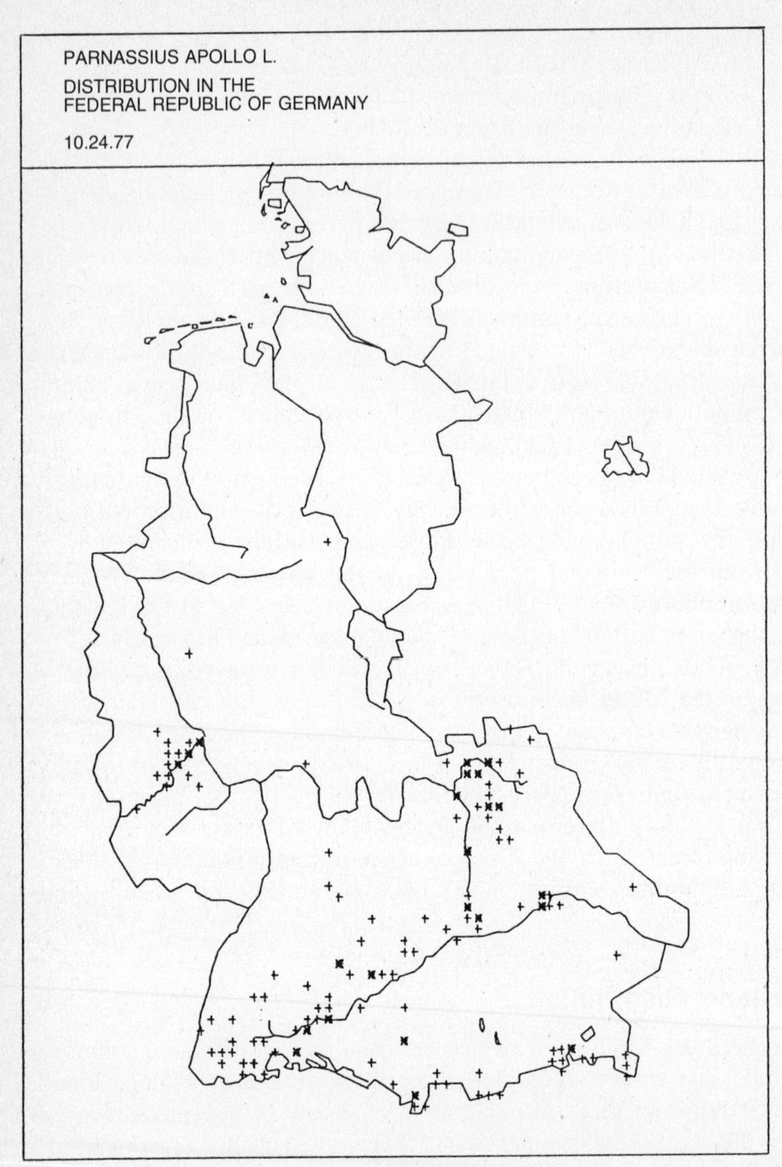

Figure 19 Former (prior to 1960; indicated by a cross) and present (after 1960; indicated by a star) distribution of the appollo (*Parnassius apollo*) within the Federal Republic of Germany. A registry of collaborators checks each individual observation point. The isolated populations north of the Danube are in part characterized as a special sub-species.

Valerianaceae, and Zosteraceae. *Viola, Papaver,* and *Empetrum,* along with a further 60 plant genera, also belong to this distribution type.

A similar distribution type exists in the pelagic and benthic fauna of the ocean; it has, however, been determined that some bipolar distributions are due to the equatorial submergence. The equatorial submergence is a vertical displacement of cold loving fauna in tropical areas, where the surface waters

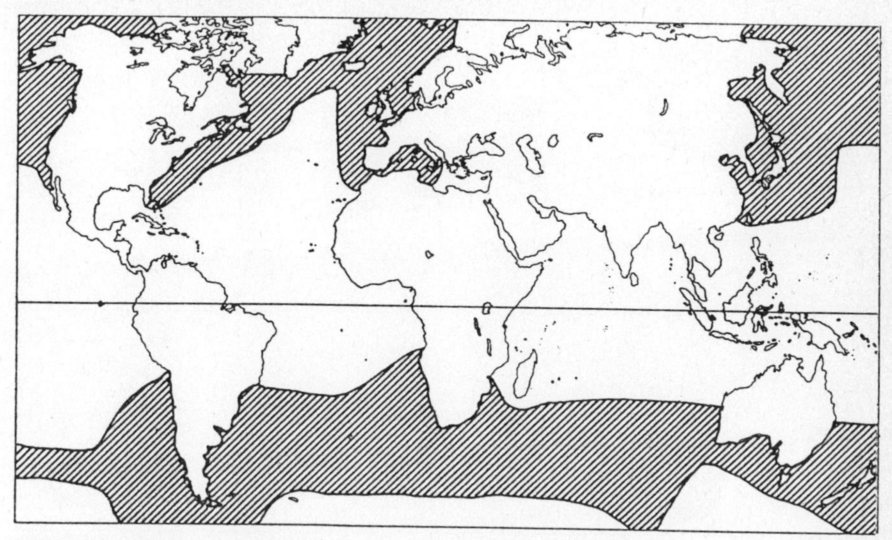

Figure 20 Bipolar distribution of the whale family Balaenidae (according to Rice, 1967).

normally preferred in the higher latitudes are avoided in favour of the deeper, colder water strata at the equator. The copepod *Rhincalanus nasutus* occupies the Atlantic surface waters, at 40°N and 30°S. Between 10°N and 10°S, it occurs only below 1,000 m. The species *Priapulus caudatus* (Priapulida), *Sabella pavonia* (Polychaeta), *Retusa truncatula, Puncturella noachina, Limocina helicina, Clione limacine* (Gastropoda), *Rossia glaucopis, Ommatostrephes sagittatus* (Cephalopoda), *Didemnum albidum, Botryllus schlosseri* (Tunicata), *Lamna cornubica* (shark species), and *Balaenoptera physalus*

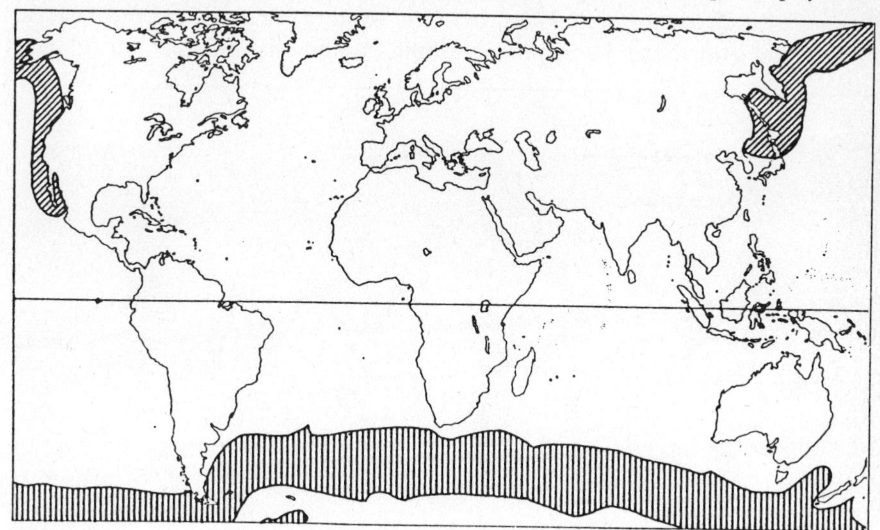

Figure 21 Bipolar distribution of the whale super species *Berardius bairdi* (according to Marcuzzi and Pilleri, 1971). The closely related species *B. arnouxi* lives in the northern hemisphere.

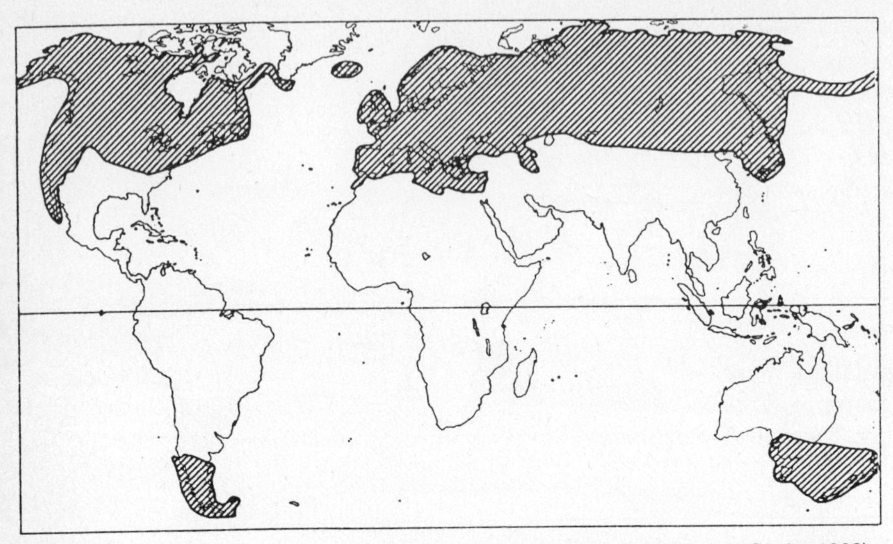

Figure 22 Bipolar distribution of the fish family Petromyzonidae (according to Sterba 1962).

(Cetacea) are further examples of bipolar disjunctions (Figs. 20–23).

Genera with a bipolar distribution include the octopus of the genus *Bathypolypus* and the feather star genus *Promachocrinus*. Up to now, *Sardina* had always been treated as a bipolar genus. It has become apparent, however, that *Sardinella aurita* as well as *Sardinella melanura* and *Sardinops* do cross the equator.

A continental disjunction often observed with plants as well as with animals is the *amphi-Pacific* one. Here, the genus covers an area generally formed by individual isolated species in the western Nearctic and the eastern Palaearctic (northern hemispheric–amphi-Pacific disjunction), or in the

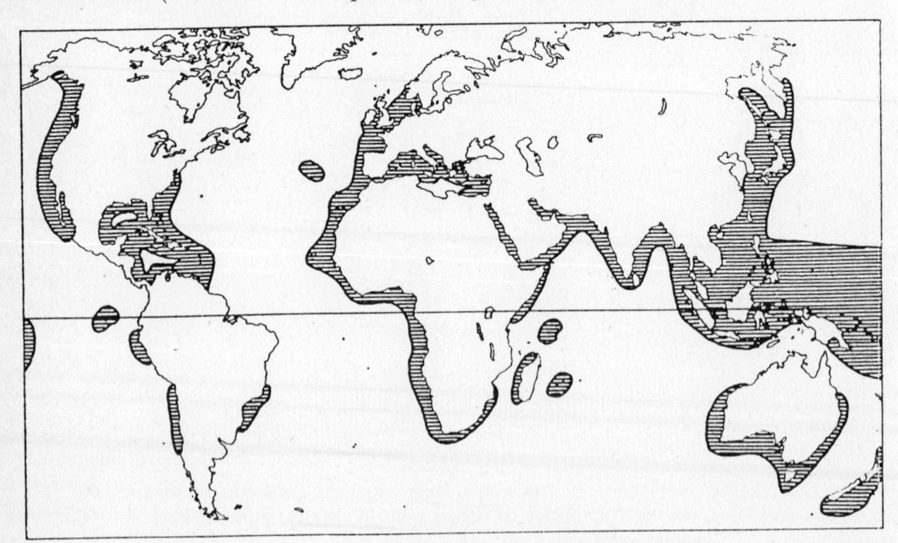

Figure 23 Distribution of the sardine family (according to Müller 1977).

southwestern part of South America, New Zealand and/or Australia (southern hemispheric–amphi-Pacific disjunction). Northern hemispheric–amphi-Pacific areas contain, for instance, the cerambycide (Coleoptera) genera *Plecrura, Callidiellum, Megasemum,* and the species *Leptura obliterata.* An example of a southern hemispheric–amphi-Pacific area is provided by the southern beech genus *Nothofagus.*

3.1.3 Spatial Structure and Differentiation of Populations

The distribution of populations and individuals, the area size of the individual population units, the distribution of the alleles, the chromosomes and genomes of the species endemic to a given area and the intraspecific structure resulting therefrom is of the utmost importance for the population-genetically controlled structural elements of a propagation area.

3.1.3.1 Genetic Differentiation In the final analysis, any change in the occurrence of alleles results in evolution, and, strictly speaking, all biogeographical work deals with the geographical dispersion of genes. Consequently, de Lattin (1967) distinguished between an allele, a chromosome, and a genome geography. It is the task of allele geography to classify the distribution of alleles of one or several gene loci within the dispersal area of a taxon (as an example, see the distribution of the recessive "simplex" allele in *Microtus arvalis;* also cf. Jones 1973). Jain (1975), for instance, showed the importance of such analyses in connection with six prairie plant species, whereby he pointed out the close ties between spatial relationships, population structure, and evolution.

Chromosome geography follows the dispersal of chromosomal mutations, whereas genome geography deals with the dispersal of entire chromosomes (aneuploidy) or of entire sets of chromosomes (polyploidy). Within polyploidy, two types can be distinguished:

Autopolyploidy = duplication or multiplication of the same set of chromosomes.

Allopolyploidy = two different species contribute sets of chromosomes by hybridization; fertility is restored by duplication.

Obviously, the occurrence of polyploid animals within an area must be due to parthenogenetic populations. The beetle *Otiorhynchus dubius,* in its Alpine sub-area, is represented by a diploid, bisexual population, whereas in its northern European sub-area, it occurs in the form of a tetraploid-parthenogenetic population. Further known examples are isopods (*Trichoniscus elisabethae*), butterflies (*Solenobia triquetrella*) (Fig. 23a) planarians (*Dugesia benazzi, Dendrocoleum infernale*), grasshoppers (*Sago pedo*) and the brine shrimp (*Artemia salina*).

Occurrences of polyploidy — particularly in connection with hybridization of species or genera — are of considerably greater importance in plants

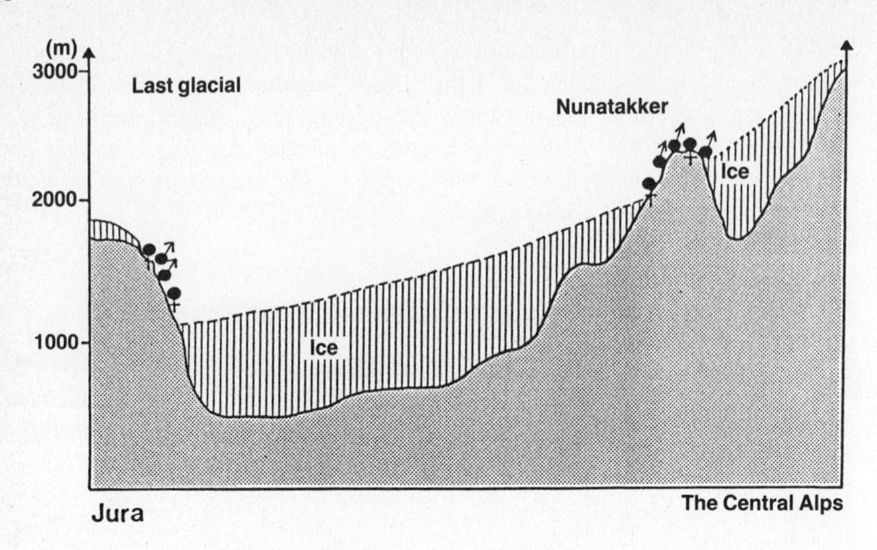

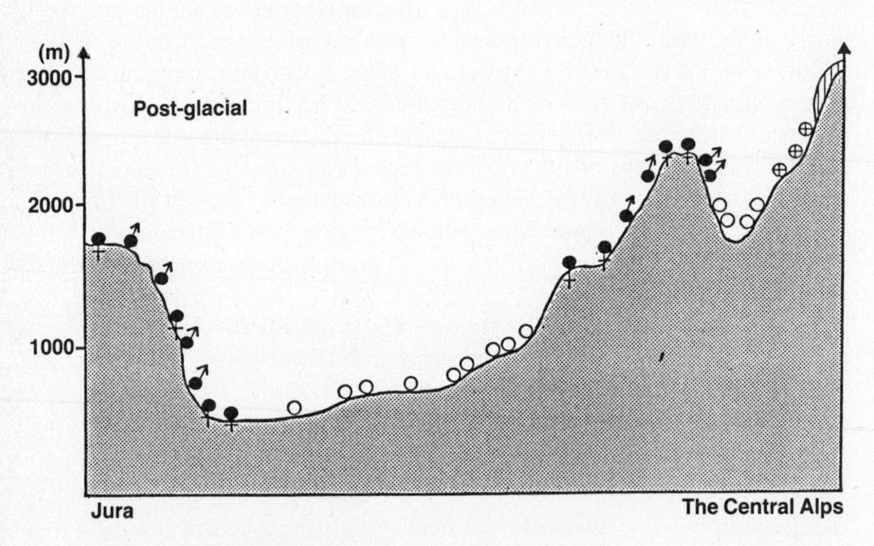

● Bisexual populations

○ Diploid parthenogenetic populations

⊕ Tetraploid parthenogenetic populations

Figure 23a Dispersal of the bisexual, diploid parthenogenetic and tetraploid parthenogenetic populations of *Solenobia triquetrella* during the last glaciation and postglacial periods.

than in animals. The phylogenetic effectiveness of chromosome mutations is particularly evident in some plant genera (e.g. *Oenothera, Godetia, Rhoeo*) (Ehrendorfer 1962, Gottschalk 1971, 1976, Vida 1972).

Within the same genus, polyploid species often inhabit a larger area than do the diploid ones. In the case of *Capsella bursapastoris, Fumaria*

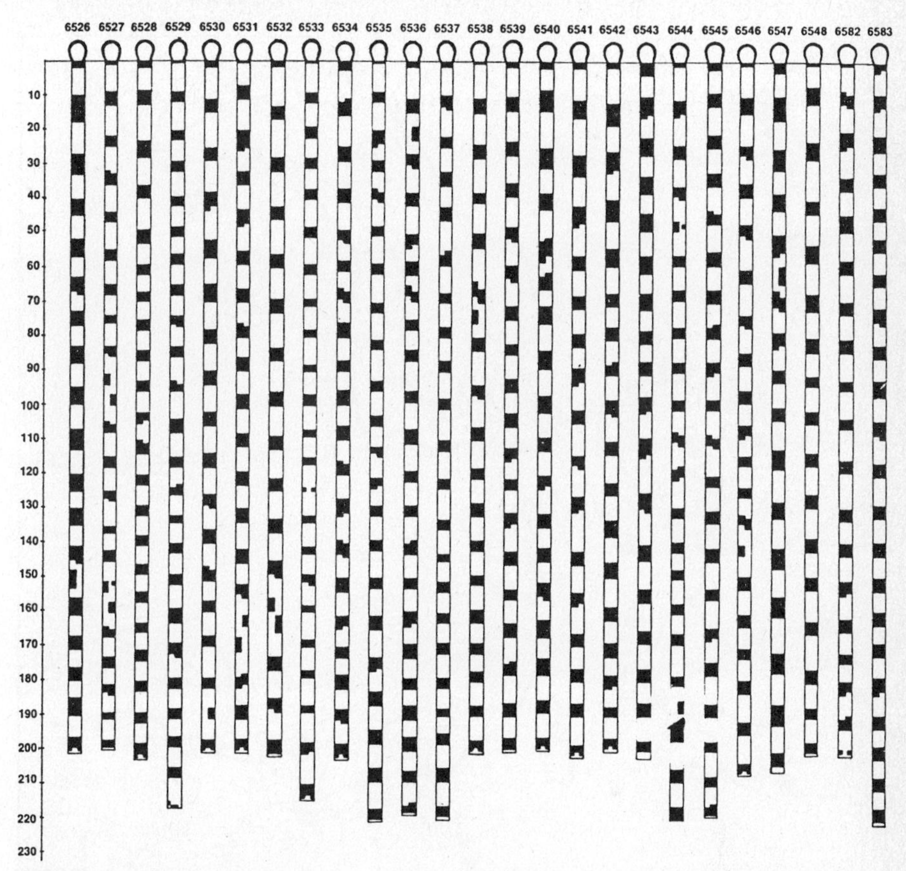

Figure 24 Polymorphism in the South American harlequin snake (*Micrurus corallinus*) on the island of Florianopolis (Brazil). Narrow, black bands alternate with brilliantly red ones. The number of ventral scales is indicated on the vertical axis, and the reference numbers appear along the horizontal axis (University of the Saarland).

officinalis, and *Erodium cicutarium,* the diploid races are only locally significant, whereas tetraploid plants of this type are cosmopolitan.

A close correlation also exists between altitude, latitude, and polyploidy. On the Japanese island of Yaku below 480 m, the liverwort *Dumortiella hirsuta* exists only as 2n or 4n individuals, whereas above 540 m, it occurs only with chromosome sets of 3n or 6n. In the higher regions of the Rocky Mountains, the Portulacaceae species *Claytonia lanceolata* occurs only in tetraploid form. In the case of *Poa alpina,* the populations with the highest numbers of chromosomes occur in the northern part of Sweden and in the high Alpine regions. In tropical India, the diploid representatives of the genus *Commelina* grow as diploids, whereas in the Himalayas, they grow in the polyploid form.

The percentage of polyploid forms often increases with an increase in latitude. The more northern the floral distribution, the more extreme the site, the higher the location of the area, the higher is the proportion of polyploids.

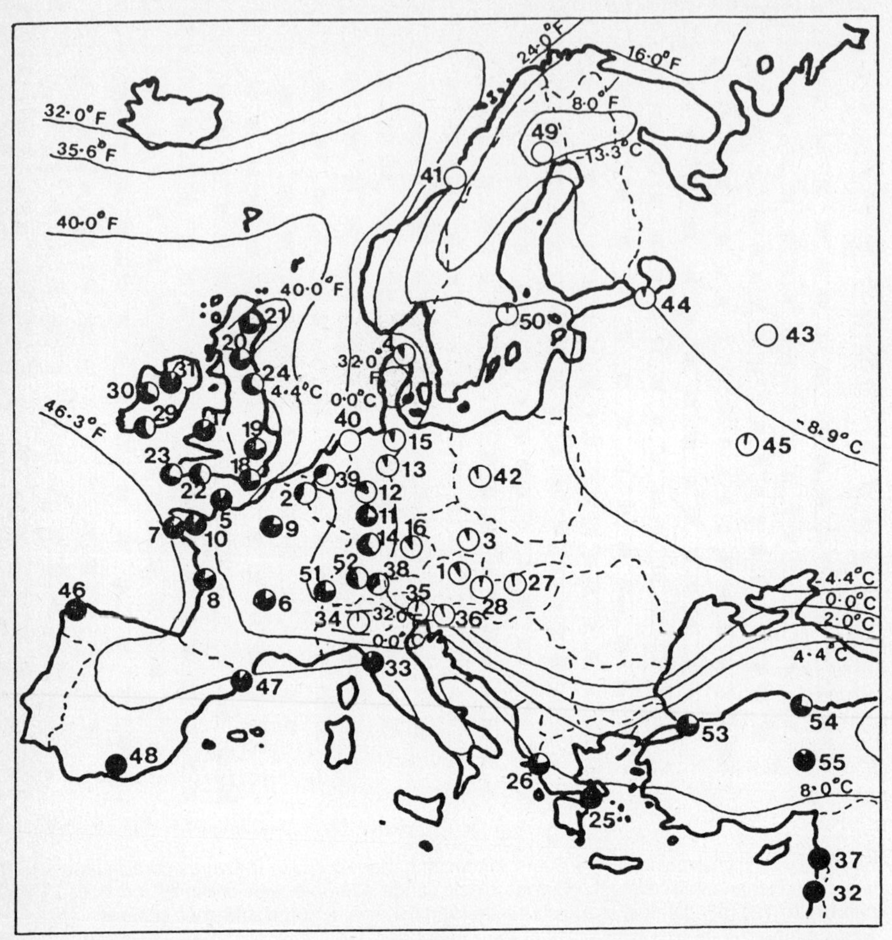

Figure 25 Relationship between the relative frequency of cyanogenic clover populations and January isotherms (according to Jones, 1972).

There are, however, some noteworthy exceptions which point to the fact that, generally speaking, the percentage of polyploids decreases as the age of the flora increases. The younger the flora, the higher the proportion of polyploids (Fig. 26).

Frequently, polyploid plants have different ecological requirements of their habitats than have the diploid ones. In Europe, the *Acorus calamus* is triploid and sterile, whereas in North America, fertile diploid populations can be found, and in eastern Asia, tetraploid and hypertetraploid populations occur. The ecological parameters of the three polyploid states vary greatly, the water supply being the most important factor. Thus, diploid plants on dry sites in the vicinity of Copenhagen were not able to produce spadices for 8 years; after having been moved to a wet site, however, the formation of a spadix did occur.

In the case of *Orchis maculatus,* the polyploid types have a greater

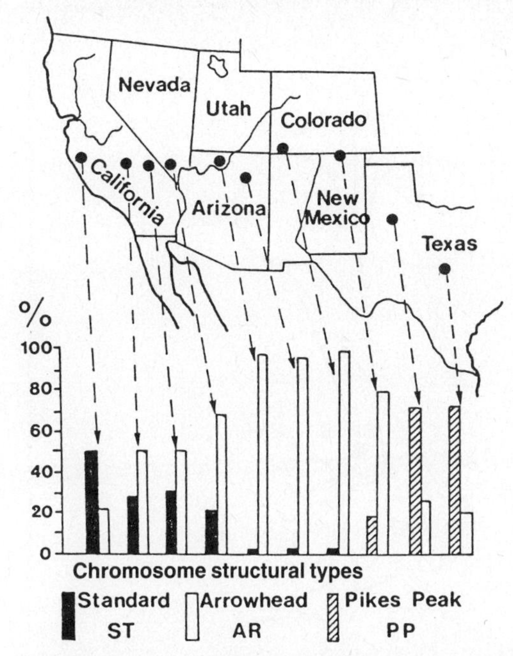

Figure 26 Chromosome structural types of North American populations of *Drosophila pseudoobscura* (according to Sperlich, 1973). The relationship of these structures to their ecological potential might have as a consequence the fact that ecological studies of Texas populations would show results which deviate from those arrived at from a study of Californian populations. Thus, the definition of ecological potential should be made only in conjunction with a genetically defined population.

tolerance than the diploid ones of aridity, wetness, cold and higher salinity of the soil.

3.1.3.2 Population Dynamics In addition to the genetic structure of a population, the biological capabilities of populations directly affect the dynamics of propagation areas (by evolutionary processes at the population level). These include: population growth, population regulation, distribution and dispersal of population elements (Bullock 1976, Myers 1976), the population structure (composition with regard to age and sex), r- or K- strategy population behaviour (Immelmann 1976, Luscher 1976), seasonal and daily rhythms (Saunders 1976) and inter-specific behaviour vis-à-vis other populations (Fig. 28).

Rarely are populations evenly distributed over an area. Their distribution is often influenced by coincidence. Generally, however, it is in part due to the pattern of distribution of habitat islands, to competing species, and to individual assortative and disassortative mating. The habitat islands is dependent on the ecological requirements and capabilities of any one species. The first organisms to arrive obviously choose the optimum habitats within a region. Food, population density, territorial behaviour, sexual partners, enemy pressures, among other things, then result in settlement in sub-opti-

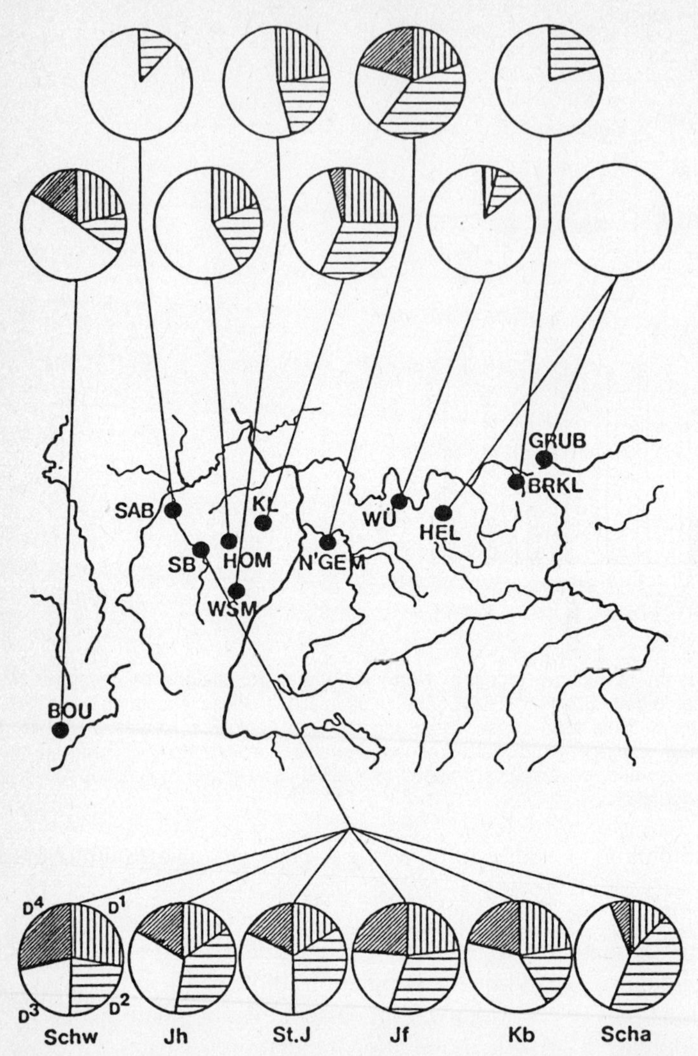

Figure 27 Geographical location of the frequencies of Est-D, -D², -D³ and -D⁴-alleles of *Abax ater* within Central Europe (according to Steiniger 1978). Legend: BOU = Boussenois; SAB = Saarburg; WSM = Wingen-sur-Moder; HOM = Homburg; KL = Kaiserslautern; N'GEM = Neckargemünd; WÜ = Würzburg; HEL = Hellnitzheim; BRKL = Brücklein; GRUB = Grube near Weissenstadt; SB = Saarbrücken.

mum or at least pessimistic habitats. The size of the habitat is specific to the species. Taking into consideration this specific size of habitat and the yearly rate of reproduction, Brüll calculated (most recently in 1977) what he termed "ordinal numbers". For *Microtus arvalis,* for instance, the ordinal number is

$$1 : 5 \text{ to } 100 \text{ m}^2 : (3 \text{ to } 7) \times (4 \text{ to } 12)$$

This means that in the case of the field vole, the female, in a habitat area of 5 to 100 m², will produce, 3 to 7 times annually, 4 to 12 offspring. Brüll (1977) assigned the following ordinal numbers to predatory animals which are

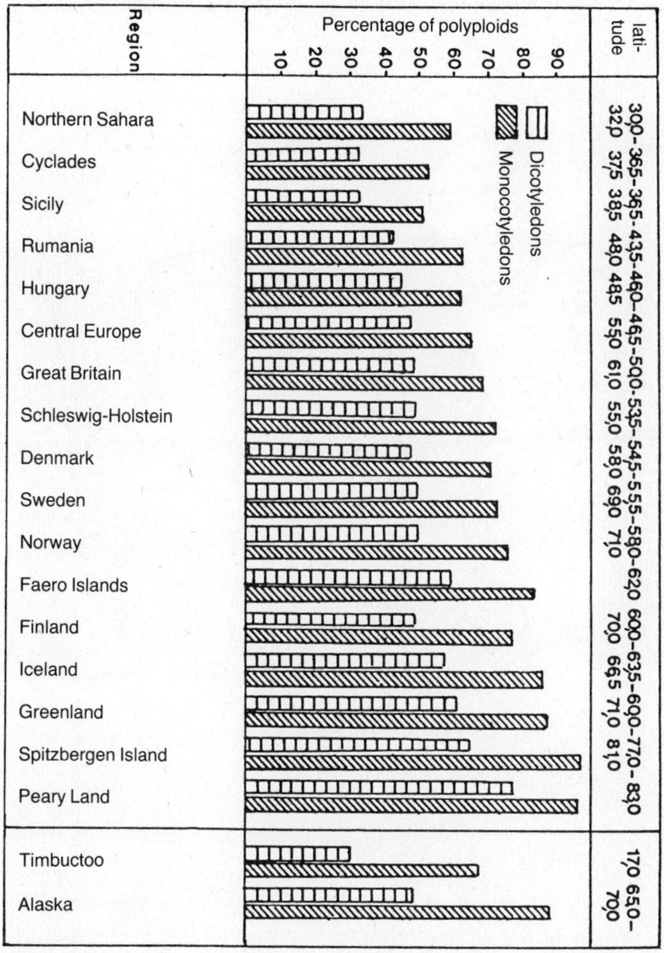

Figure 28 Degree of polyploidy in plants at various global localities (according to Gottschalk, 1976).

related to the high reproductive rate of the small mammals which form their prey:

Mustela nivalis	1 : 4–6 ha : 5–7
Mustela erminea	1 : 8–12 ha : 4–7
Buteo buteo	2 : 80–130–200 ha : 1–3
Martes foina	1 : 100–300 ha : 3–5 (2–7)
Martes martes	1 : 100–800 ha : 3–5 (2–7)
Vulpes vulpes	1 : 500–1500 ha : 3–6 (12)
Accipiter nisus	2 : 700–1200 ha : 3–6
Accipiter gentilis	2 : 3000–5000 ha : 3–4
Bubo bubo	2 : 6000–8000 ha : 2–3
Aquila chrysaetos	2 : 8000–14000 ha : 1–2

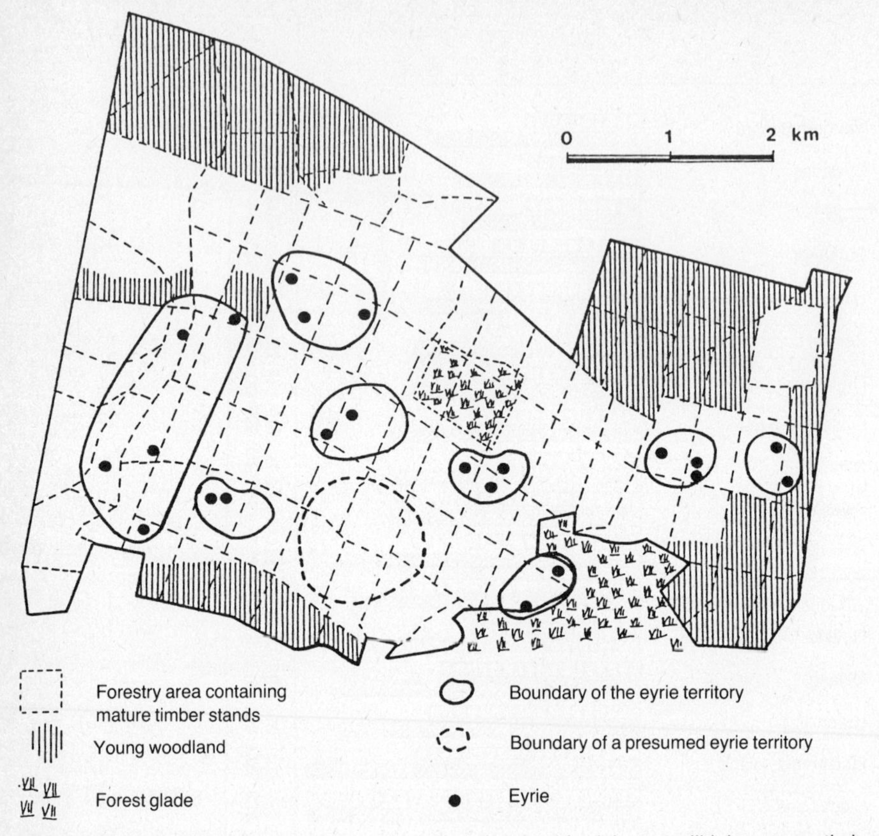

Figure 29 Distribution and territory sizes of goshawks (*Accipiter gentilis*) in mature timber stands of a Polish forest (according to Pieschoki 1971).

Changes in the biotic elements during the course of the year lead to seasonal changes in the structure of the habitat. Related to this is the availability of nutrients for plants and food supply for animals. During the course of their phylogeny, parasites adapted — through specific evolutionary strategies — to these fluctuations. The change in biotic and abiotic parameters is of great significance, since — for example — the seed production of a plant, the breeding success of a predatory bird (Fig. 29) or the survival of young mammals directly depends upon these factors. It is of decisive importance to clarify the population-genetic characteristics and the fitness measures required for the survival of a population. Types of natural selection, fitness of individual genotypes, or the affiliation of populations to the r- or K-strategy of population growth (Pianka 1974, among others) are significant key words in this connection. With increasing size of population, the population density can become the most effective environmental factor (leading to the problem of stable and unstable population sizes).

The description of population growth is usually expressed by means of the Lotka-Volterra equation. It describes constant age-group structures, whereby it is presumed that population growth is determined only by the

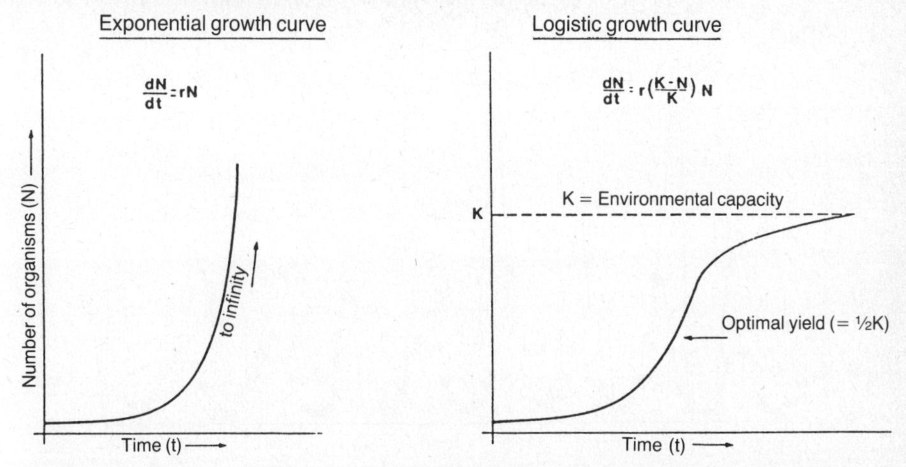

Figure 30 r- and K-strategies are characterized by varying reproductive capabilities. The r-strategy organisms possess exponential, the K-strategies logistic growth curves. Depending on exterior factors, however, certain species fit into the r- and well as the K-strategy group.

parameters r and K (r = growth rate, K = maximum attainable size of population) (Fig. 30). The r-strategies are selectively favored in ecosystems with frequent successions, whereas in ecosystems which settle down to a stable point, K-strategy organisms are preferred. The frequency of a species (i.e. whether it is rare or common) thus becomes an adaptation strategy to its habitat, and therefore that the term 'endangered species' (cf. discussion of the "red lists") can be nonsense.

The connection between polymorphism in populations and certain spatial parameters can only be ascertained by means of population genetics. Polymorphism in a species can express itself both in its structure and in its function (Schmidt 1974). The correlation between population-genetic and ecological factors has been confirmed in recent years by both autecological (Neumann 1974, Thiele 1974) and synecological studies (Ellenberg 1973, Funke 1977). These, however, have neglected to include urban ecosystems, although their analysis clearly shows the physiological limitations and adaptive capabilities of living systems (Gill and Bennett 1973, Li 1969, Müller 1974, 1975, 1977, Stearns 1967, 1971, Sudia 1972, Sukopp, *et al.* 1974, Swink 1974).

3.1.3.3 Species and Subspecies Basic relationships exist between spatial distributions and species or subspecies. Groups of species constitute populations of mutually fertile individuals and represent the largest potential of reproducing communities, which are specific to only their own communities and can therefore be distributed sympatrically (within the same area) with other communities without losing their identity (Fig. 31).

This biological species definition emphasizes the reproductive isolation as the most important criterion, and largely neglects morphological characteristics. In practice, morphological, chromosome-structural, hybridological,

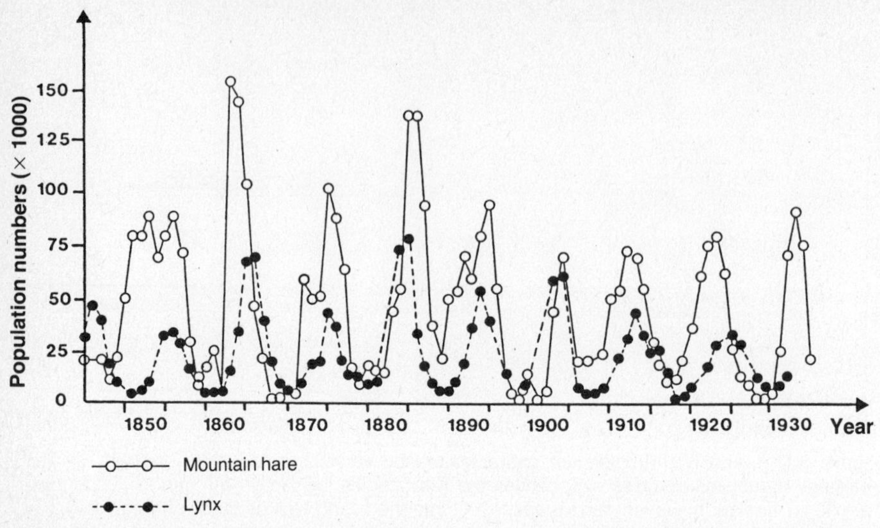

Figure 31 Relationship between mountain hare and lynx populations in Canada (according to MacLulich, 1937).

cytological, genetic, and perhaps also serological and other biochemical criteria must also be taken into consideration, "which, in their entirety, are certainly more likely to limit species as a biological system than is the dogmatic application of individual criteria" (Gottschalk 1971).

In contrast to monotypic species, polytypic species often occupy an area composed of at least two allopatric subspecies areas. Subspecies (geographic races) are populations of a polytypic species, geographically limited to a specific area; these populations must, by definition, differ from other allopatrically (mutually exclusive) distributed populations of the same species in 70% of a characteristic feature.

Since, in contrast to species, subspecies are fertile between each other, hybridization zones are frequently present within the contact region of subspecies areas. Many closely related populations are, however, sharply separated from each other by mountains, seas or other ecological barriers (ecogeographical separation). In these cases it is useful to deal chorologically with subspecies propagation areas and species propagation areas. Semispecies which belong to a superspecies complex are genetically allopatrically distributed. Superspecies are groups which have arisen monophyletically from essentially allopatrically distributed species (semispecies), which are morphologically different and whose cross-fertilization is often severely restricted, so that they cannot be viewed as a single species (Keast 1961, Mayr 1967). *Platanus occidentalis* (eastern U.S.A.) and *Platanus orientalis* (Mediterranean) interbreed and form the vigorous hybrid *Platanus acerifolia.* In natural conditions the two populations do not come together. It is the same with *Catalpa ovata* (China) and *C. bignonioides* (eastern U.S.A.) whose hybrid is fertile. Temporal (for example, seasonal) isolation can give the same result. Hence *Lactuca graminifolia* flowers in spring and *Lactuca canadensis* in summer. If atypical climatic conditions result in overlapping of flowering periods, hybrid swarms develop.

Many animal species which are morphologically, ethologically, or ecologically similar (they can replace each other) may exhibit allopatric distribution patterns.

Subspeciation can be the result of differing ecological, environmental conditions in various regions in the world, of spatial isolation (cf. dispersal center), or often also of both factors. After the long duration of an isolation phase, subspeciation can lead to speciation; this is the reason why subspecies can be interpreted as being species in *statu nascendi.* It must be pointed out, however, that, for the zoogeographical interpretation of an area, the mode of differentiation of the population relevant to that area must be known (cf. dispersal centers).

Races represent a particularly interesting type of specifically differentiated population. This often concerns cryptically colored, subspecifically differentiated populations of a polytypic species which are, for example, dependent on a specific type of soil. The concentration of a particular phenotype at a certain soil location is thus interpreted as the result of selective processes. The formation of a local color congruent with a certain background is an occurrence which can be observed with most varying groups of animals and plants. Lewis (1949) described substrate races of lizards in New Mexico. Benson (1933), Blair (1943), Hoffmeister (1956), and Baker (1960) concerned themselves with the lava races of individual species, Neithammer (1959) and Vaurie (1951) were occupied with the races of bird species. Lorkovic (1974) showed that the light grey *lorkovici* race of the Satyridae *Hypparchia statilinus* in Yugoslavia represents a form of adaptation to the light-colored karst ground. "*Hypparchia statilinus,* therefore, because of the coloration of the under surface of its wings, would be a good ground indicator."

In this connection, the spatially-dependent parallel developments between parasites and their hosts (cf. among others, distribution and differentiation of *Myrsidea* parasites on the different crow races *Corvus macrorhynchos*), as well as the distribution of mimicry complexes (for instance, neotropic Heliconides and their imitators; *Papilio dardanus; Micrurus*) are particularly noteworthy. Certain plant viruses owe their dispersal to the nematodes which function as the transmitters (Lamberti, Taylor and Seinhorst 1974). In the case of brood parasitic bird species, such parallel differentiations have been well examined (cf. among others, Nicolay 1977). The whydahs of the genus *Steganura,* whose males are long-tailed, are brood parasites of the finches (genus *Pytilia*) of which five species inhabit the savannas of Africa (*melba, afra, phoenicoptera, hypgrammica,* and *lineata*). *P. hypogrammica, phoenicoptera* and *lineata* form a superspecies complex. *P. melba* can be divided into 14 subspecies, which in turn can be separated into two groups (one in which the facial mask of the male is divided by a grey strip; and one with a uniformly red head), which, on the basis of the difference in their singing, behave like species in a contact zone. Each *Pytilia* species has its own brood-parasitic whydahs. Distribution patterns of *Pytilia* and *Steganura* species and their differentiation are explained by parallel evolution.

Along the boundaries, border races often occur. In the case of the butterfly of the Euro-Siberian distribution area, there actually are border race centers within the western Palaearctic. The same applies to small mammals and reptiles.

The intraspecific variability and modes of differentiation of the Central European wall lizard populations described as "subspecies" have not been explained satisfactorily, in spite of numerous examinations, particularly along the boundaries. The question as to whether the subspecific differentiation is primarily a result of refugial isolation or of location-dependent selection conditions — or the question as to how far both of these factors have an impact — is still open to investigation.

In the case of the Breton lizard populations, for example, these population-genetic questions can only be answered by detailed cross-breeding experiments. Such experiments proved particularly interesting in the region of northwestern France, since varying ecological conditions as well as border conditions favorable to the isolation of populations are present there and the tendency towards the development of border races is quite distinct for vertebrate as well as invertebrate populations (Richards 1935, Zimmerman 1950). Attempts at explaining the materialization of these border races extend from an indirect dependence on the recent climate to "Postdispersal Isolation" (Müller 1977).

Experiments have shown that populations occur along the northwestern boundary of the limit of *Lacerta muralis* which deviate from each other in their characteristics. They can be summarized in three similarity groups:

Group A *L. m. calbia* Blanchard 1891
Group B *L. m. oyensis* Blanchard 1891
Group C *L. m. muralis* (Laurenti) 1768.

Whether we agree with such a classification or whether we consider the phenotypic distances too short to separate *calbia* and *oyensis,* in the final analysis depends on the question as to what morphological level we apply within one species in order to define the subspecies (Mayr, Linsley, and Usinger 1953). Since criteria of relationship can only be determined in relation to time and not to the phenotypic similarity, an opinion on this question seems to be mainly of taxonomic interest. It should be recognized, however, that a "subspecies" developed in refugial isolation can have a completely different genetic population from a peripheral border race, characterized under certain circumstances by an extreme elimination of the allele (Müller 1973, p. 232).

3.1.4 Co-Evolution of Plants and Animals

Just as the evolution of numerous parasites can only be understood in terms of the evolution of their hosts (cf. among others, Franke 1976), the examination of symbionts is also required. For example, the ringed *Boletus* is dependent on birch trees and the pungent *Russula* dependent on beech trees. The phylogeny of numerous plants and animals can, therefore, only be understood in terms of their co-evolution.

The types of adaption of flowers visited by insects, birds, or bats are numerous. Many Neotropic Amaryllidaceae genera are characterized by frequently recurring divergence into sphingophile-ornithophile species. *Hippeastrum calyptratum* found in the mountainous forests of the Serra de Mar is chiropterophile as are many species of the genera *Burmeistera, Centropogon,* and *Siphocampylus* (cf. Vogel 1969). Chiropterophile species also occur among the Bromeliaceae (among others, *Vriesia bituminosa, Vriesia norrenii*). Some plants avoid being eaten (cf. among others, Morrow *et al.* 1976; cf. life forms), while others have developed specific mechanisms to react against certain enemies. For example, as a consequence of the infestation of the carrot by *Heterodera schachtii,* the carrot grows an excessive number of roots, the so-called carrot beard, which, however, absorbs insufficient nutrients; the plants therefore turn yellow. In addition to the carrot, the turnip and beet, etc., are also prone to infestation. The actual existence of hostile species is important to the presence of the defence mechanism. The roots activate the larvae resting within the cysts, but they are unable to develop any further. Lucerne and rye are other examples of hostile plants.

3.1.5 The Dynamics of Propagation Areas

3.1.5.1 Basic Problems All the elements within a propagation area are in the process of constant change. Changes within the population-genetic structure lead to spatial changes, for instance, in the form of area expansions or regressions. We have applied the terminology proposed by Sukopp (1972, 1976), concerning the temporal and spatial features of area expansions or regressions of plants, also to animals, in spite of group-specific peculiarities (Müller 1976). The following terminology for the features of spatial changes has, therefore, proved appropriate:

1. Period of immigration.
1.1 Native (prior to the period when man started interfering with the landscape).
1.2 Immigrated prior to 1500 A.D. (Archaeophytes, etc.).
1.3 Immigrated after 1500 A.D. (Neophytes, etc.).
2. Country of origin (categorizing organisms by their centers of dispersal).
3. Direction of immigration.

   The country of origin and the direction of immigration are closely related. This applies in particular to numerous Siberian, Atlanto- and Pontomediterranean faunal and floral elements. The direction of immigration, however, does not always necessarily indicate the country of origin.

4. Speed of immigration.

5. Method of immigration.
5.1 Without human interference.
5.2 With human interference.
5.2.1 Accidentally transported.
5.2.2 Deliberately introduced species.
5.2.3 Species which invade because of human activity.
6.1. Limited to natural or semi-natural ecosystems.
nited to natural or semi-natural ecosystems.
6.2 Occurring in ecosystems either totally or partially created by man.
6.2.1 Epecophytes and epezoes. Weeds and pests.
6.2.2 Ephemeral plants and animals.
6.2.3 Useful species.
7. Period of decline.
8. Direction of decline.
9. Speed of decline.
10. Method of decline.
10.1 Without human interference.
10.2 With human interference.
10.2.1 Direct interference.
10.2.2 Indirect interference.
11. Degree of decline.
11.1 Extinction of a species.
11.2 Extinction of the populations within a country.
11.3 Extinction of the populations within certain parts of a country (for example of the Federal Republic of Germany).
11.4 Extinction of or change in the allele diversity of a population (for example, in the Federal Republic of Germany).
11.5 Relict areas.

Dispersal can be described on the basis of the dispersal speed (by determining the size of a newly-inhabited area in km^2 within the period of, say, one year, and expansion of the population by determining the average size of the newly-inhabited area per year in relation to 100 km of the actual boundary of the expanding area, i.e. dispersal front; cf. Nowak 1975). Recent spatial expansions of a great number of plant and animal species are known (among others, in Europe: serin finch, redstart, woodpecker, green wood warbler,raccoon dog, burnished brass moth, fritillary; and numerous Campo species in South America, cf. Müller 1970), although the causes differ widely. Population structure, dispersion (distribution of the individuals within the area), dispersal (modes of movement leading to dispersion), ecology and changes within the genetic structure of the population, are of prime importance. Dispersal capability can be easily studied, by recording the time taken for newly-created areas to be settled (den Boer 1973).

Species whose areas are known to have been influenced by man are termed hemerochores (cf. Sukopp 1972, 1976). A further differentiation can be made according to the date of their immigration, the manner in which man contributed to the immigration, and/or the degree of their integration into the present ecosystem. Of the 2,338 plant species existing in Germany, 16% are considered to be hemerochorous; the percentage for the 1,908 plant species present in the British Isles is also 16%, and for the 1,227 occurring in Finland, 18% (cf. urban ecosystems).

3.1.5.2 Animal Migrations Animal migrations are changes in location by animal populations which have ecological as well as genetic causes. From an ecological point of view, this seasonal change in the ecosystem or region is of great importance not only to the migrating species but also to the ecosystem serving as their location at any one time. Thus, varying enemy pressures on young birds of dominant tropical species, together with a changing food supply, probably control the species diversity of tropical forest birds to a greater extent than is the case in extra-tropical areas. Migrations from the tropical and extra-tropical areas, therefore, may also be due, in part, to an evasion of enemy pressures during the breeding period.

The orientation and navigation ability of some species during their migrations is one of the most amazing performances of living organisms.

Plankton also perform regular, vertical migrations; they frequently rise to the ocean surface at midnight, and then descend into the depths during the morning hours. The Pacific palolo worm (*Eunice viridis*) undertakes migrations which are coupled to the lunar phases (lunar periodicity). The Florida palolo (*Eunice furcata*) appears regularly three days prior to the last moon quarter around July 28. Similar seasonally dependent vertical migrations are also known in the case of mountain animals (for example, the Virginian deer in North America; the chamois and the red deer in the Alps; and the Andean hummingbird). Savanna (gnus and zebras), steppe (bison), and tundra (reindeer) animals also undertake long migrations. In South Africa, the passage of the springboks used to be anticipated with great fear.

Numerous marine organisms have long migrations. This also applies to the nocturnal Caribbean lobster (*Panulirus argus*), which, by forming long chains, can pass through shallow water (Emmel 1976). The herring, tuna, haddock, and cod belong to a group of fish which regularly migrate. Eels are catadromous fish, i.e., initially, they live in fresh water and only visit the ocean (Sargasso Sea) during the spawning period. Conversely, salmon are anadromous fish which ascend the rivers from the ocean to spawn. With its sense of smell, each salmon finds the same river in which it itself hatched from an egg.

Numerous additional fish species are considered anadromous (for example, *Petromyzon marinus, Caspiomyzon wagneri, Lampetra fluviatilis*). Among the sturgeons, *Huso huso* which supplies the famous Beluga caviar; *Acipenser gueldenstaedti,* which ascends the Danube as far as Bratislava; *Acipenser naccari, Acipenser nudiventris* and *Acipenser stellatus* migrating from the Adriatic Sea, all belong to this group, just as much as the true sturgeon, *A. sturio*. (In 1890, 2,800 of these fish were still caught in the Elbe River but the number caught in 1918 had dropped to 34.) The Cyprinidae also include anadromous species (among others: *Rutilus frisii, Barbus capito*), as

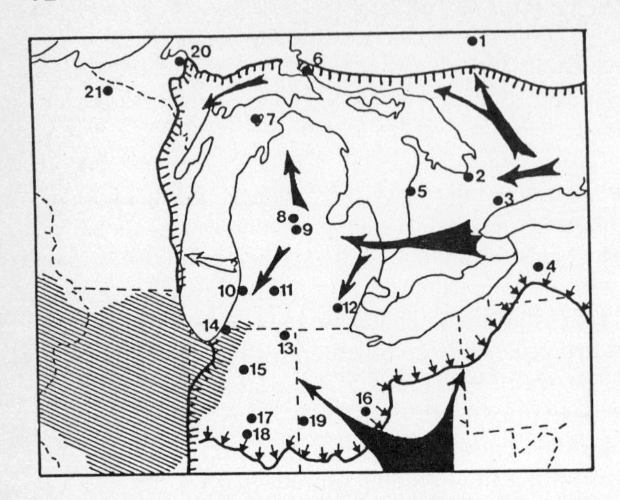

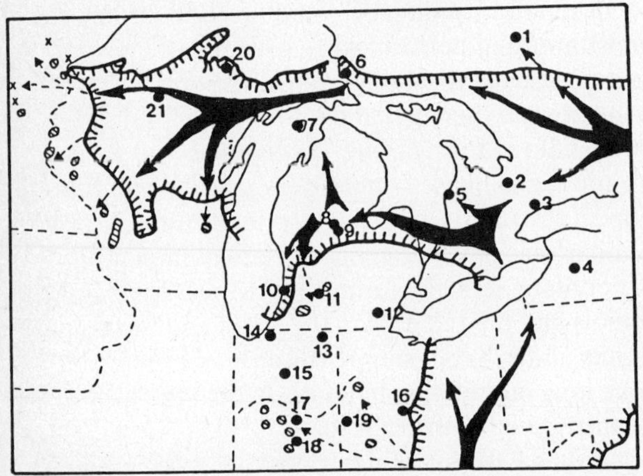

Figure 32 Postglacial migration paths of the American beech (*Fagus grandifolia*) and the hemlock fire (*Tsuga* canadensis; according to Fowells 1965).

do the Clupeidae (*Alosa alosa, A. fallax, A. pontica, Clupeonella delicatula*), the Salmonidae (*Salmo salar, S. trutta labrax, S. trutta caspis, Salvelinus alpinus*), the Coregonidae (*Coregonus oxyrhynchus, Stenodus leucichthys*), and the Osmeridae (*Osmersu eperlanus*). In the case of the Brazilian mugilides (which in Brazilian is called "Tainha"), Hans Staden had already observed their anadromous migratory behaviour in 1548 and described the effect upon the mobility of Indian tribes. Stationary as well as anadromous populations may occur within a species.

Migrations may involve a major change in habitat. Seals, penguins, and sea turtles go ashore for reproductive purposes, whereas land crabs (*Gecarcinus, Birgus latro,* and others) migrate into the sea. The amphibians' passage leads from the land to fresh water. In this connection, strict loyalty to a particular location, i.e., to spawning pools, for example by toads (*Bufo bufo,* among others) has been observed. North American newts of the genus *Taricha* find the river section where they were born, year after year.

In antiquity, great attention was paid to the migration of birds. Bird migrations must certainly have existed since the early Tertiary. The migration paths of the recent migratory birds, however, are in all likelihood extremely young. "All such adaptations must be the product of evolution in something like the last 10,000 years, a conclusion shattering to much current evolutionary theory" (Moreau 1972).

In the case of the Norway lemming (*Lemmus lemmus*), as well as the forest lemming (*Myopus schisticolor*), a distinct seasonal change in biotope occurs, which is closely related to the seasonal change in their most important food plants. The mountain lemming is thus known by its famous long-distance migrations (Kalela 1963; cf. Tundra biome, Chapter 5.1.8).

The monarch butterfly (*Danaus plexippus*) which reproduces in spring in northern North America, migrates in fall in enormous droves to its wintering territory along the Gulf of Mexico. Trees serving as a resting place for the animals are revisited with great regularity year after year, and in many states, are protected by wildlife conservationists. In the following spring, many animals return for reproduction to the U.S.A. and Canada. Similar migrations are also known in bats. The migrations of European butterflies are usually spread over several generations. These include, among others, the painted lady (*Vanessa cardui*), the red admiral (*Vanessa atalanta*), the lambda moth (*Autographa gamma*), the clouded yellow (*Colias hyale*), the death's-head moth (*Acherontia atropos*), and the sweet-potato hawkmoth (*Herse convolvuli*). Regular migrations which may be due to various causes are also undertaken by dragonflies (among others: *Sympetrum fonscolombei, Hermianax ephippiger, Aeschna affinis*), and millipedes (for example, *Schizophyllum sabulosum*).

The mass migrations of the Pallas sandgrouse (*Syrrhabtes paradoxus*), the nutbreaker (*Nucifraga caryocatactes macrorhynchus*), and the crossbill (*Loxia*) occur at irregular time intervals with expansions into Central Europe. During invasion years, immense flocks of *Nucifraga caryocatactes macrorhynchus* migrate westwards from Siberia. The cause for these migrations primarily lies in the availability of a particularly favorable seed crop of the coniferous trees in the invasion areas; the population density is also a contributory factor. Irregular precipitation results in mass reproduction and the consequent invasion by migratory locusts into savanna and steppe regions (cf. Savanna Biome, Chapter 5.1.2).

3.1.5.3 Drifting A general characteristic of migrations is the fact that only in rare cases is the actual reproduction area expanded considerably. Spatial changes due to passive dispersal mechanisms are, however, a completely different matter. Numerous plants and animals use this method to expand their area.

Two types of drifting can be identified in plants (Ridley 1930): autochory (formation of shoots, scattering of seeds; propagation through falling due to gravity: barochory), and allochory (seeds and fruit passively dispersed by other means).

In the case of steppe plants, the wind tears loose entire bushes and then blows them across the steppes in the form of tumbleweeds (for example, *Crambe, Eryngium, Falcaria, Seseli, Phlomis, Centaurea*). The same is true of the ball-like pieces of thallus of the manna lichen (*Lecanora esculenta*) which roll over the ground, again blown by the wind. *Lecanora atra* and *L. calcarea,* can also be dispersed by the wind (Bailey 1976). Collembola often carry away the soredia of *Pertusaria amara.* Many of the littoral lichen species live epiphytically on animals and are thus carried off. For instance, *Arthropyrenia halodotytes, Verrucaria maura, V. microspora, V. mucosa* and *V. striatula* are frequently found on *Littorina* and *Patella.* Giant turtles of the Galapagos often transport lichen of the species *Dirinaria picta.* Numerous dust-like, small seeds and sporules (mosses, ferns, basidiospores of the pileus fungi, conidia and ascospores of the ascomycetes, seeds of orchids and the tobacco plant), or seeds with special drifting mechanisms are dispersed through the air. Others have adapted to water as a means of transportation. Their ability to float is attained by air-conducting cells (water plantain and arrowhead, for example), by large intercellular spaces (marsh marigolds), or by special flotation vesicles (water lily and sedges, among others). Darwin showed that the seeds or fruit of various plant species can float in fresh as well as saltwater for long periods of time without losing their germination capacity. Rode (1913) arrived at similar conclusions.

Of particular interest are plant seeds which have adapted to animals for the purpose of dispersal. Similar observations were made by Theophrastus, Pliny and Darwin. Griffin vultures (*Gyps fulvus*), via their prey (for example, Rodentia), become the transmitters of seeds which were eaten by the prey prior to being killed. Balgooyen and Moe (1973) proved this for the grass species *Calamagrostis canadensis* whose seeds -- ingested by rodents — were transported by *Falco sparverius* for distances of up to 440 km.

This method of dispersal includes such seeds as the burdock fruit which hooks into the hairy coat of animals by means of barbs or claws (corn crowfoot, burdock, avens). A particular form of these epizoitic species are the burr grasses of steppe and desert regions which are transported by hanging onto the legs of mammals. Other seeds or fruits adhere to animals by means of a stickly slime and are thus dispersed (Ridley 1930; Pijl 1957). A synzoitic method of dispersal is attained by those animals which carry away fruit in order to eat it undisturbed, or to store it away as food provision (such as hamsters, squirrels, mice, jays, nutbreakers, woodpeckers, etc., cf. Holtmeier 1966). The myrmecochores (common celandine, violet, and others) always carry light, oily appendages (elaioplasts) on their seeds which are eagerly eaten by ants which then act as the dispersers. Endozoochorously dispersed fruit is usually quite conspicuous because of a tempting colour or juicy fruit pulp, while a hard shell protects the germ against digestion (berries, stone fruit, mistletoe, and others).

Passive dispersal of animals is primarily accomplished by means of the wind, running water, transport by other animals (phoresy, zoochory), or by man (Fig. 33). Book scorpions of the genus *Chelifer,* for example, are regularly transported by flies. Swamp or water fowl (ducks, for instance) often carry away the spawn or the developmental stages of aquatic organisms. Usually, it is only the small and light plants and animals that are suited for transportation through the air (aeroplankton). Air currents are of special importance to passive dispersal of insects, because normally, the greater the air speed the further insects can be carried. Aeroplankton, drifting in the wind can reach a height of 4,000 m. The vertical dispersal of aeroplankton can be studied using samples taken with a net from an aeroplane. Dispersal directions (passive as well as active ones) have been identified by means of samples collected from oil drilling rigs and lightships in the ocean. In the collecting trays set up on the lightships in the North and Baltic Seas, up to 95% of diptera found had come predominantly from ecosystems close to the coasts. The flight distance correlated with the various wind directions and speeds (Heydemann 1967). The syrphid fly (Syrphidae; for example, *Syrphus ribesii, S. balteatus, Tubifera trivittata*) were caught over the open sea (Weidner 1958, among others).

The aphid *Cinara abieticola* regularly appears with the appropriate weather situation at Spitzbergen more than 1,000 km from its closest occurrence on the Kola peninsula. Down currents finally force the aeroplankton to the surface and in many regions cause both a daily and hourly prolonged bombardment of the ground with small insects. For many insects, long-distance travel between seas is known (Visher 1925; Bowden and Johnson 1976). Even larger, generally flightless species can be carried by the wind far beyond their own area. Of the 361 bird species confirmed as living on Heligoland (409 subspecies, cf. Vauk 1972) only 18 are recorded as regular breeding birds, whereas more than double that number of species reach the island during certain weather conditions as stray guests. Passive drift through

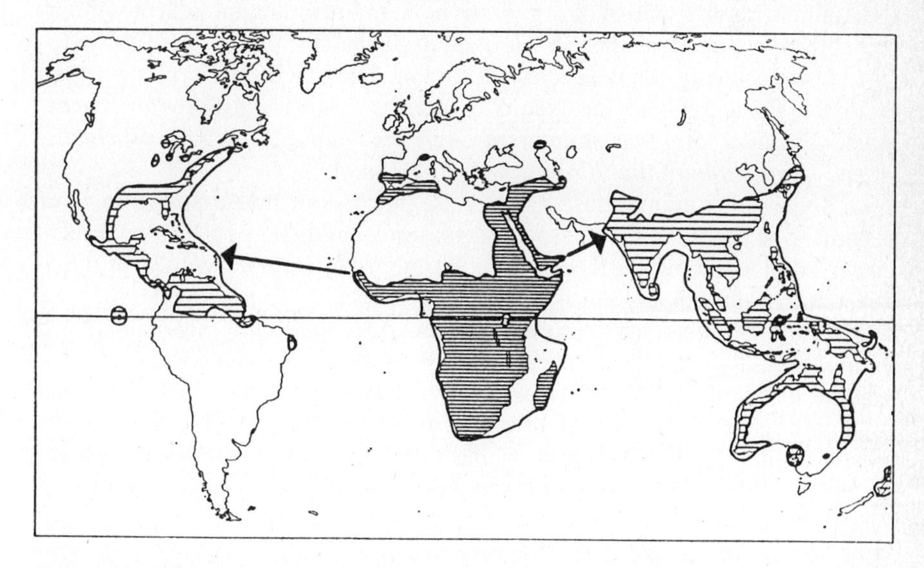

Figure 33 Present distribution of *Ardeola ibis.* The original Ethiopian area has been close hatched, and the area settled through expansion during historical times has been broad hatched.

the use of wind and water is called anemohydrochory. Palmer (1944) studied the phenomenon in detail. He examined the margins of rivers and lakes which had insects washed-up (density of the washed-up insects: about 4,000 individuals/m^2) and was able to demonstrate that they were insects of the Baltic area from southern Finland (100 km) some of which had survived several days in saltwater. The regularity of the anemohydrochorous dispersal in this region is the reason for the so-called "Baltic direction of immigration of many Finnish insects."

Generally only animals which live in water are dispersed by the hydrochorous method. Floating materials serve as a means of transportation for terrestrial species (flotation theory). In many cases, however, there have been very few experiments to test the effectiveness of passive drift over water by terrestrial vertebrates, though there have been some chance observations. A series of experiments with lizards performed both in the laboratory and in the open field (Mediterranean) confirm the existence of considerable differences in the swimming ability and orientation capacity of individual species. At water temperatures of 21°C, *Lacerta muralis* (wall lizard) as well as *Lacerta viridis* (green lizard) were able to remain afloat in the water for more than 60 minutes without ill-effect. In the case of the larger *Lacerta galoti* living in the Caribbean Islands, however, the experiments had to be stopped after 15 minutes to keep the animals from drowning. The large South American land tortoise *Testudo carbonaria* which can grow to a length of 70 cm, is also capable of swimming in water for 24 hours. This fact is all the more noteworthy as this species' closest relative is the Galapagos giant tortoise.

3.1.5.4 Introduction and Dispersal by Man Under dispersal we understand the unintentional dispersal of organisms by man; introduction indicates intentional dispersal. Often, there is an overlap of the two processes (Table 8).

The following plant species (excluding Dicotyledons) were dispersed from the Palaearctic to the Nearctic: *Acorus calamus, Agropyron repens, Agrostis canina, Agrostis stolonifera, Agrostis tenuis, Asparagus officinalis, Avena fatua, Lolium multiflorum,* and *Poa annua.*

Because of man, house rats, house mice, and house sparrows have become cosmopolitan. Numerous reptile and amphibian species have also dispersed. The large South American marine toad, *Bufo marinus,* was introduced into Cuba, Haiti, Eastern Australia, and New Guinea for the purpose of controlling insects, whereas the small gecko *Hemidactylus mabouia* arrived unintentionally in South America on the first slave ships from Africa; it is now a domestic animal in South America. Five additional geckos were dispersed by man beyond their original area into various continents. The gecko *Cnemaspis kendalli,* endemic to Indo-China and to the Indo-Australian archipelago, was introduced into New Zealand. *Gehyra multilata* of Eastern Madagascar, Sri Lanka, and Oceania came to Mexico (via the sea ports of Nayarit and Sinaloa); *Gonatodes albogularis,* native to the northern Neotropic, today also occurs in Florida; *Hemidactylus frenatus,* originally from India and Oceania, now belongs to the Herpetofauna of South Africa, and *Hemidactylus turcicus,*

TABLE 8 NUMBER OF FLOWERING PLANTS INTRODUCED INTO CENTRAL EUROPE (according to Sukopp 1976)

Trees and shrubs	2650
Herbaceous ornamentals, including bulbous plants	2000
Agricultural plants	100
Field and garden weeds	150
Grass seed — newcomers	52
Bird food mixtures	230
Grain mixtures	several hundred
Tropical fruit mixtures	
Adventitious plants	1600
Others	?

often found in the Mediterranean region, has been introduced into North America and Mexico. *Testudo marginata,* common to some regions of Sardinia, arrived on the island from southern Greece. The brook trout, originally limited to the western part of Eurasia, was released for economic reasons in North America, Chile, Argentina, South and East Africa, Madagascar, Australia and New Zealand.

Nearctic fish species, such as the rainbow trout (*Salmo gairdneri*) belonging to the Salmonids, and the brook trout (*Salvelinus fontinalis*), the largemouth black bass (*Micropterus salmoides,* of the Woerthersee, for example), and the sun bass (*Lepomis gibbosus*) assigned to the Centrarchidae (sunfish), the catfish (*Ameiurus nebulosus*) and the American mud fish (*Umbra pygmea* of the ponds of Schleswig-Holstein and Lower Saxony, for example) have been introduced into many European waters (Fig. 34).

For the same reason, man also dispersed the oyster (*Ostrea edulis*)

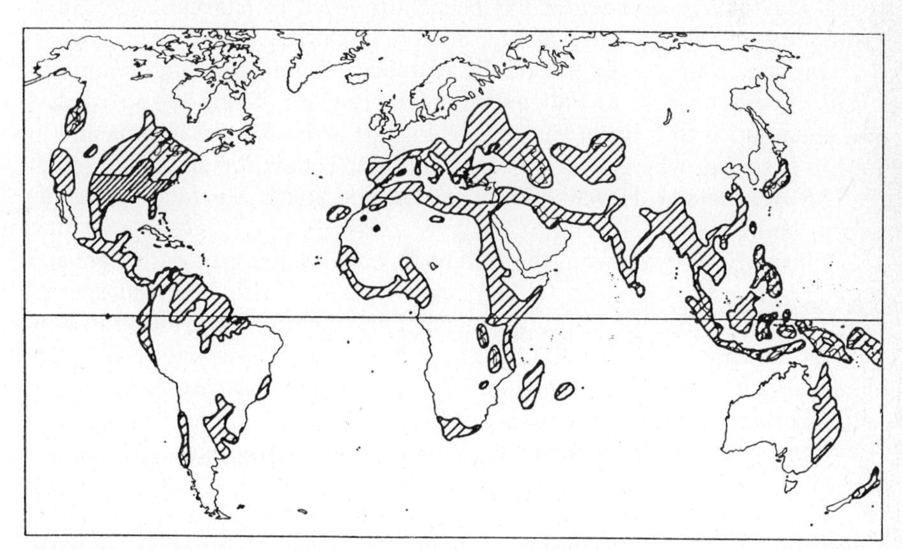

Figure 34 Original (close hatched) and present areas of the mosquito fish *Gambusia affinis.* This species was introduced into many tropical and subtropical waters in an attempt to control mosquitoes.

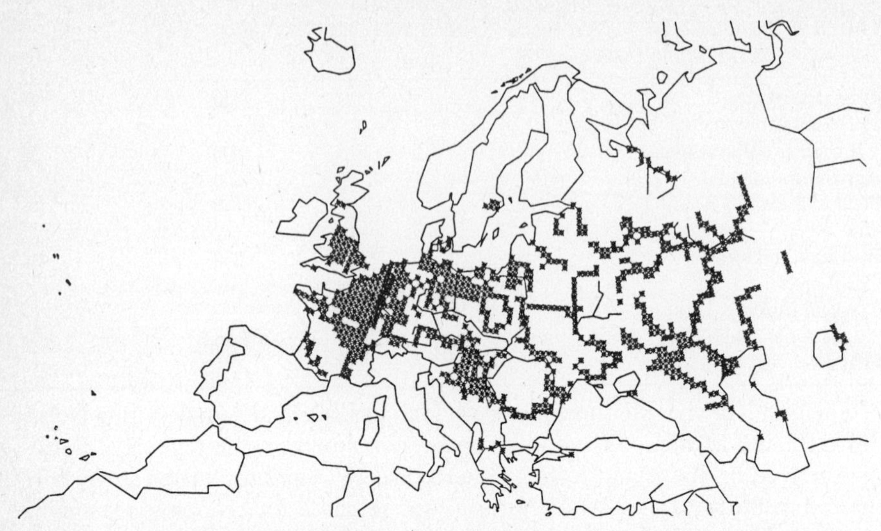

Figure 35 Distribution of the migratory mussel, *Dreissena polymorpha* in Europe.

beyond its original area. Due to the crayfish disease, caused by the fungus *Aphanomyces astaci* originally native in North America, many European brooks and rivers became devoid of crayfish in the second half of the last century. While North American crayfish species are largely resistant to the fungus, the European, Asian and Japanese populations quickly perished when attacked. For this reason, nearly one hundred years ago the North American *Cambarus affinis* was introduced, which replaced the Central European *Astacus fluviatilis* in many areas. For approximately the last ten years, *Pacifastacus leniusculus* has been introduced in Europe, where it is being bred, in Sweden for example, in large quantities.

In Roman times, the eastern Palaearctic common pheasant (*Phasianus colchicus*) was brought to Italy and southern England. Since 800 A.D., it has also been native to Central Europe. In 1523, it arrived in St. Helena and in 1667 in Madeira, where it later became extinct. Today, it is being hunted in the U.S.A., Canada, Hawaii, New Zealand, and South Australia, as well as Cyprus and Chile.

Of the European fauna, the wild boar, brown hare, grey partridge, and starling were exported to North America. Originally North American species which were able to settle in Europe included the potato beetle, muskrat, grey squirrel, and raccoon (Fig. 37). Multiple dispersion is also known in many animal species. Although the number of mammals and birds with which naturalization was attempted in Europe was large (Niethammer 1963), only a few of them have actually been able to adapt to the ecosystems there (Table 9).

Modern transportation also contributes to dispersal. Dipterous insects and pathogenic micro-organisms constantly hitch rides over the Atlantic on board jets.

Lepidoptera, particularly Noctuidae, can often be found in railway cars

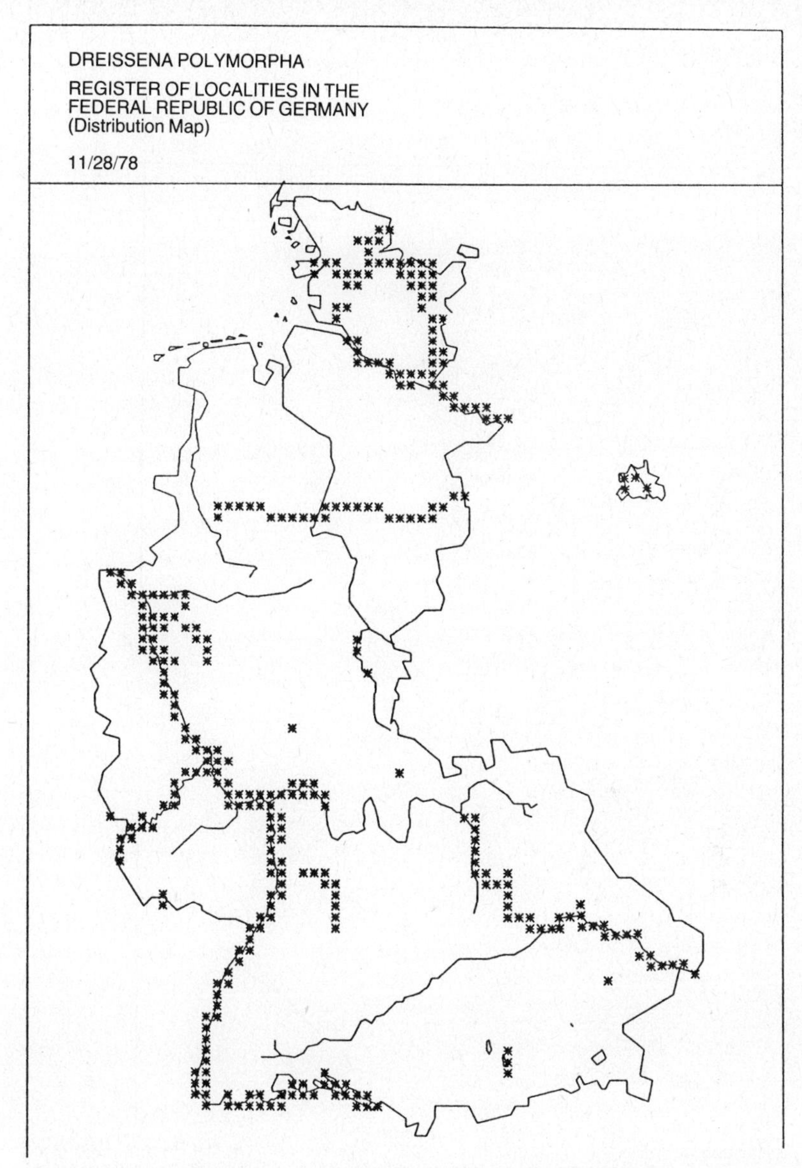

Figure 36 Distribution of the migratory mussel, *Dreissena polymorpha,* in the Federal Republic of Germany.

where they have been attracted by the illuminated windows.

Introductions of animal and plant species have led to a 'foreignization' of the New Zealand fauna and flora (e.g., the Australian opossum; from Europe, the red deer, fallow deer, wild boar, brown hare, polecat, stoat, weasel, hedgehog and partridge). Today, some of these species are used commercially (e.g., red deer). The Australian opossum (*Trichesurus vulpecula*) lives, in New Zealand, in the treetops of laurel and coniferous forests

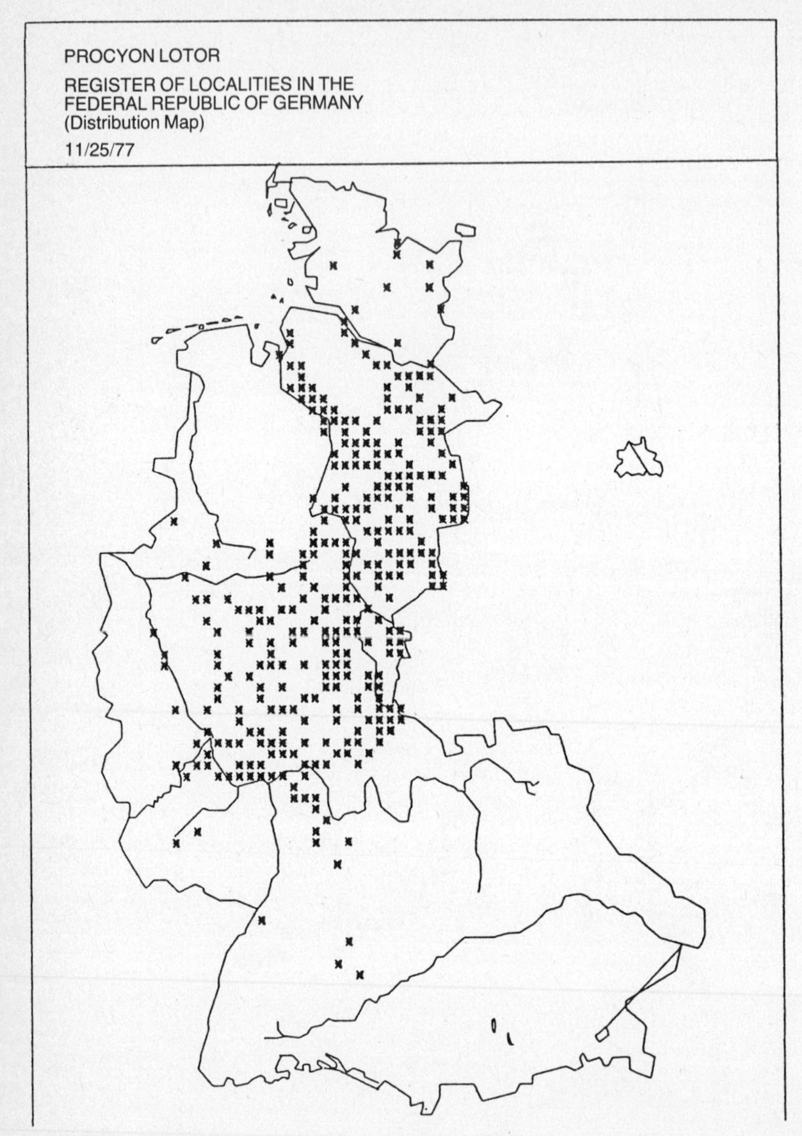

Figure 37 Distribution of the North American raccoon (*Procyon lotor*) in the Federal Republic of Germany (Status as of 11/25/1977). Today, this species can be considered an integral part of the German fauna.

(preferred tree species: *Metrosideros*), and, by chewing the treetops, it breaks up the dense rainforests. Red deer which cannot find favorable living conditions in the dense rainforests, penetrate the opossum woods. The result of the combined effect of opossum and red deer is the devastation of these forests.

The New Zealand flora also includes many species which have been introduced by man (e.g. *Ulex europaeus, Lupinus arboreus, Rosa canina, Rubus fruticosus, Thymus serpyllum, Digitalis purpurea*). The ecological consequences of introduced species are often not predictable as is seen in the

TABLE 9 NUMBER OF BIRDS AND MAMMALS WHICH HAVE SUCCESSFULLY INTEGRATED INTO THE EUROPEAN FAUNA (according to Niethammer 1963)

Origin	Europe	Africa	North America	South America	Asia	Australia	extra-European
Mammals	16	2	7	1	7	0	17=52%
Birds	7	1	2	0	3	0	6=46%
	23	3	9	1	10	0	23=50%
	NUMBER OF BIRDS AND MAMMALS WHOSE INTEGRATION INTO EUROPE WAS NOT SUCCESSFUL						
Mammals	4	2	3	0	4	1	10=72%
Birds	13	7	6	7	14	4	38=75%
	17	9	9	7	18	5	48=73%

history of the naturalization of the muskrat (*Ondatra zibethicus zibethicus;* cf. Pietsch 1970), the potato beetle in Central Europe, the rabbit in Australia, and the agate snail (*Achatina fulica*) (Fig. 38) in Southeast Asia. Within the past 200 years, the east African *Achatina* has reached the cultivated zones of Asia, the Pacific islands (including Hawaii), California, and Florida, and due to its high rate of reproduction, its high rate of consumption of food and its capacity as a carrier of plant diseases, is now very dangerous to its new areas of colonization. Due to the absence of effective enemies, various species were able to disperse into the plantation regions. Today, they are controlled with viruses, poisonous bait, and predator snails of the genera *Gonaxis* and *Eugladina.*

A similarly negative experience occurred with certain introductions of fish species. In 1967, the Amazonian cichlid *Cichla ocellaris* was released in

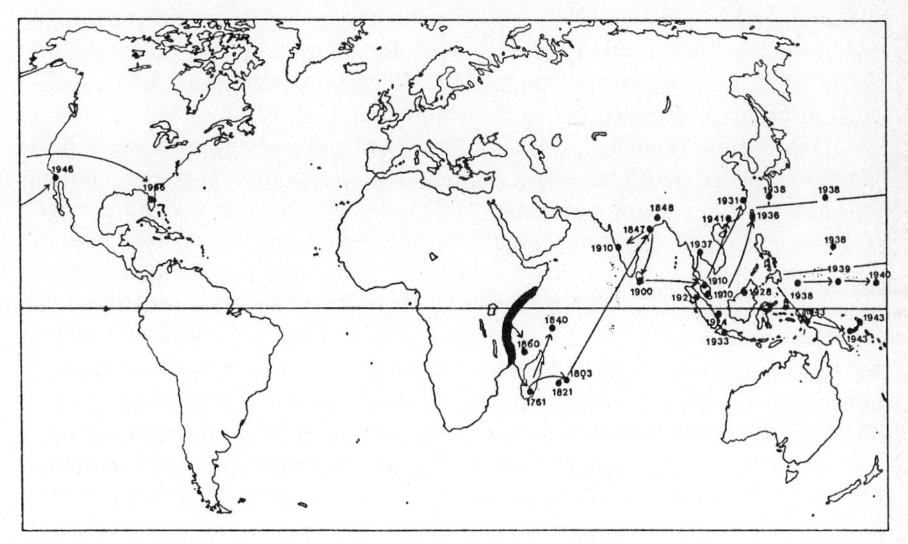

Figure 38 Dispersal routes of the agate snail (*Achatina fulica*) introduced by man (according to Müller, 1977).

the Gatun Lake (47,315 ha) in the Panama Canal (Zaret and Paine 1973). Fish of this species can grow up to 50 cm long and 2 kg in weight. They live predaciously. The impact of this introduction on the local fish fauna was dramatic. The plankton eater *Melaniris* was considerably reduced in numbers which in turn affected the planktonic composition. Bird species (kingfisher, heron), living on smaller fish disappeared from the vicinity. The reduction of insect eaters among the fish led to an increase in the mosquito populations and to an epidemic spread of malaria.

3.1.6 Recording and Mapping of Propagation Areas

For more than 100 years, systematists and biogeographers have demanded distribution records of the various animal and plant groups, a task which, in part, was accomplished successfully by our scientific fathers and grandfathers, but which, in part — depending on the specific biological characteristics of the taxa and the number of experts available — is still awaiting a satisfactory answer to this day. The present situation, however, is in basic contrast with that which previously existed. It is no longer exclusively systematists, faunists and biogeographers who, out of love for their research, undertake to establish records, but rather an increasing number of people responsible for the environmental planning of our countries. It is basically desirable to be completely aware of the spatial distribution of each plant and animal group. A look at the present number of species described of the most important animal groups (more than 1 million), however, illustrates the zoological problems very clearly.

These obvious difficulties are confronted by the urgent necessity for record-keeping. On the one hand, the task is made more difficult by world-wide anthropogenic changes, and the burden on the ecosystems and individual animal species; on the other hand, it is gaining increasing importance due to the fact that living organisms, once their ecological potential has been assessed, can be used as indicators of environmental quality.

Distribution boundaries, unless they can be delineated by natural barriers to dispersal (such as water, mountains, competitor species), can be indicated only in varying percentages of probability for species and subspecies. The ecological challenge which the population face along their boundaries, together with endogenous population dynamics, results in boundary fluctuations of the area. A representation of the facts concerning the distribution thus requires a representation of the localities and environments in which the organisms live in certain countries or on other organisms (in the case of parasites, for example). This raises the question of which chorological recording methods should be used.

The recording of animal groups with an abundance of species presents difficulties which are dependent on the recording methods as well as the quality and number of employees engaged in this task.

The speed of spatial changes must be assessed by an organized method which, because of the amount of work involved, must be generalized in such a manner as to apply it to more than just a single problem. Udvardy (1969) supplied a list of the various spatial mapping methods (cf. Klein 1976). It goes without saying that the use of certain maps is determined by the information desired. We must keep in mind, however, that a map showing only the boundary of a distribution, however precise, simulates a fact which, in reality, often exists for a short while only. Each map which goes beyond an exact recording of an occurrence at a specific time and place, therefore, reflects the intention of the cartographer in question. Without questioning the significance of such maps which, after all, we all use, we should be aware of the problems they pose. They do not fulfill the requirements of being generalizable; they possess nearly infinite possibilities of localities, and, generally speaking, can only be used to answer specific questions. The conclusions, therefore, are that each map must have as a basis a producible statement of the locality . Localities defined by coordinates, or by a grid are probably the best. The advantages and disadvantages of both of these methods (boundaries and grids) have been sufficiently discussed and it is clear that they are complementary rather than contradictory. Grid mapping of defined localities offers a good approach as far as recording and general presentation are concerned. This mapping method has been used for a long time in connection with biogeographical maps and is also generally used in the establishment of distribution maps. The establishment of distribution maps requires extensive recording work. Since the density of information is necessarily great and the information is recorded over time, it is being processed by computer (Figures 39 to 42).

These requirements are fulfilled to a large degree, for example, by the European Invertebrate Survey. This project represents the zoogeographical supplement to the Flora of Europe (Perring 1963, Niklefield 1971, Haeupler and Schönfelder 1973). The considerable advantages of this record-keeping project are evidenced by the abundance of available information (Heaht and Leclerq 1970, Müller 1976). Computer processing is certainly of special significance. Furthermore, it is of importance that the status of work within a particular area can be recognized and gaps pinpointed precisely. The grids can be examined at regular intervals and fluctuations in population recognized more precisely and more systematically. Since this grid reference system is based on being able to locate a specific area exactly on a standard EDP index card, any argument about the accuracy of grid-mapping is futile. Each entry depends not only on the organizational structure and its financial backing, but also on the specialists participating and their abilities.

As has already been done for plants, intensive work is being carried out towards the establishment of distribution maps for approximately 50,000 animal species within the Federal Republic of Germany (cf. Müller 1976, for example). This work is being speeded up because of the threatened extinctions of certain populations (Table 10, Table 11).

For systematic, topographical population-genetic, and regional reasons, 28 animal groups have been included in this recording programme (1977). A grid map (2,712 grid squares, each 10×10 km) of the Federal Republic of

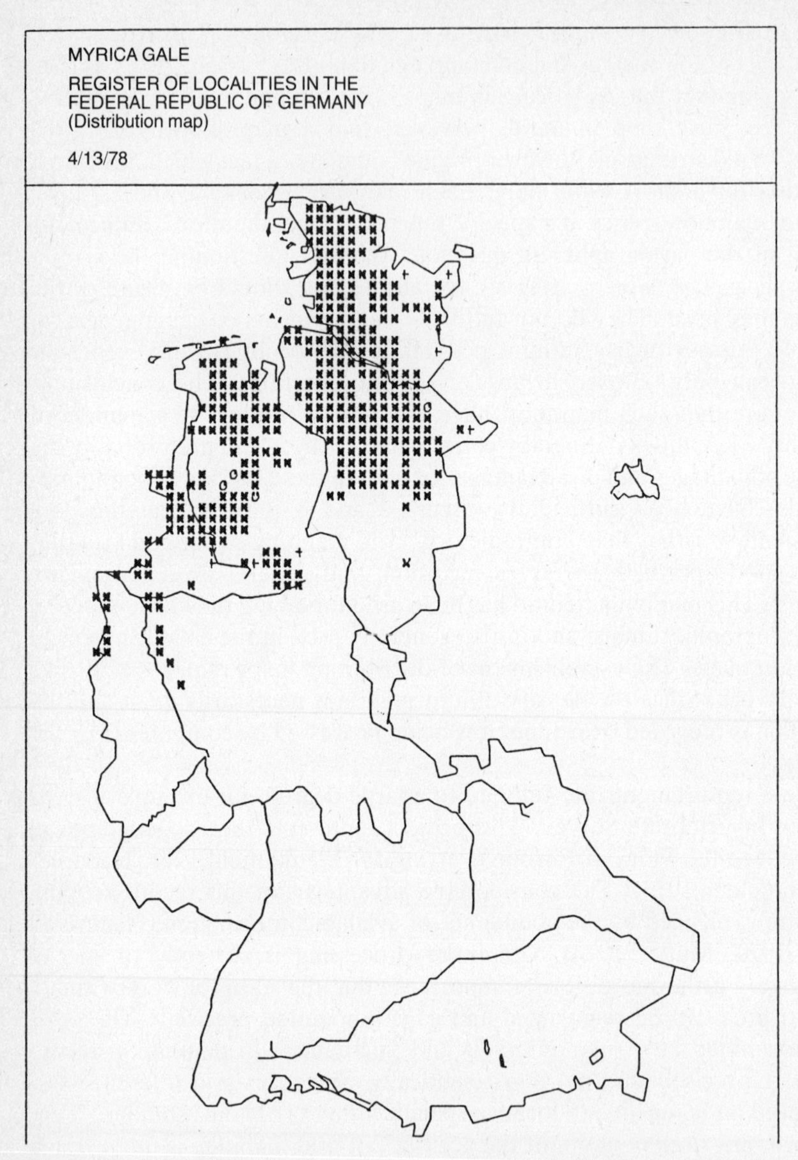

Figure 39 Distribution of the bog myrtle, *Myrica gale* within the Federal Republic of Germany (according to a Central European Mapping Survey) (Distribution map).

Germany is used as the base map. Volunteers provide distribution information either directly onto index cards which, after having been punched, can be immediately transferred to the computer for processing or onto lists of species or localities (Fig. 43).

Each locality is determined exactly, from entries made over a period of time, on a locality index card. Using the computer programs for the Federal Republic of Germany, the computer-controlled plotter will print the information onto a 10×10 km grid.

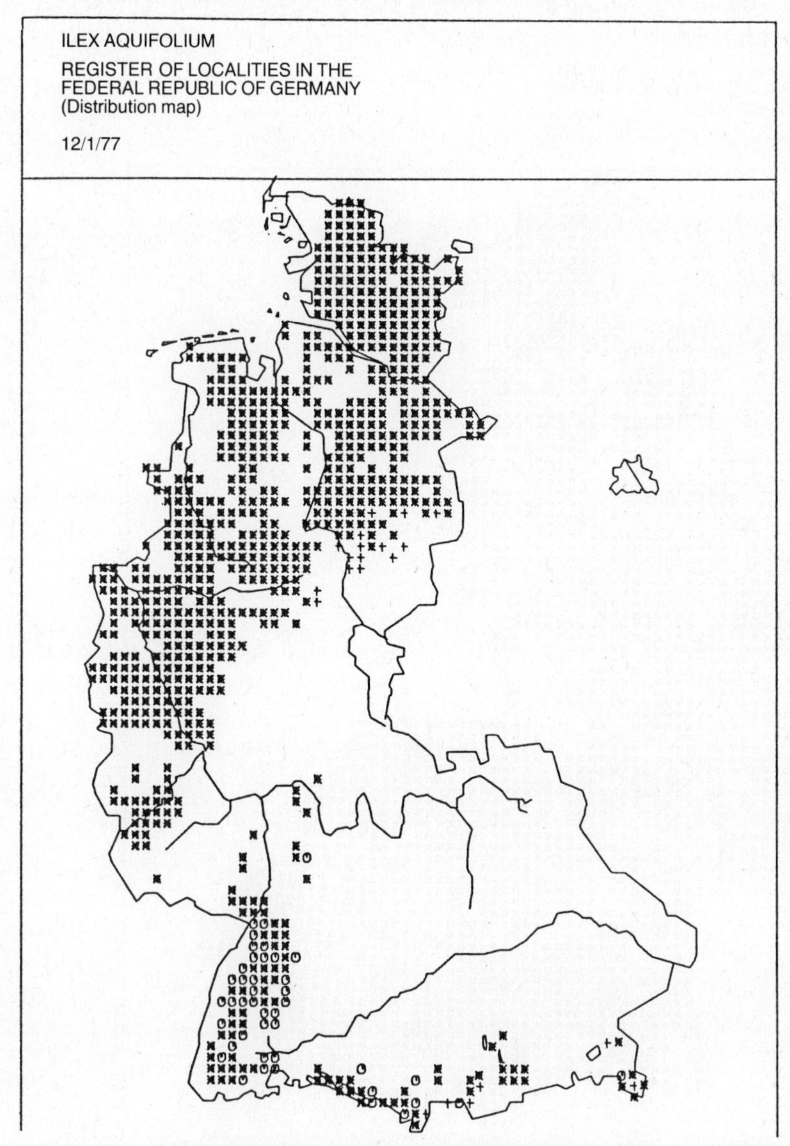

Figure 40 Distribution of the holly, *Ilex aquifolium,* as an indicator of moderate Atlantic conditions within the Federal Republic of Germany. Planted individuals are indicated by circles (transferred onto UTM from a Central European Mapping Survey)

At the European scale, 50×50 km grid fields are used. For smaller regions, the area size is reduced to 1×1 km or even 500×500 m. A causal analysis of the distribution, however, is only possible by associating the representative localities with an ecosystem analysis.

Each locality card of a species is checked by the recorders — as well as by the information register. It is the task of these two persons to ensure the uniformity of treatment of the larger areas. Particular attention is centred

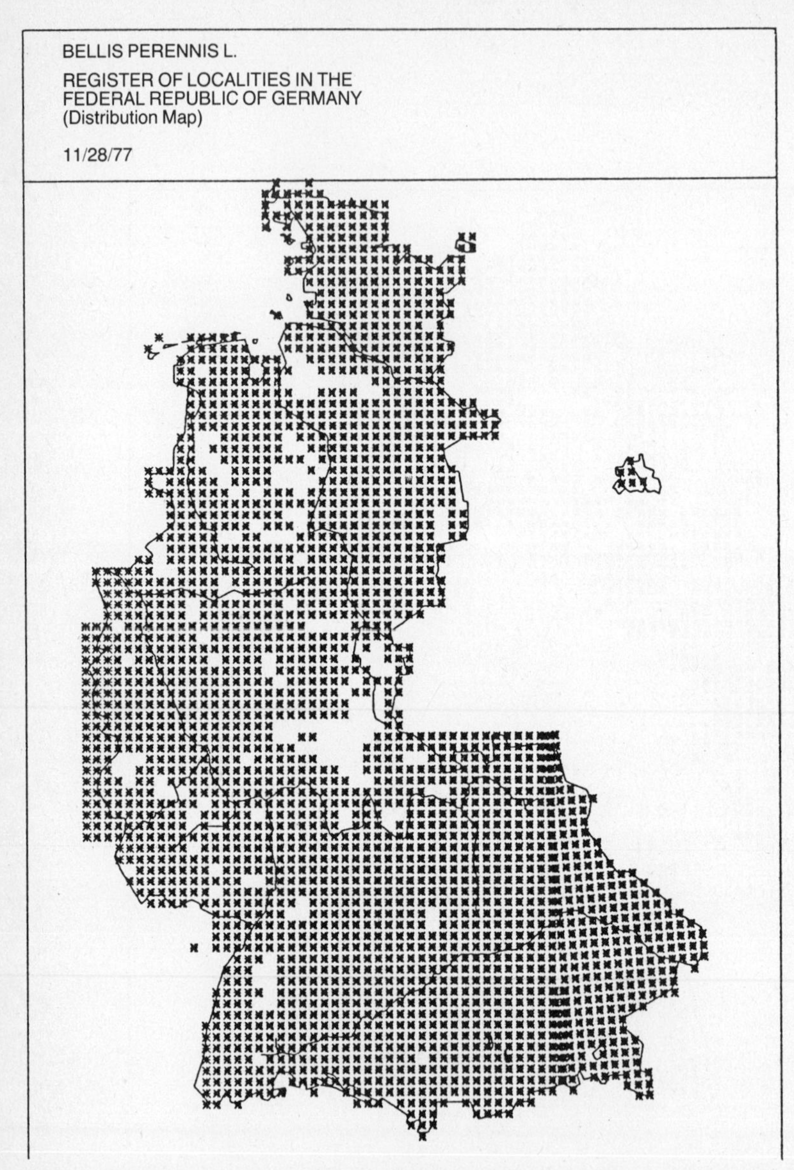

Figure 41 Distribution of the daisy (*Bellis perennis*), within the Federal Republic of Germany (transferred onto UTM from a Central European Mapping Survey).

upon the smaller and larger spatial discontinuities; rare and introduced species; the control of boundaries, coincidences with natural barriers, competing species and food plants as well as the various pressures on our landscapes. The surface covered by the available information, for example, prevents the scientist from concluding — without the corresponding back-up — that an observation made concerning a population decline of a taxon in Bavaria, also applies to a similar population in Lower Saxony (Fig. 44).

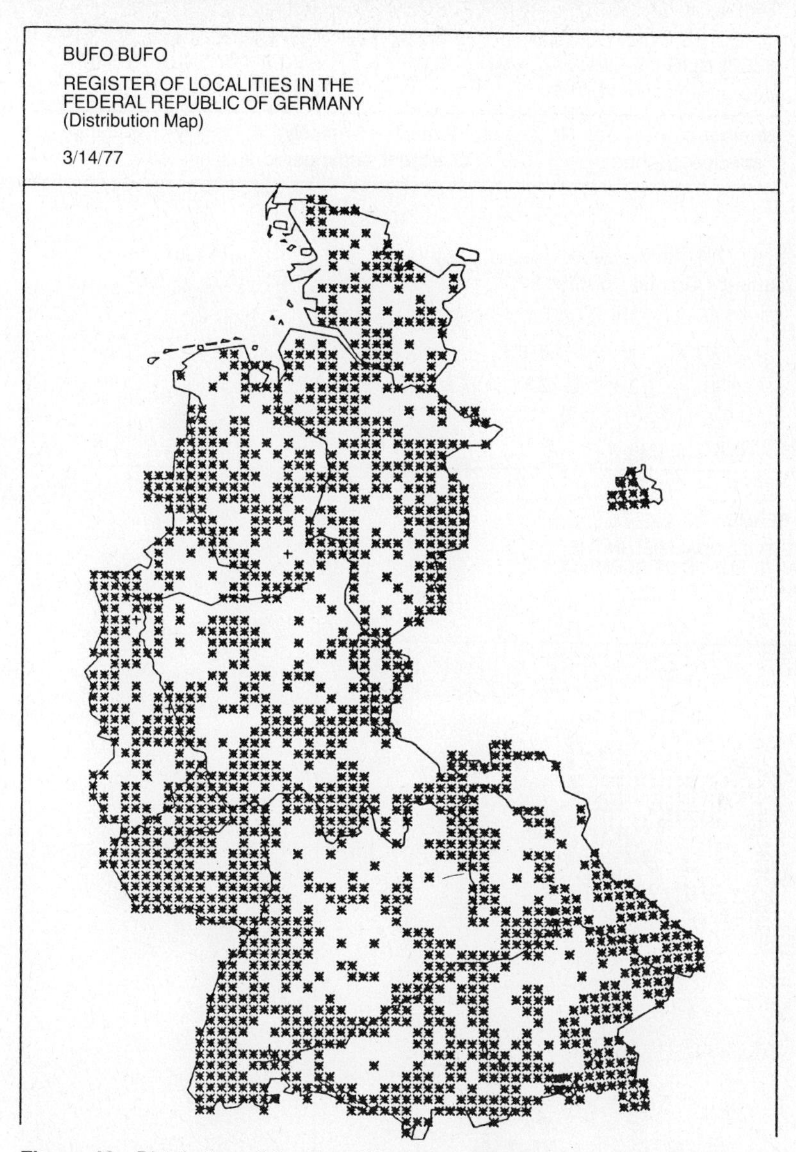

Figure 42 Distribution of the Euro-Siberian toad (*Bufo bufo*) within the Federal Republic of Germany. While in the case of species such as the striped field mouse or *Myrica gale,* the degree of extinction of the species within an area can be determined on the basis of spatial changes, this is not possible for species occurring everywhere, such as the toad. Here, the exclusive determining factor is contained within the population itself.

3.2 DEPENDENCE OF ORGANISMS AND POPULATIONS ON AREA

Organisms which make up propagation areas can be integrated into either similar or different ecosystems. Therefore, changes in ecosystems have long-lasting effects on propagation areas. In those cases where ecosystems and propagation areas are identical, the destruction of an ecosystem can cause the

TABLE 10 SPECIES COMPOSITION (%) OF SOME PLANT GROUPS IN THE FEDERAL REPUBLIC OF GERMANY WITH NUMBERS EXTINCT OR ENDANGERED (from: Sukopp and Müller 1976)

	Number of species	Hemero-chores	Neophytes	Extinct or absent	Acutely endangered	Highly endangered	Endangered
Sinut fungi:							
FRG	72						
Berlin	(23)	–	–	66 (48)	?	16 (26)	82 (74)
(in parentheses Gramineae smuts)							
Mosses	1200	3–4	0.2	1	?	?	10
Lichens	2000	?	0–0.1	0.6	?	?	50
Ferns	80	2.5	2.5	2.5	12.5	10	41
Flowering plants	2282	16	9	2.4	7.5	21.6	38.5

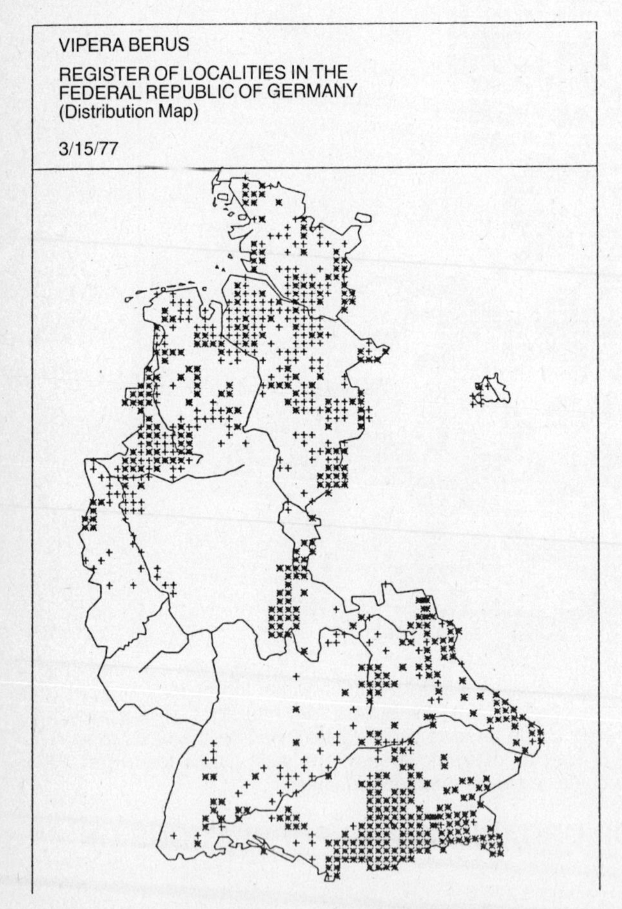

Figure 43 Past (crosses) and recent (stars) distribution of the Euro-Siberian dispersed adder (*Vipera berus*) within the Federal Republic of Germany. The species has always been absent from the Rhineland-Palatinate and the Saarland. The older information pertaining to the Eifel, which can no longer be checked today, is doubtful.

TABLE 11 NUMBER OF SPECIES, WITH NUMBERS EXTINCT OR ENDANGERED, OF THE VERTEBRATE FAUNA IN THE FEDERAL REPUBLIC OF GERMANY (from: Sukopp and Müller 1976)

	Number of species	Extinct	Endangered
Fish	128	2	45
Amphibians	19	–	2
Reptiles	12	–	8
Birds	240	17	130
Mammals	90	8	48
	489	27	243

destruction of a species; where a particular species is dominant in an ecosystem, the destruction of the ecosystem is more often than not the result of the extinction of that species.

3.2.1 Ecosystems

Biogeography has to concern itself with research on ecosystems in order to be able to resolve satisfactorily its basic problems.

Ecosystems are spatially arranged dynamic structures composed of biotic (including man) and abiotic elements with the capacity for self-regulation. The living content of the ecosystem is the biocenosis. By biocenosis (Möbius 1877), we understand a symbiosis which constitutes the dynamic part of an ecosystem and which, together with its biotope, forms a mutually dependent unit with its own dynamics. Although individual animals may change within this environment, the population , once it has adapted to the environmental conditions, survives in its balance of species' characteristics resulting in a population equilibrium. Möbius recognized these self-regulating abilities of the biocenosis.

The structure of the ecosystem was first recognized by Woltereck (1928). He noticed "that the pelagic Cladocera, as with nearly all the other organisms (including plant organisms) in the water, by no means inhabits all of the strata and zones of a lake evenly, but rather these species depend on a part of the available space, limited in a vertical as well as a horizontal direction."

In his case, the experiments performed in connection with the above statement led to the recognition of "morphological and ecological systems" as a result of synthetic analytical research. Woltereck's "ecological systems concept" was changed by Tansley (1935) to the present terminology of the ecosystem.

In natural landscapes, where man has not as yet interfered significantly with the land, the hierarchical organization and spatial limitation of terrestrial ecosystems often correlates with what we would expect at any given location. The world-wide pressure of cumulative, summative, and concentrative poisons has nullified to a great extent the traditional distinction between natural and cultivated landscapes. At present, there are only secondary

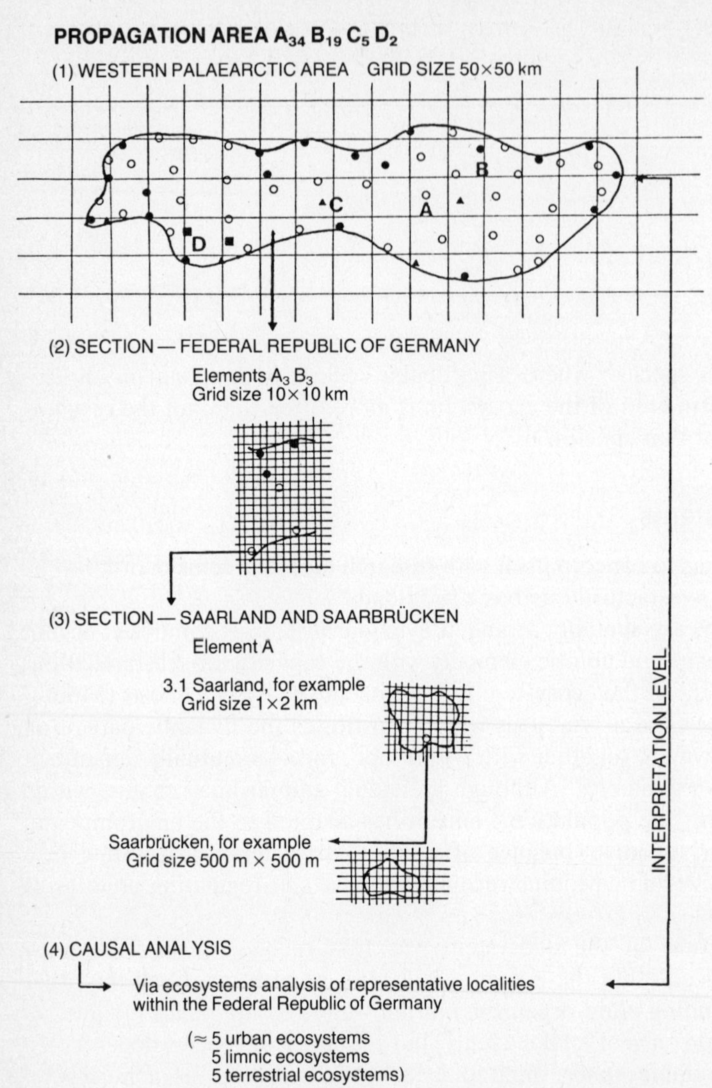

Figure 44 Schematic processing steps for the recording and analysis of propagation areas. (1) is based on a Western Palaearctic propagation area containing the species A B C D with allele frequencies A 34, B 19, C 5, and D 2. Its recording could only be ascertained on the basis of 50×50 km grid fields. This rough data needs to be regionally improved. This is done in section 2, for the Federal Republic of Germany. The size of the grid field can be reduced to 10×10 km. One must remember, however, that one is probably left with only a portion of the allele diversity (therefore, A3 B3) contained in the total propagation area. Where locally possible (cf. 3.1 and 3.2), individual areas can be covered by narrower grid fields (1×1 km, 500×500 m, for example). These are checked at regular intervals for the presence of elements A B C D in the propagation area. The distribution of the elements can be correlated with abiotic or biotic spatial parameters. Thereby, the reasons for the presence can be narrowed down. A causal analysis (work section 4), however, generally speaking is only possible via an ecosystems analysis, which must be undertaken at least for representative localities within large areas (e.g. the Federal Republic of Germany).

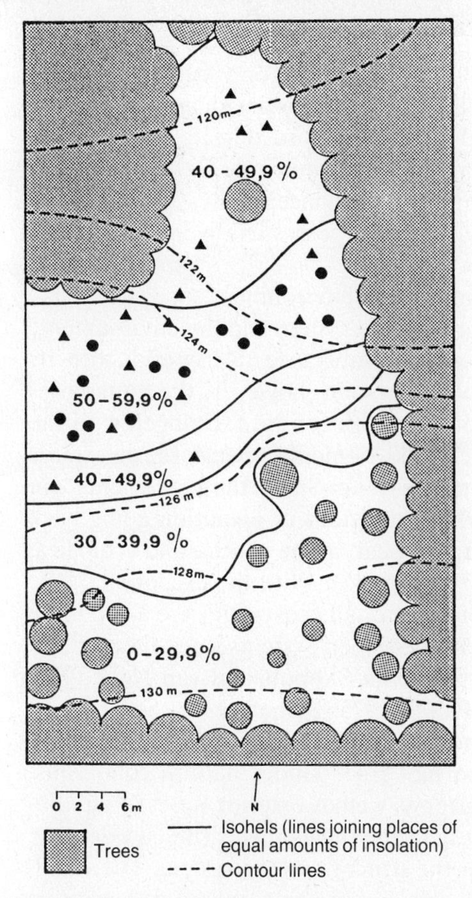

Figure 45 Distribution of the grasshopper species *Pholidoptera griseoaptera* (▲) and *Metrioptera roeseli* (●) which depends upon the sun's irradiation in a forest glen (Mucher Wiesental) in the Siebengebirge. Apart from insolation, the environmental factors recorded show relative constancy over the entire surface. Irradiation (insolation) decisively controls the distribution of both species (according to Brocksieper, 1978).

ecosystems remaining with differing extents of anthropogenic influence. Man, in his capacity as the decisive species, has been changing the structure, dynamics, and spatial distribution of ecosystems ever since the Neolithic period. By no means do the ecosystems of today have to coincide with the physical geography of a landscape. Urban ecosystems are sitting — with a completely new "naturalness" — on top of the former semi-natural systems. Ellenberg (1973) submitted a classification of the semi-natural ecosystems (group 1) of the earth, according to their functional aspects (marine, terrestrial, semi-terrestrial). Although a scientific, theoretical formulation does exist for the second group of ecosystems, which, from the human point of view, must be viewed as urban-industrial ecosystems, a consistent classification has so far not been published.

3.2.2 Ecosystems, Pollution Limits and Stability

Pollution limits as well as stability of ecosystems can only be meaningfully defined in the light of certain relationships. As is the case with any system, the ecosystem can be demarcated towards the margins so that its elements are more closely related to each other than their environment. Generally, it can be said of a system that its boundary, which conceptually surrounds the system and separates it from the environment, possesses diverse properties and can serve the following purposes:

(a) to completely isolate the system from its environment, or

(b) to allow a specific interaction between the system and the environment.

An isolated system exchanges neither substance nor energy with its environment. In the case of a non-isolated system, however, the boundaries are such that either energy or substance or both can be exchanged with the environment. The relationships between the elements, therefore, determine the structure and also the size of the system. Since these elements are effectively organisms which, in part, have a history of evolution going back for thousands of years, it is appropriate to speak of the genetic and ecological structure of an ecosystem (Müller 1977) (Fig. 46). With the exception of the highest-ranking ecosystem, i.e., the biosphere, all ecosystems are acted upon by external factors. These should be termed open systems since they submit to outside influences as well as energy inputs (Abbott and van Nees 1976, Cernusca 1971, Hall and Day 1977, Patten 1975, Solberg 1972, Stuart *et al.* 1970, etc.). As in thermodynamics, the theoretically possible state of equilibrium of the ecosystem plays an important role. Under natural conditions, ecosystemns are characterized by interior as well as exterior states of equilibrium. The stability of the relationships between the elements (linkage density, linkage, and cross-linkage) determines the structure of the system. The stability of an ecosystem, therefore, can be assessed by the reaction of the system to interferences caused by limited inputs and the amount of compensation required to avert permanent changes in structure.

Webster, Waide, and Patten (1975) proposed to replace this concept of stability by a concept of relative stability. "The argument that ecosystems are asymptotically stable focuses attention on the critical area of relative stability." Their "asymptotic stability" is dependent on an equilibrium existing between all structural elements of the system and the cycling and output of material. This theory leads on to the fact that pollution limits and elasticity of the system are considered from structural as well as from energetic viewpoints (cf. Hall and Day 1977, Holling 1973, Lewontin 1970, MacArthur 1955, May 1972, 1973, 1977, Patten 1974, Smith 1972).

In ecosystems abundant in species, the individual elements generally possess a higher linkage density than in those which are poor in species. It can be assumed therefore, that these systems possess a greater internal stability. They may, on the other hand, be more mobile and unstable where exterior

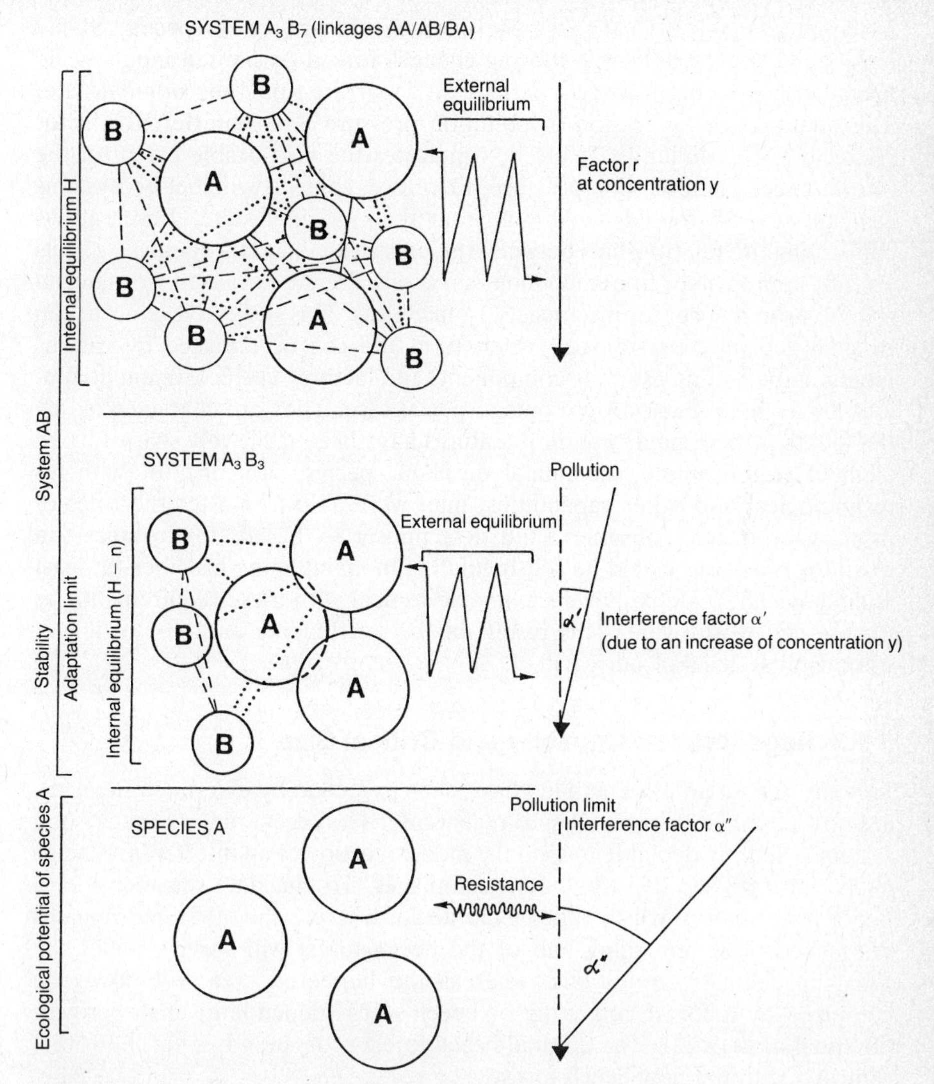

Figure 46 Functional diagram of an ecosystem with elements A and B for the interpretation of the concepts of area stability, equilibrium, and pollution limits (according to Müller, 1977). Between the individual elements, system A_3 B_7 possesses the linkages AA, AB, and BA. The relationships between these elements are (theoretically) in an interior state of equilibrium (H), which in turn is in an external equilibrium with external factors (factor r at concentration y). If, by an increase in the concentration of y, we introduce interference factor α' (say, pollution), the external as well as internal equilibrium changes (for example, by a reduction of element B which has a sensitive reaction to α'). After the removal of interference factor α', the system — due to the reproductive potential of element B — has the possibility of returning to the original state (regeneration capacity). If, on the other hand, element B is eliminated due to an increase in interference factor α'', any regeneration is out of the question. The adaptation limit or the stability of system AB has, therefore, been exceeded. Only element A then remains; it is resistant to interference factor α''.

influences are strong, than are those which are poorer in species. Stable ecosystems are capable of buffering changes forced by limited inputs while unstable ecosystems undergo a permanent departure from their original state. The quantitative expression of pollution pressure is an interference factor (Fränzle 1977, Müller 1977, etc.). It indicates the measurable extent of the interference. The pollution pressure, therefore, controls whether ecosystems can still reverse forced deviations from their original state. Reservations concerning the relationships between the pressures and anthropogenic effects are not appropriate, in our opinion, since any pressures caused by man can also occur in nature (for instance, SO_2, nuclides). This subjective concept can be analyzed by ecosystematic observation and can be replaced by various large numbers of measurable components and factors. The ecosystem dominated by a single species represents a special case. They are characterized by the fact that their main structural features have been decisively shaped by an element (for example, an animal or plant species, or man). If, with his technological and other capabilities, man were able to master the roles of primary producer, consumer, and decomposer — based, for instance, on fossil energy — he would be self-regulating in maintaining his cities (at least technologically). He is, however, unable to deal with these requirements, as is observed world-wide. This results in the overloading and destruction of some vital systems of our earth.

3.2.3 Regeneration Capacity and Critical Size

Systems, whose areas of stability have been exceeded by external influences, are now no longer in a position to regenerate. The regeneration capacity of a system is directly dependent upon the mode of action of an interference factor and on the linkage density and cross-linkages. To elucidate this, consider a beech forest half of which is being cut down, whereas the other half remains untouched. The remaining half of the beech forest will survive under the prevailing climatic conditions, whereas the lumbered area will slowly be colonized by a forest consisting of beech trees (depending on the growth potential of the site). The original beech forest ecosystem has not, however, returned with the new beech forest.

It is particularly experimental biogeography which during the past few years has shown that close relationships exist between area size and genetic structure (Lack 1976, MacArthur 1972, MacArthur and Wilson 1971, Simberloff 1976). Through numerous experiments, it has been possible to substantiate the dependence of the extinction rate within a system on area (Müller 1977, Simberloff 1976, among others). These findings led to the establishment of area requirements in the setting up of reserves (Diamond 1975, among others). They led to the conclusion that only systems with large areas can retain their genetic structure over longer periods of time. Perhaps, the large population declines in animal and plant species within the Federal Republic of Germany are due partly to area effects. Further arguments which speak against the simple regeneration of a destroyed system can be derived from the differing evolution capabilities of individual species. Finally, it should be pointed out that regeneration capacity does, of course, not only depend on the loss of area, but also on the distribution of harmful substances within the system as a whole (Blau and Neely 1975, Smith *et al.* 1966), its chemical properties and its relationships to other systems.

3.2.4 Ecosystems and Species Diversity

In ecosystems abundant with species, the individual elements generally have a higher linkage density than those which are poor in species. We can conclude, therefore, that these systems have a higher internal stability. They can, however, be more mobile and unstable where external influences are strong, than those which are poor in species. The type of cross-linkage serves as a measure of the ratio between inputs and outputs of similar elements, in similar and also alien systems. Linkage density and cross-linkage type clarify a system's elements and/or subsystems in terms of the total system. Their elucidation makes it possible to predict the system's reaction to certain pressures. Organisms and populations are parts of mutually dependent biotic communities which are located at a specific site. Although populations and biotic communities react in accordance with their own rules, they can only exist if they constantly store and process information concerning other components within their propagation areas. Thus, the structural and/or energetic analysis of biotic communities can provide significant clues to the space inhabited by them. The biotic community structure is dependent on the diversity of its individual elements.

Numerous methods and mathematical models have been developed to ascertain the informational content of biotic communities by the analysis of the group diversities constituting it (Wiener 1948, Shannon 1948, Shannon and Weaver 1949, Peilou 1975, 1977, Murdoch and Oaten 1975, Patten 1975, Whittaker 1975). Each abstract or concrete system which can be divided into its individual elements and subunits can be characterized by the specific diversity of its elements (for example, species diversity, biomass diversity, ecotype diversity). The species diversity of a system is dependent on the frequency of the occurring species and evenness of the distribution of individuals among the individual species (= equitability in the sense of Stugren 1974; evenness or equitability in the sense of Pielou 1969, cf. also de Benedictis 1973). Species diversity can be compared to the degree of probability associated with making a random "grasp" into a system and obtaining a certain informational element. Consequently, the diversity will be the greatest whenever — with the largest possible number of species — all species âre equally frequent. The limits of this statement have been shown by Scherner, 1977. Number of species and quantity of individuals can be related to species diversity as follows:

$$H_s = -\sum_{i=1}^{s} p_i \log p_i.$$

where H_s represents the species diversity and p_i the relative abundance of species i. A system containing two equally frequent species thus possesses a greater species diversity than does an eleven species system, where one species provides 90%, and each of the remaining species only 1% of the individuals (Wilson and Bossert 1973). The possibility of two systems having the same H_s-values at different species compositions can be shown by the diversity difference (H_{diff}) equation:

$$H_{\text{diff}} = H_t - (H_1 + H_2)/2.$$

H_1 indicates the species diversity of the first, H_2 that of the second and H_t that of both sites (as a unit).

The frequency of species at site 1 is indicated by p_i, whereas that in site 2 is denoted by p_i'. *Should a species be missing in one of the two systems,* p_i or p_i', respectively, equals zero.

To illustrate these relations, consider the following two extreme examples from Nagel (1976):

(a) If, at two sites, there are no differences either in the number of species or their abundance and species composition, $H_{\text{diff}} = 0$. As proof, let's assume two sites, each containing the same species a and b, each of whose individuals dominance amounts to 50%, i.e. $p_a = p_b = p_a' = p_b' = ½$, then $H_1 = -2\,(½ \log ½)$

$= -\log ½ = \log 2$

If we insert the values in the formula for H_{diff}, the result is zero.

(b) A completely different species composition at two sites with the same species diversity must result in the highest possible diversity value. As proof, let's assume species a and b at site 1, A and B at site 2, p_i being ½ for each species:

$$H_s = -\log ¼ = 2 \log 2$$

$$H_{\text{diff}} = \log 2$$

This means that the maximum possible difference in diversity between two sites cannot exceed log 2.

Discussions during the Intecol-Congress in 1974 in The Hague, as well as our own findings (Müller *et al.* 1975, Schäfer 1975, Thomé 1976) point out the possibilities and limits of diversity analyses. In the case of biogeography, the species diversity and diversity difference value show the possibility of determining the degree of relationship between biotic communities and the sites they populate. Diversity indices are generally used for measuring anthropogenic pressures, since the species diversity of a system will generally diminish with changes in the relative abundance of organisms within a system. The degree of change, however, is dependent on the structural diversity of the habitat. Thus, the diversity and distribution of benthic communities in the Saar is determined not only by the chemical and physical pressures, but also by the structural features of the banks and hydrography of the various river sections (Müller and Schäfer 1976). Diversity and information are thus closely related. It is to the credit of MacArthur (1955), Margalef (1957, 1958, 1975), Peilou (1979) and Stugren (1974), etc., that the necessary prerequisites have been created to integrate information theory and communication theory into ecology. The abundance of species in a habitat is dependent on the diversity of its living conditions and its history. The area of a habitat is of importance if numerous species compete with each other within the community. Competitive pressure and the number of niches within a biotope also control the species abundance and are closely related to diversity. The term ecological niche indicates the role which a species assumes based on its demands on its environment and its utilization of the environmental conditions within an ecosystem, i.e., within the relationships of a symbiotic link between themselves as well as in their common environment. Niche describes any relationship between diferent species and environmental conditions. The more uniform the environmental conditions, the less species competing among themselves can live there in the long term. The more diverse the environment, the larger the number of species that can exist alongside each other.

3.2.5 Plant and Animal Sociology

The structure of biotic communities depends on the daily and seasonal fluctuations of the species-specific niche variables. Since biotic communities are associated with a certain type of site, they often change where the combined effects of terrain factors change. This can mean that within a natural landscape containing various biotic communities areas of the same ecosystem contain the same biotic community. The spatial classification of the potential natural vegetation which explains the present growth potential of a habitat thus becomes the indicator for the physiography of a landscape (cf. Smithüsen 1967). The mapping of certain associations, for instance in weed communities, not only allows a determination of the age of the cleared land surface, but also (by knowing the course of succession), of the future development of the landscape (Tüxen 1950, 1956, 1970). There is a close correlation between growth performance and substrate characteristics (Seibert 1968, Pfadenhauer 1975, among others).

Often, close relationships exist between the available nutrients in the soil and their concentration within individual parts of the plant (for example, leaves), although such connections often show a seasonal and age-dependent periodicity (cf. Höhne and Nebe 1964). Troll (1968) was correct when he pointed out that for a full understanding of the ecological dynamics of an area, community analysis is indispensable. Experiments made during the past few years have been able to demonstrate convincingly the value of plant- and animal-sociological methods in assessing spatial qualities. Here, too, however, it has become clear that the consistent application of plant-sociological methods to zoological communities poses problems not only from the theoretical point of view, but also in practice.

3.2.6 Ecoclines

Many populations show characteristic gradients which correlate with environmental parameters existing in biomes or regions already studied. They often show a depedence upon geographic-climatic factors (Kurten 1973). Various zoogeographic rules makes these close relationships between geographic-climatic factors and the differentiation of the taxa clear. Regularities of the geographical variability as they apply to a climatic gradient are expressed in the following rules:

1. Bermann's rule which implies an increase in body size of warm-blooded animals in colder climates.
2. Allen's rule assumes a decrease in the length of body appendages (extremities, tail, ears) in colder climates.
3. Renn's hair rule which, at rising temperatures, presumes a reduction in the amount of hair on mammals.
4. Gloger's rule, whereby subspecies in warm and humid regions are more strongly pigmented than those in cooler and drier climates.
5. Hesse's rule, whereby the relative weight of the heart increases in cooler climates.

3.2.7 Life Forms

"The characteristics described as adaptations to the environment, be they structures or modes of behaviour, ...constitute the life form of an organism" (Koepcke 1971, p.6). By first principles, it is clear that vital behaviour by organisms, such as swimming, flying, running, fleeing, hunting, collecting, grazing, or protection by means of giant physique, requires different habits and are designated by Koepcke following Remane (1951, p.354), as types of habit. Habit types, however, are reflected in the structure and form of organisms. These types of life form are understood as complex individual adaptations to the environment.

"The life form of plants determines the appearance of vegetation, and thus, to a large extent, the physiognomy of a landscape... As long as its arrangement is based solely on taxonomy, the abundance of forms remains indeterminable" (Schmithüsen 1968).

Raunkiaer (1903, 1905) distinguished 30 life forms for higher plants which he classified into five main groups:

1. Phanerophytes = plants whose regeneration buds are located more than 1 m above ground level;
2. Chamaephytes = plants whose buds lie above soil level but below 1 m;
3. Hemicryptophytes = plants whose buds are close to the ground surface;
4. Geophytes = plants, whose buds survive in the ground, whereas their parts located above ground level die off.
5. Therophytes = annual plants which survive the unfavorable conditions as dormant seeds.

Schmithüsen (1968) distinguishes 30 classes of growth forms:

1. Crown trees, mega- and meso-phanerophytes
2. Microphanerophytes
3. Savanna grasses
4. Creeping and hanging plants
5. Liana
6. Shrubs, nanophanerophytes
7. Dwarf trees
8. Stalked succulents
9. Herbaceous stalked plants
10. Epiphytes
11. Dwarf shrubs
12. Subshrubs
13. Dwarf succulents
14. Chamaephytic herbaceous plants
15. Hemicryptophytic woody plants
16. Hemicryptophytic herbaceous plants
17. Winter annual plants
18. Geophytic herbaceous plants
19. Therophytic herbaceous plants
20. Floating aquatic plants
21. Submerged herbal plants
22. Thalloid epiphytes
23. Thalloid chamaephytes
24. Thalloid hemicryptophytes
25. Thalloid geophytes
26. Thalloid therophytes
27. Underground plants
28. Thalloid hydrophytes
29. Plankton
30. Endophytes

Life forms are the ecological answer of a living system to the challenges of its environment. They have been formed not only by climatic and pedological influences, but also by a wealth of biotic elements. Their spatial distribution can provide insight into important seasonal climatic fluctuations or soil properties.

Chapter 4

Propagation Areas and Ecosystems

Propagation areas are related to ecosystems in many different ways. Knowledge concerning the structure and function of lakes, moorlands, rivers, or cities — to name but a few ecosystems integrated into the biomes (large habitats of the the earth) — thus must be recognized as one of the prerequisites for a causal interpretation of their spatial distribution.

4.1 THE LAKE ECOSYSTEM

4.1.1 Structure and Dynamics

Although seemingly isolated, a lake is still related to its environment in many different ways. Its hydrological cycle is directly influenced by its geographical location. Thus, it is justified, when classifying lakes, to include frozen waters within the polar region (the surface being frozen during several months; with the warmest water at depth), the warm water lakes in the tropics (or subtropics), and lakes of the middle latitudes. It should be recognized, however, that such a classification cannot include regional peculiarities. Precipitation, evaporation and a multitude of influences directed by the terrestrial systems located close to the banks of the lakes modify water balance. The subsurface profile, too, is of considerable importance (Fig. 46a).

The quantities of water present in standing or flowing water bodies are relatively small when compared with those in the lithosphere or oceans. Approximately 1.8% of the land surface (= 2.5 million km^2) are covered by lakes, the largest lakes (from the point of view of surface) being: the Caspian Sea (424,300 km^2), Lake Tanganyika (35,000 km^2), and Lake Baikal (33,000 km^2). The varying ecological factors around a lake (climate, for example)

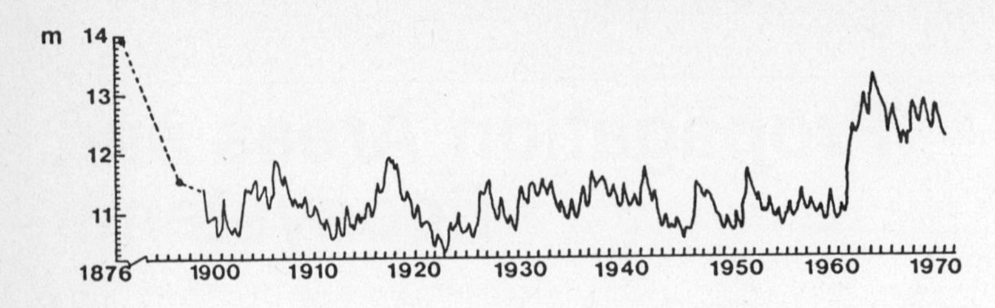

Fluctuations in the water level of Lake Victoria

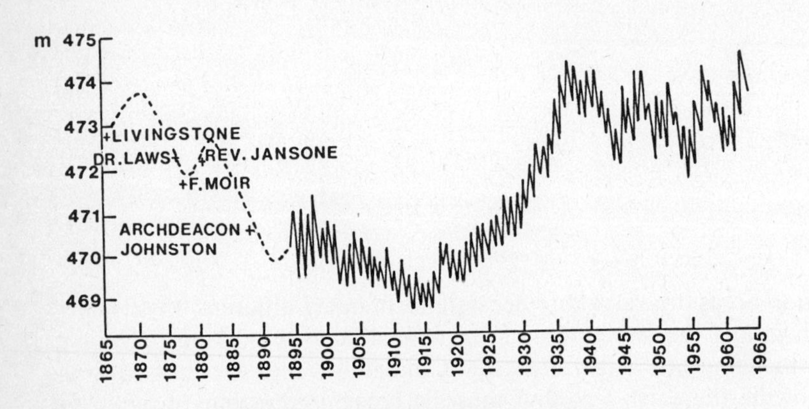

Fluctuations in the water level of Lake Nyasa

Figure 46a Fluctuations in the water levels (in m) of Lakes Victoria and Nyasa as indicators of changes in the water balance of their catchment areas.

affect the lake which, among other things, can lead to varying rates of sedimentation, the extent of which can be determined by analyses of lake sediments. Structural changes of the lake ecosystem, therefore, must be viewed in conjunction with the surrounding biomes. Information concerning changes which have already taken place over the centuries during winter freezing of ice over the Central European lakes are a dependable testimony of past temperature fluctuations. Some of the Antarctic lakes possess impressive frozen surfaces all year long and, due to the lack of intermixing, their chemical stratification is practically uniform (Goldmann 1970).

The specific characteristics of water as a medium (for example, maximum density occurs at 3.94°C; boiling point = 100°C; dielectric constant = 78.54; latent heat of vaporization in kcal = 545.10 (= 2282.22 kJ); thermal conductivity at 25°C in cal (cm×sec×°C) = 0.00136 all affect the living conditions of organisms in the community.

According to Henry's law (concentration of a saturated solution of a gas = temperature-dependent solubility coefficient × partial pressure of the gas), the solubility of a gas in water diminishes with increase in temperature and decrease in pressure. At a higher temperature, therefore, the oxygen content of a lake decreases while, simultaneously, oxygen demand by water fauna living in the lake increases. The vertical lake gradients (hydrostatic pressure, temperature, light, and chemical factors) arrange the water of the lake into a mosaic of spatial areas of varying attractiveness to water organisms. There is a causal relationship between the above and the specific distribution of organisms and migrations between the bodies of water of varying quality (Goltermann 1975, Halbach 1975). Two large habitats can be distinguished in every lake: the pelagic habitat and the benthos.

The pelagic habitat or free water zone is inhabited by organisms whose development cycles occur in open water. The inhabitants of the benthos, on the other hand, are limited to the bottom of the lake. The benthic region can be subdivided into the bank, or littoral zone, and the deep, or hypolimnetic zone. The boundary between littoral and hypolimnetic is the compensation level which can be characterized by the fact that below this level, most photoautotrophic organisms no longer possess a positive photosynthetic balance. Thus, the compensation level divides the lake into a photic zone with mostly photoautotrophic production and an aphotic zone. The light-flooded photic zone of the pelagic habitat is designated the epipelagic, while its aphotic zone is called the bathypelagic habitat. The upper part of Central European lakes is characterized by reed belts including emergent plants such as *Scirpus lacustris, Phragmites communis, Typha, Alisma*. This zone is adjoined, particularly in sheltered areas of the lake, by a belt of floating plants, including Nymphaeaceae and *Potamogeton* species which, in the lake centre, are usually superseded by submerged water plants such as *Myriophyllum, Ceratophyllum, Elodea, Vallisneria* and then by underwater meadows predominantly of Characeae. Symbiosis below the lake's compensation level and in the bathypelagic habitat is based on the biomass production of organisms within the littoral and epipelagic habitats. The sediments resulting from these production processes are deposited in the bathypelagic habitat in the form of gyttya or dystrophic matter.

Dystrophic matter develops in lakes poor in plant nutrients (dystrophic lakes) by means of an intensive supply of allochthonous material, such as fallen leaves, while gyttya consists of the microscopic remains of organisms and is characteristic of oligotrophic and eutrophic lakes. Based on the amount of floating debris (particularly human acids), standing waters can easily be classified into dystrophic or brownwater lakes (with high humic matter content) with poor visibility (moorland waters for example) and into clear water lakes (with little turbidity).

In most lakes, substances dissolved in the water, physical factors and the biomass maintain complicated interrelationships which are subject to seasonal fluctuations. Definition of individual lake types, therefore, can be undertaken in detail on the basis of their structure during any given season. Generally, sufficient oxygen is supplied to the upper water strata (epilimnion) and, due to constant circulation, has layers of even temperature. In the metalimnion (= middle strata) located below the epilimnion, the temperature falls rapidly and, in the deepest lake zone (hypolimnion), it is close to 4°C. At this temperature, water is at its maximum density. In holomictic lakes, this stratification can be destroyed, at least in the short-term, through the mixing of surface and deeper water. Meromictic lakes are never completely intermixed; in many cases, this can be attributed to a difference in the specific gravity of the individual water layers or to a wind-sheltered location (Fig. 46b).

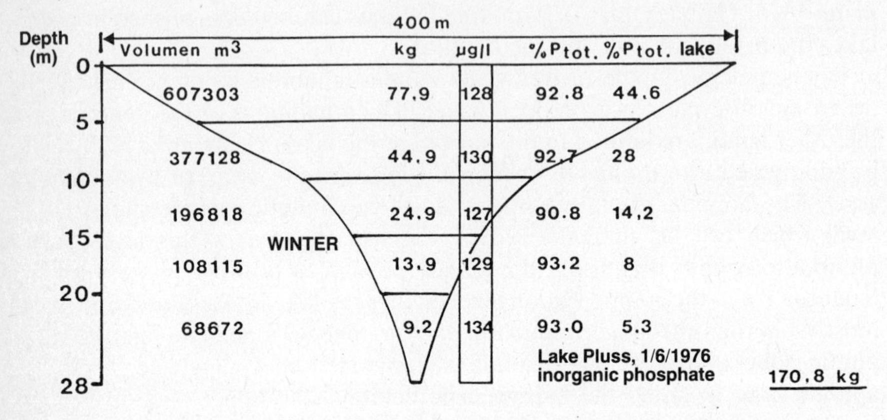

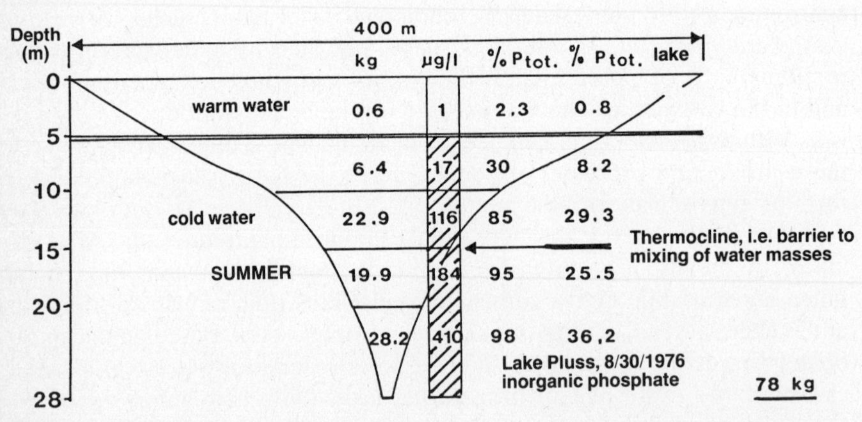

Figure 46b Cross-section showing phosphorus balance in Lake Pluss (eastern Holstein). With its diameter of 400 m, Lake Pluss is one of the smaller lakes in this area; however, its depth of 28 m and its sheltered location and protecting moraines result in a stable stratification during summer. During winter, phosphorus is evenly distributed; during summer, the formation of a stratified structure due to heating, greatly limits the vertical exchange of water and results in a considerable concentration of phosphorus at depth. µg/l = microgram phosphate/liter; $\%P_{tot}$ = percentage of inorganic phosphate shown in the figure in relation to the total phosphorus content per liter. Whereas in winter, nearly all the phosphorus is present in the form of inorganic phosphate, in summer, it is only a small percentage. $\%P_{tot}$ lake = percentage of phosphate quantities present in the individual water volumes as compared with the total inorganic phosphate content of the lake. The difference in inorganic phosphate content of the lake between winter and summer means that in summer, a large portion of inorganic phosphate is absorbed by the biomass. This leads to a considerable decrease in phosphate in the surface water (according to Overbeck, 1980).

TABLE 12 PRODUCTION IN ETHIOPIAN AND PALAEARCTIC LAKES (according to Beadle 1974)

Lake	Altitude (m)	Euphotic zone (m)	Chlorophyll (mg/m^2)	Production ($g\ C.m^{-2}.a^{-1}$)
Lake Victoria (Uganda)	1230	to 20	35–100	950
Windermere (England)	<100	to 10	5–100	20.4
Bunyoni (Uganda)	1970	4		
Kivu (Zaire)	1500			540
Mulehe (Uganda)	1750			
George (Uganda)	913	0.7	70–280	1980
Chad	283			
Mariut (Egypt)	<100	0.5		
Aranguadi (Ethiopia)	1910	0.14	221–235	
Lungby Sø (Denmark)	<100	1.0	–	660
Esrom Sø (Denmark)	<100	10	–	144–204
Lunzer Untersee	608	20	–	30

The pelagic organisms, autotrophic algae and heterotrophic macro- and micro-consumers are just as dependent on these dynamics as the benthal macrophytes and the bacteria, a significant part of which goes into secondary production (Overbeck 1974). During the summer due to stratification of the water in Central European lakes, the phosphate present in the epilimnion is generally used up. A renewal of the phosphate level from the hypolimnion whose phosphate content is much higher, is cut off by the thermocline in smaller lakes and can only be replenished in winter with the resumption of full circulation. The phosphate concentration in the hypolimnion is attributed to iron in the form of $Fe(OH)_3$, present in the oxygen-rich part of the lake which binds the inorganic phosphorus and transports the latter into the deoxygenated hypolimnion. Through a reduction of iron, or the frequent formation of iron sulfide, the bond is broken and the phosphorous liberated. This mechanism is utilized in the biological purification of water in that, through the addition of iron salts during the third purification phase, close to 90% of the dissolved phosphates are removed from the waste water. The phosphate content directly controls the growth of plankton and since this is determined by the adsorption by $Fe(OH)_3$, it is closely dependent on the oxygen condition of the lake.

During the summer, stratification in the hypolimnion (which registers an oxygen deficit all year round) results in the oxygen content usually being completely depleted. At the same time, an intensified production of hydrogen sulfide takes place within the same layer. Since H_2S is being depleted by bacterial photosynthesis, the hydrogen sulfide concentration shows a vertical day/night rhythm. The phosphate, oxygen, and hydrogen sulfide levels show how the structure of a lake is controlled by metabolic processes. An increase in the populated part of the epilimnion can lead to active catabolism by the floor organisms and thus to a lack of oxygen and the formation of the mud characteristic of eutrophic lakes (marked by the presence of the chironomid *Chironimus* and therefore termed Chironimus lakes).

Oligotrophic lakes (characterized by the chironomid *Tanytarsus,* therefore, termed Tanytarsus lakes), on the other hand, have a much reduced production of biomass and, therefore, smaller mud deposits and a sufficient oxygen supply for the hypolimnion. Since, during eutrophication, the yearly production of phytomass and the associated growing period are of importance, the primary production in transparent lakes from middle latitudes to the poles generally decreases.

4.1.2 Pollution in Lakes

Pollution inputs into lakes have to be degraded by the lake sediments. In addition to their pollution pressure, cities located on the banks of lakes and agricultural enterprises in their catchment basins cause eutrophication processes to be set into motion.

Radionuclides introduced into lakes and running waters either naturally or under anthropogenic influences, are gaining in importance, since these radionuclides may be utilized in the course of food chains.

Since the concentration of plutonium (^{239}Pu) and caesium is highest in the benthal substrate, benthal populations characteristically exhibit higher concentrations. This is evidenced by ground fish in Lake Michigan (Table 13). Miettinen (1976), during the examination of marine sediments in the Gulf of Finland, arrived at comparable results.

At one time, Lake Constance was a typical *Tanytarsus* lake. The most conspicuous changes in Lake Constance during the past decades correlate with an increase in phosphorus content, an increase in plankton production, and an oxygen decrease within and below the thermoclime cause by a decrease in the plankton biomass. While this lake during the past centuries was used by man for fish and to carry goods traffic and ferries between lakeside communities, today it serves man from near and far as a recreation area and, at the same time, is the largest reservoir in Germany. From 1950 to 1970, the population of Baden-Württemberg grew by 38.7%, rising by 44.3% within the immediate vicinity of Lake Constance. Studies concerning the intensity of production of organic substances by plankton by measuring radioactive carbon showed that maximum production occurred during bright sunlight and usually at a depth of 1–2 m; the largest amount of organic material is produced within the top 5 m of the lake. The total production per unit of surface area exceeds that of oligotrophic lakes and is closer to the value of an eutrophic lake. The winter production of organic material is impeded by the circulation of the water; during the winter of 1963, under the ice, however, it nearly reached the summer levels. The total yearly production of living material within Lake Constance amounts to approx. 2 million tons which, for their total catabolism, would consume about 130,000 tons of oxygen. About 21% of the 36,000 tons of oxygen-consuming organic matter, about 9% of that consuming nitrogen (17,900 tons of total nitrogen), and 20% phosphate (1,750 tons of phosphate) originate in the lakeside communities of Lake Constance. 12,640 tons of nitrogen, i.e., 74% of the total nitrogen content, as

TABLE 13 PLUTONIUM IN THE FISH OF LAKE MICHIGAN (according to Miller and Stannard 1976)

Species	No. of specimens examined	% Ash residue as % of wet fish weight	^{239}Pu pCi/kg wet fish weight	^{239}Pu Concentration factor
Cottus cognatus	4	0.033	0.205±0.076	273±100
Coregonus sp.	6	0.021	0.025±0.003	33± 4
Alosa pseudoharengus	3	0.031	0.018±0.003	24± 4
Osmerus mordax	4	0.021	0.013±0.003	16± 4
Coregonus clupeaformis	2	0.023	0.010	14
Oncorhynchus kisutch	1	0.020	0.005	7
Oncorhynchus tschawytscha	1	0.017	0.003	4
Salvelinus namaycush	2	0.022	0.001	1.5

well as 650 tons of phosphorus (37% of the total phosphorus content) in the tributaries originate from flushed-out fertilizing agents from agricultural areas. 63% of the total amount of phosphorus originates from waste waters from the lakeside communities and tributaries. Comparing these numbers with the overall quality of Lake Constance, the inferior quality of tributary waters of the lake in the industrial locations and cities becomes understandable (Buchwald *et al.* 1973, Lang 1969, Lehn 1976).

Because of the formation of sapropel, the habitats of the mesopsammon (cf. Kloft 1978) in Lake Constance are nearly free of turbellarians. Density of individuals, and species composition of Tubificideae populations are dependent on the depth of water, quantity of nitrogen, and on the kind and actual availability of sediment materials. In the shore zone, *Euilyodrilus* and *Limnodrilus* species are dominant to a depth of about 30 m, whereas at 60 to 90 m, *Tubifex tubifex* is dominant. With the increase in sedimentation of putrefactive organic matter, maximum colonization shifts towards the deeper areas of the lake. *Euilyodrilus* and *Limnodrilus* migrate downwards, by 40 m; *Tubifex tubifex* in some areas can be found more than 100 m deeper than normal for this species within Lake Constance. Within the area of output of predominantly domestic waste water, mainly *Euilyodrilus* and *Limnodrilus* develop whereas where the influx of industrial waster water occurs, *Tubifex tubifex* becomes dominant.

Fishing yields have increased dramatically from 90,000 kg per year in 1925 to 340,000 kg in 1968. The whitefish of Lake Constance (*Coregonus wartmanni,* among others) have grown well over their normal size since 1960. At two years of age (before spawning maturity), they have been caught in nets whose mesh density had previously been designed to catch four year old or older fish. Today, the mesh density has been changed accordingly. The increase in the fish population caused a rise in fish predators. During the summer, about 1,200, in winter nearly 9,000 tippet grebes (*Podiceps cristata*) live on the lake; efforts are made to control their population by means of mass

TABLE 14 AVERAGE CONCENTRATIONS OF AMMONIA, NITRATE, AND PHOSPHATE IN THE TRIBUTARIES OF LAKE CONSTANCE (1962 and 1963)

	m^3/S	Ammonia-N	Nitrate-N	Phosphate-N
New and old Rhine, Bregenzer Aach, and Argon	313	3.8 t/d 140mg/m^3	14 t/d 520 mg/m^3	0.32 t/d 12 mg/m^3
Other tributaries	30	1.6 t/d 600 mg/m^3	2.3 t/d 900 mg/m^3	0.39 t/d 150 mg/m^3
Total	343	5.4 t/d 180 mg/m^3	16.3 t/d 550 mg/m^3	0.71 t/d 24 mg/m^3

shootings. Other bird species, such as the red-crested pochard (*Netta rufina*) feeding on Characeae — which at one time visited Lake Constance by the thousand during moulting time — disappeared together with the Characeae, from the Basin of Ermatingen (Table 14).

4.1.3 "Old" Lakes

Only in rare cases does the evolutionary history of lakes cover a very long time in geological terms. Exceptions include Lakes Baikal, Tanganyika, Nyassa, Victoria, Titicaca, and Ochrid; Lake Baikal, for example, has been an isolated fresh water lake since the Cretaceous period and was also an open lake even through the Pleistocene. This has resulted in the development of numerous endemics (for example, the fungus family Lubornirskiidae, the fish families Comephoridae and Cottocomephoridae, and the snail families Baikaliidae and Benedictiidae). According to Kozhov (1963), 583 of the 652 animal species occurring in Lake Baikal are endemic. The 286 m deep Lake Ochrid on the south Yugoslavian–Albanian border is a typical karst lake dating back to the Pliocene. Of the endemics, sponges (*Spongilla fragilis, Spongilla stankovici, Ochridaspongia rotunda*) turbellarians (for instance, the genera *Promacrostomum, Mesovortex, Proamphibolella, Jovanella, Vranjella, Archopistomum*), gastropods (90% = 34 species are endemic; Stankovic 1960), ostracods (66% = 24 species are endemic) and fish (60% = 10 species are endemic, including *Salmothymus ochridanus, Salmo letnica, Pachychilon pictus, Phoxinellus minutus*) are dominant. In contrast with Lake Baikal, on the other hand, the amphipods play a subordinate role.

As is the case with islands or isolated caves, isolated lakes offer an evolutionary experimentation field for insights into biogenetic processes.

The South American lake, Titicaca, at an elevation of 3,812 m and with a surface area of 8,100 km^2, is the largest mountain lake and the highest navigable lake in the world. It originated at the end of the Plio-Pleistocene. The full circulation during the summer and partial circulation during the winter affect the plankton of the lake. The zooplankton consists predominantly of Copepoda and Cladocera. The original fish fauna is represented by *Orestias* species and the silurid *Trichomycterus rivulatus* which have been considerably reduced since the rainbow trout (*Salmo irideus*) and brook trout (*Cristivomer namaycush*) were introduced in 1939 from the U.S.A. *Orestias agassii* is the most widely distributed fish although *O. cuvieri* most probably became extinct in 1960. In the reed zone in the outer shallow waters, representatives of another *Orestias* species, *luteus,* regularly occur. This species is a predator living close to the bottom of the lake and eating mainly molluscs and ostracods.

The open bays of Lake Titicaca are inhabited by two more species, easily distinguishable from each other: *Orestias cuvieri,* with a length of 30 cm is the largest of the *Orestias* species, and an obvious predator, whereas *Orestias pentlandi,* at 20 cm, is somewhat smaller and lives mainly off plankton. The ecological and geographic differentiation of Lake Titicaca was probably as polymorphic throughout its evolutionary history, particularly at the time of the Plio-Pleistocene Lake Bolivian as it is today. Possible separations existed several times during the course of its history.

The Philippine lake, Lake Lanao and East African lakes with their species-abundant fish fauna have provided us with a significant insight into the species found in lakes.

Considerably fewer species are known in the Central European waters. Furthermore, their species composition has, in part, been changed fundamentally by anthropogenic influences.

4.1.4 Relationships and History of the Lakes

The relationships between the limnic fauna and flora — due to their ecological ties and the history of their habitat — often lead to different proposals for biogeographic classifications when compared with classifications of terrestrial organisms (cf. among others, Banarescu 1969, Illies 1967, Thienemann 1950). While there are numerous lake dwellers that are cosmopolitan, others are limited to individual lakes and types of lakes and exhibit an island pattern of distribution (the fish *Salmothymus ochridanus* and *Phoxinellus minutus,* for instance, are endemic to Lake Ochrid; the barbel *Barbus albanicus* only occurs in Lake Yanina). A representative of the phytoplankton, *Aphanizomenon flosaquae,* is to be considered cosmopolitan whereas a closely related species, *A. gracile,* so far has only been found in Northern, Central, and Eastern Europe. In addition, there are seasonal migrations of lake fauna to the land to spawn and also vice-versa. Pond turtles (*Emys orbicularis*) lay their eggs on firm ground on the shore, while the Central European anurans and newt species (*Triturus*) go to standing waters to spawn.

Seasonally dependent fluctuations in the presence of individual species (spring animals, fall animals, etc.), in the phenotypical form of individuals (polymorphoses, for example), and in the vertical and horizontal distribution of individuals and species within a lake also detract from a generally valid classification. Furthermore, numerous lakes, during the course of their development, passed through marine stages or, at least, had contact with marine

TABLE 15 ENDEMIC SPECIES OF GASTROPODS AND LAMELLIBRANCHIAE IN LAKES TANGANYIKA AND MALAWI

	Gastropods				Lamellibranchiae			
	Families	Genera	Species	Endemics	Families	Genera	Species	Endemics
Tanganyika	9	36	60	37	4	12	14	5
Malawi	6	8	19	11	3	6	12	8

ecosystems (the Caspian Sea, for example). This is probably also true for Lake Nicaragua in Central America, the only freshwater lake inhabited by sharks.

The Baltic Sea passed through a variable history of alternating limnic and marine stages. Its post-glacial developmental phases were characterized by a change in the fresh- and saltwater conditions, and occurred during the early post-glacial when a peripheral lake, of melted snow and ice was trapped south of the Scandinavian ice cap. When the ice receded northwards, the sea expanded to its present size.

Between 14,000 and 8,000 B.C., similar peripheral lakes were formed south of the ice caps in the region of the Holarctic (Magaard and Rheinheimer 1974, Thienemann 1928, 1950, Segerstrale 1957). Colonization of the Northern American ice barrage lakes occurred, in part, from eastern Siberia. The North American barrage lakes extended from the St. Lawrence River to the foothills of the Rocky Mountains (Saskatchewan). Around the Baltic Sea — a further regression of the southern ice periphery via Lake Väner, Lake Vätter, and Lake Mälar — revealed an overflow channel through which freshwater flowed into the ocean so that the level of the lake sank to sea level, making a direct marine connection. Central Sweden was formed, therefore, around 7,500 to 7,000 B.C., the peripheral lake developing into a peripheral sea called the Yoldia Sea after its predominant fossil *Yoldia* (today: *Portlandia*). Due to a further isostatic rise of Scandinavia, the marine connection was again interrupted, and, once more, a freshwater lake formed which persisted from 7,000 to 5,000 B.C., called the Ancylus Lake, again after its predominant fossil *Ancylus fluviatilis,* the freshwater mussel, which still occurs in unpolluted trout rivers in Central Europe. A prolonged sinking of the land between Jutland and South Sweden, created a connection to the North Sea, which still exists today, and the Baltic Sea again received saltwater. The salt content at first was higher than that found today, and this stage of the Baltic Sea is, therefore, termed the Littorina Sea (climatic optimum) after its predominant fossil, the periwinkle *Littorina littorea* (Fig. 47). While *L. littorea* at present is limited to the western part of the Baltic, freshwater fauna is again penetrating the brackish waters of the central and eastern areas. The present stage of the Baltic Sea is called the Lymnaea Sea (Segerstrale 1954) after the snail *Radix peregra* (previously *Lymnaea ovata baltica*), which has immigrated from the freshwater region (Fig. 48).

The post-glacial history of the North Sea is important for an understanding of the distribution of numerous British fresh-water organisms. During the

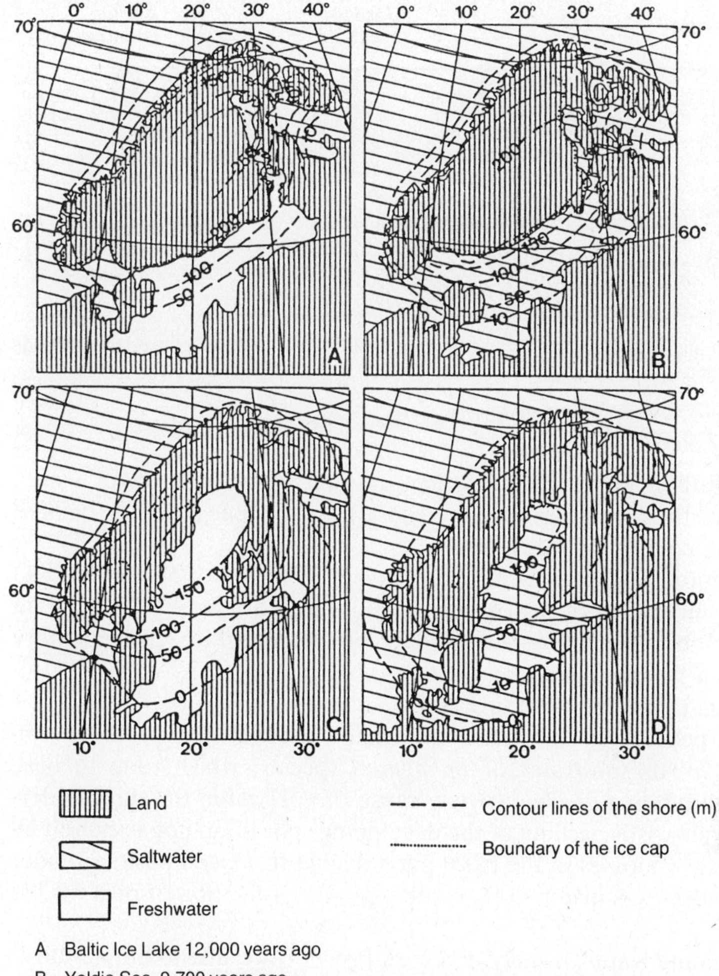

A Baltic Ice Lake 12,000 years ago
B Yoldia Sea, 9,700 years ago
C Ancylus Lake, 8,000 years ago
D Littorina Sea, 7,000 years ago

Figure 47 The post-glacial developmental stages of the Baltic Sea.

last ice age, significant parts of the North Sea were dry land. The sea level was between 50 and 60 m below present mean sea level and the ice boundaries ran through Jutland and Schleswig-Holstein. As during the Tertiary, a land connection existed at this time between the European mainland and Britain. Today's coastlines developed during the formation of the Littorina Sea and Great Britain became an island between 5,500 and 3,000 B.C.

The history of the Mediterranean and the Black Sea (cf. history of the Sarmatian continental lake) shows how many other continental lakes passed through the terrestrial and limnic stages even before the Pleistocene (de Lattin 1967, Banarescu 1971).

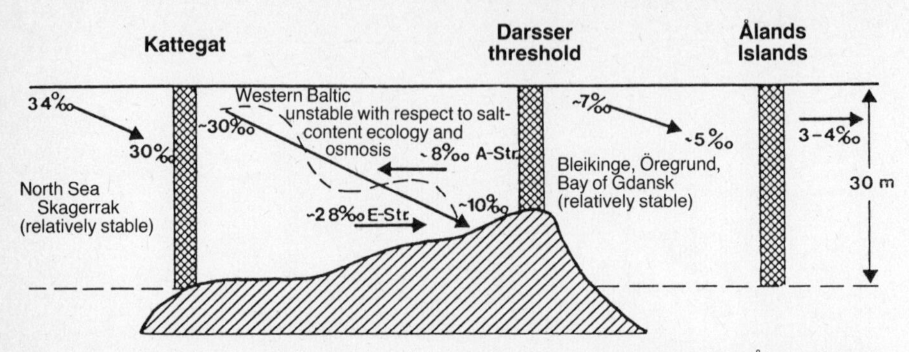

Figure 48 Fluctuations in the salt concentration between the Kattegat and the Ålands Islands in the western Baltic.

4.1.5 Bog Ecosystems

The process of filling up by sedimentation generally heralds the end of a lake, and the start of the formation of bogs whose largest expansion is associated with the boreal coniferous forest biome.

By definition, bogs require for their existence a layer of peat at least 30 cm thick or a succession of peat layers. Peat, whose dry mass must contain 30% organic substance, is formed by layered plant residues (a sedentary process), which do not completely decay. Mud is formed by deposition in standing waters (a sedimentary process).

The flat peat surface, when appropriately wet (all year round), can become the growth substrate for *Sphagnum* species, which, due to base absorption and binding of electrolytes, cause the pH-value to fall considerably. The characteristic arching of the developing sphagnum bog is caused by apical growth, the mosses in the basal parts dying off. Decomposition under anaerobic conditions is incomplete, resulting in the formation of peat (Table 16).

Free-standing water, usually of a dark brown color due to humic substances, occurs as bog pools between moss hummocks and sedge tussocks. The small water bodies in sphagnum bogs usually possess lower pH-values than those in other bogs. Starting along the edges, they can become overgrown by mosses and grasses and can eventually be dominated by grasses. Typical bog plants are calcifuge.

The following invertebrate species, in particular, occur regularly in bogs. Thecamoeba, Cladocera (*Streblocercus, Acantholeberis, Scapholeberis, Holopedium gibberum*), Copepoda, Rotatoria, and Odonata (Schwoerbel 1977). In Central Europe, numerous Lepidoptera (*Colias palaeno,* for example), are dependent on bogs for their food plants. Thiele (1977) listed *Pterostichus nigrita* and *Pterostichus vernalis* as the dominant carabids of the eutrophic western Palaearctic bogs.

TABLE 16 CHEMICAL PROPERTIES OF SOME SMALL WATER BODIES IN THE MOORLANDS OF AUSTRIA (according to Loub 1958)

Moorland area	Sphagnum bog	Transitional bog	Flat heathland
pH	3.4 – 5.0	5.0– 6.2	5.8 – 7.2
Alkalinity*	0.07– 0.023	0.2– 0.45	0.36– 2.0
Cl^-	0.4	0.4	2 – 8 plus
SO_4^{2-}	5 – 15	5 – 15	5 – 20 plus
NO_3^-	0	0	0 – 4 plus
NH_4^+	0 – 3	0 – 3.5	2 – 10 plus
Ca^{2+} (in mg/l)	1 – 6	5 – 9	7 – 40
Mg^{2+} (in mg/l)	1 – 6	3 – 8	12 – 40
Na^+	1 – 5 (30)	5 – 30	17 – 30
K^+	3 – 28	4 – 34	30 – 80
$Fe^{2+/3+}$	0 –100	0 –200	0 –100

*Bicarbonate content

4.2 FLOWING WATER ECOSYSTEMS

Flowing water bodies are flowing systems whose function, under normal conditions, is determined by their seasonally changing structure (speed of flow, volume of water, temperature, oxygen balance, amount of floating material, geological and pedological substrate) and whose history is closely related to the development of the landscapes through which they flow. The interrelationships between running water and the surrounding land are so close and so varied that predictions with respect to the future of a river or brook are often possible only by means of multivariate mathematical models (Herrmann 1977, Rump, Symader and Herrmann 1937, Schrimpff 1975, Whitton 1975), which take into account the erosion capacity of a river, its discharge, the precipitation occurring in its catchment area, the existing groundwater supply, the different forms of utilization, and the vegetation along its riverbanks. Running water creates a landscape by alternating erosion and deposition. It forms different shaped valleys, and extends the seaward boundaries of river mouths — which are often wide deltas — far into the ocean. Running water connects and creates various terrestrial ecosystems and thus promotes exchanges of energy. It can also be viewed as the end link in the chain of effects on a landscape — the river represents the "excretion of a landscape, so to speak" (Sioli 1968). Quantitative conclusions about the many processes occurring within a landscape can be drawn from the individual landscape elements themselves (Figs. 49 and 49a).

4.2.1 The Amazon River

The catchment area of the second largest river in the world measures 7,050,000 km^2. Its average discharge at its delta is estimated at 218,000

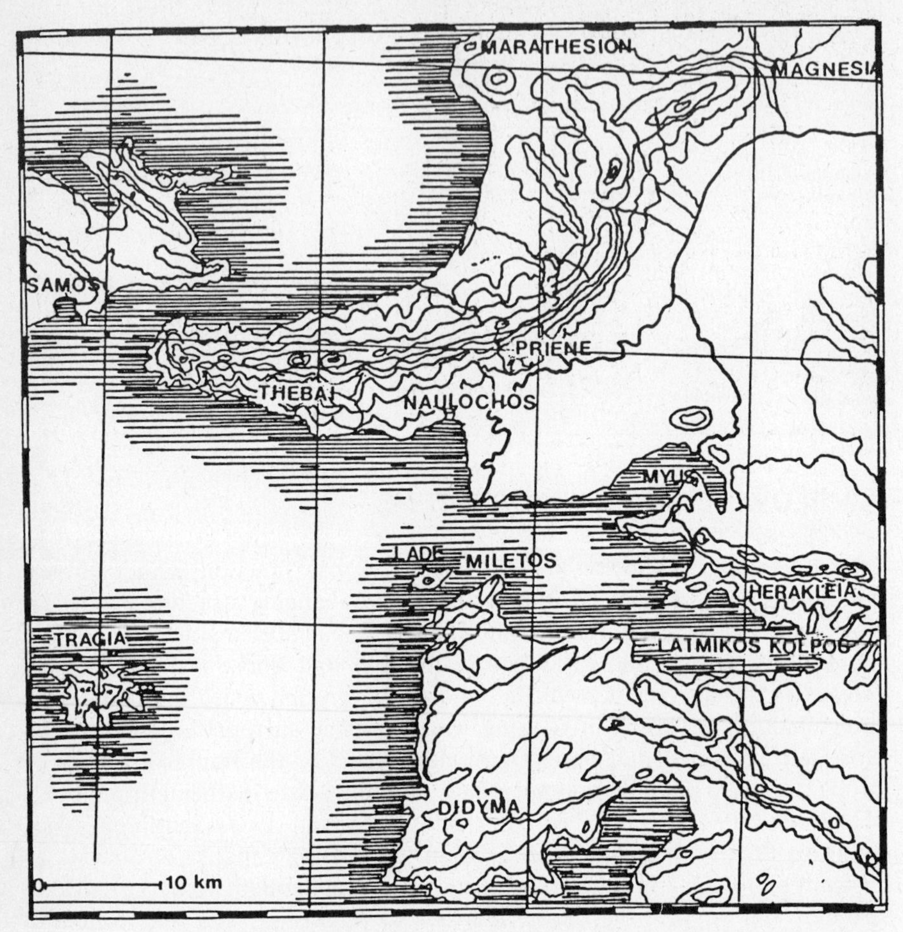

Figure 49 Coastlines of the Menderes Valley and the Latmic Gulf around 451 B.C. (above) and in the 20th century (opposite), (according to Jusatz, 1966).

m^3/sec, which amounts to 18% of the water discharged by all rivers into the oceans. The load it carries corresponds to approx. 690×10^{12} kg per year. This river and its tributaries are subjected to considerable fluctuations in water level. At high water, the aquatic fauna has a huge volume at its disposal for several months, while the terrestrial organisms are pushed back to areas not affected by the high waters. This yearly change particularly favors the fauna having short life cycles and high reproduction rates.

On the basis of the physico-chemical load, the Amazon can be divided into three areas:

(a) White water (Agua branca), i.e., loamy-yellow, turbid water. Depth of visibility between 10 and 50 cm (Example: Rio Solimões).

(b) Clear water, i.e., clear, transparent water of a yellow-greenish color and having a visibility depth of between 60 cm and 4m (Example: Rio Tapajoz).

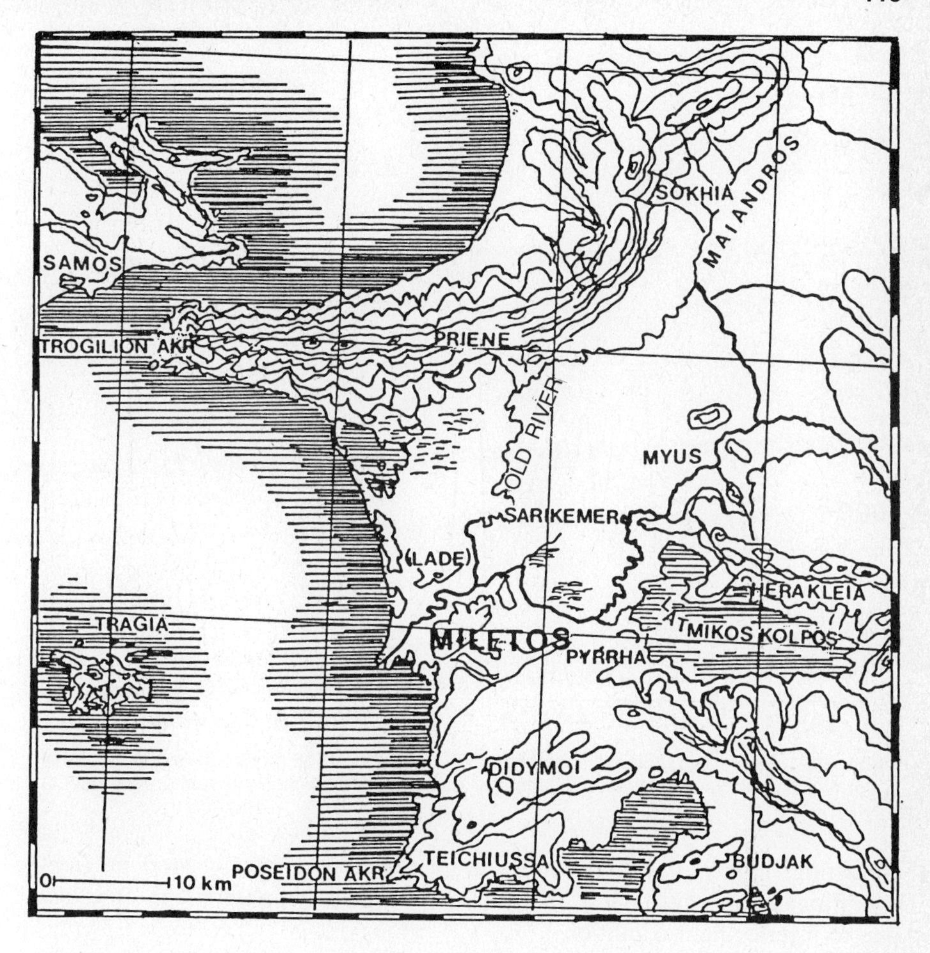

(c) Black water (Agua preta), i.e., transparent, brown to red-brown colored water whose depth of visibility varies between 1 and 1.5 m (Example: Rio Negro).

The water chemistry of these river types on the one hand is based on the different geological, mineralogical, climatic, and also topographic conditions of their spring and catchment areas, and, on the other hand, must be viewed in the light of the origins of the vast South American landscapes. While the spring areas of the white water rivers (Rio Madeira, Rio Solimões, Rio Purus) are located in the Andean mountains and other young mountains or accretion plains, the black water rivers have their source in the huge podsol areas covered by rainforests. In addition to the black water rivers in South America, such rivers, lacking electrolytes but rich in humic acid, occur in Sarawak, the Congo region, and Malaya. Due to the lack of electrolytes in Central American waters (Table 17), it is not surprising to find that they are also characterized by a low biomass production. The values obtained so far vary greatly between the river systems consisting primarily of black and those

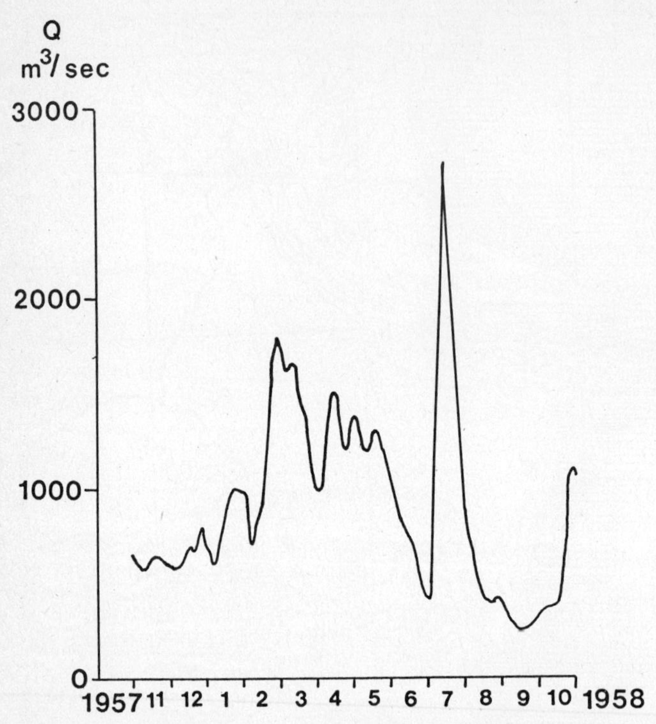

Figure 49a Flow of the Elbe at Neudarchau (534 km) during 1958 (from The Annual Report of Water and Shipping Administration, Hamburg). Summer flooding occurred during the month of July.

consisting mainly of white water. In spite of the relatively good nutrient conditions of the white water, the primary production of phytoplankton is low due to inferior light conditions.

The clear water rivers usually originate in regions with a geologically old surface and a stable relief. The history of the Amazon River is a necessary prerequisite for the understanding of the present conditions. Marine facies show that the present Amazon basin was occupied by the ocean from the Silurian to the Upper Devonian era. Tectonic movements caused the formation of sills between the Upper Devonian and the Lower Carboniferous periods, dividing the originally uniform sea basin into several smaller basins. The Mesozoic river systems of western Amazonia flowed into the Pacific since the Andes did not emerge until the Miocene. The western flow was then blocked by the rising Andes and an immense freshwater lake formed whose huge sandy-loamy sediments can be observed into the Pleistocene. From time to time, there existed a marine contact northward with the Caribbean. The present delta region of the Amazon River shows a varied history of marine, terrestrial, and lacustrine deposits (Ludwig 1966, 1968). It was only in the Pleistocene that the Tertiary Amazonian continental lakes opened up in an easterly direction. On the bottom of the lakes which had dried up in the meantime (today's terra firma of Amazonia), rainforests — which today

TABLE 17 IONIC CONTENT IN THE FLOWING WATERS OF AMAZONIA (in ppm; according to Irion and Förstner 1976)

River system	Ca	Na	K	Mg
Andean rivers	54	25	4.4	6.6
Foreland of the Andes	30	10	1.4	3.3
Solimões	12	3.4	1.0	1.7
Amazon	6.5	3.1	1.0	1.0
Acre	9.6	2.3	1.4	1.3
Madeira nr. Abuná	6.8	2.3	1.4	2.1
Pleistocene Varzea	<1.0	0.8	0.4	0.2
White water carrying kaolinite	1.5	1.6	0.8	0.8
Clear and/or black water rivers	0.5	0.6	0.2	0.2
Global mean	15	6.3	2.3	4.1

determine the physiognomy of Amazonia — were able to immigrate, together with the open Campo Cerrado formations (Müller 1973).

It is only since that time that the river carrying the largest quantity of water of all the rivers in the world has eroded into its present riverbed. The history of the Amazon basin, however, is not only the history of the Amazon River; it also explains, by way of its facies and geotectonic conditions during the Early Palaeozoic and Lower Carboniferous periods, that in the actual basin, for instance, petroleum deposits of economic proportions are highly unlikely to exist and that many of the existing soils are lacking in nutrients due to rapid leeching processes.

4.2.2 Rhine and Saar

An examination of the Central European river systems shows how strongly their present conditions have been determined by history and the natural landscape. The Rhine which is 1,320 km long and has a catchment area of 225,000 km^2 (which, in comparison with the Amazon, is small) flows through a variety of landscapes, originating in the St. Gotthard mountains and ending up in the North Sea. The river sections vary considerably depending on their origin, their hydrography, and their characteristic landscape; they only joined together during the Pleistocene. At the end of the last glaciation, the North Sea dried up eustatically and the Thames became a tributary of the Rhine (Fig. 49b).

The history of the Saar which is only 242 km long (7,421 km^2 catchment area) also started in the middle Tertiary. Until the end of the Cretaceous, marine deposits dominated large parts of the area through which the Saar flows today. It only became a flowing water system on an Oligo-Miocene plain which had a south–southeastern/north–northwestern incline. Remains of this plain are still preserved in the area of Orscholz. It was only during the Pleistocene that a significant depth of erosion by the river took place, which continued through the Quaternary and led to the formation of several terraces. The Saar whose source is on the highest peak of the Vosges (new red sandstone), the Donon (1,009 m), flows down into the Moselle, traversing

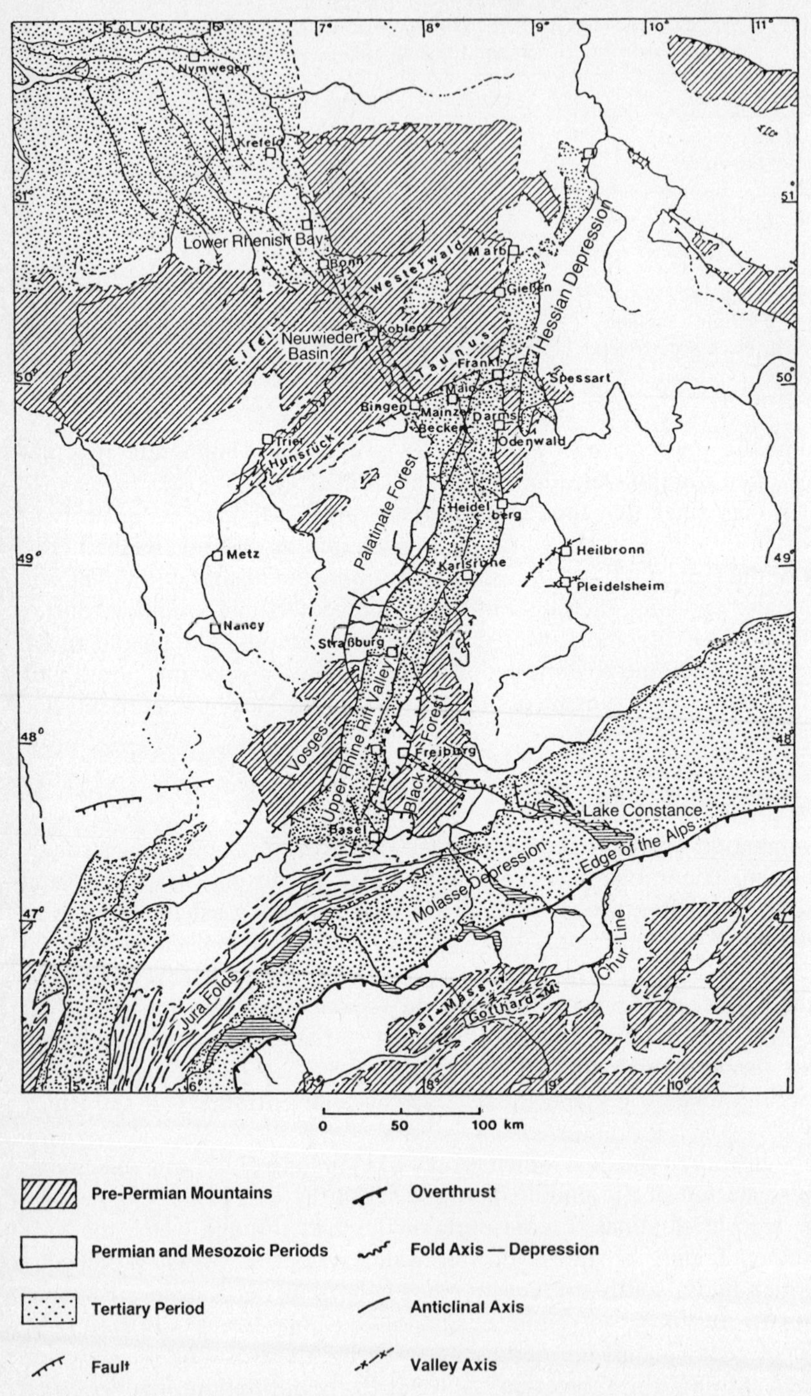

Figure 49b Tectonic structures affecting the River Rhine. The middle section of the Rhine valley developed from a Miocene flat basin, via a Pliocene valley and Pleistocene gorge to the present landscape formation.

varying geological formations which have a decisive influence on its water chemistry, its oxygen balance, and the direction and velocity of its flow. The most prominent external manifestation of the close interlocking of water bodies and landscapes is seen in the formation of characteristic zones of vegetation along the riverbanks.

4.2.3 Biogeographic Classification of Flowing Waters

Although flowing waters — as shown by the examples of the Amazon, Rhine, and Saar rivers — can be characterized, in part, by a rather young geological development and by great seasonal changes, they generally exhibit a distinct zonation from their source to their mouth. In spite of some drift losses, biotic communities survive in separate sections of the river, thereby allowing a regional arrangement and classification of the flowing water bodies. At their place of origin, they possess characteristic source fauna and flora, composed of humidity-loving land and water organisms. Therefore, the actual source region, the eucrenon, can be distinguished from the headstream, the hypocrenon (Malicky 1973).

Gill snails of the genus *Bythinella,* which require temperatures consistently below 8°C, occur in Central European source areas and form a characteristic *Bythinella* biotic community (Musmann 1970, Jungbluth 1973).

Terrestrial biomes in the source areas (Cardamino-Montion) develop not only in the mountains, but wherever seeping waters flow out from inclined slopes and valleys. The acid reaction of the springwater is of significance for the formation of these plant communities. Characteristic species of the silicate spring river banks are the mosses *Philonotis fontana* and *Mniobryum albicans,* while in calcareous spring river banks, the lime-loving mosses *Philonotis calcarea* and *Cratoneuron commutatum,* the butterwort (*Pinguicula vulgaris*) and the sedges *Carex davalliana* and *Carex pulicaris,* characteristically occur.

Depending on the type of source outlet, rheocrenes (= water flows off directly from the source), limnocrenes (= water flows first into an overflow basin), and helocrenes (= water appears in a swamp location) can be distinguished.

In contrast to the headwaters (rhithral) which are characterized by cold-stenothermal species adapted to the high velocity of flow (Fig. 49c), the lower reaches (potamal) contain species adapted to higher temperature fluctuations and lower velocities of flow. On the basis of their characteristic fish fauna, annual mean temperature fluctuations and structure of the river bottom, Central European streams and rivers can be classified into at least four regions, characterized by the dominant species. The uppermost region of a brook, the trout region (containing *Trutta fario*), is characterized by a gravel and stone base with clear, oligotrophic, oxygen-rich water which is uniformly cool all year round. Next comes the grayling zone (containing *Thymallus thymallus*) with generally warmer water and a partly sandy river bottom. In the barbel zone (containing *Barbus barbus,* cf. Illies 1958), characterized by

Figure 49c Rithral Region of the Pagsanjan River on Luzon (Philippines, 8/27/1980). The Pagsanjan Waterfalls are located 110 km southeast of Manila.

fast-flowing river courses, the water becomes turbid due to the sandy or muddy bottom (Fig. 50 and Table 18). Illies (1958) distinguished between a benthal gravel fauna and a mud fauna in this region.

In the lower reaches with low velocity and muddy bottoms, the barbel

TABLE 18 RIVER (OR STREAM) REGIONS (according to Schwoerbel 1977)

Crenal	=	Source zone	Source fauna
Rhithral	=	Mountain stream zone	Salmonidae region
Epirhithral	=	Upper mountain stream zone	Upper trout region
Metarhithral	=	Middle mountain stream zone	Lower trout region
Hyporhithral	=	Lower mountain stream zone	Grayling region
Potamal	=	Lowland river zone	River zone
Epipotamal	=	Upper lowland river zone	Barbel region
Metapotamal	=	Middle lowland river zone	Bream region
Hypotamal	=	Lower lowland river zone	Ruffle flounder region

region becomes the bream region (containing *Abramis brama*). The species occurring here are adapted to the turbid oxygen-deficient, warmer water.

Comparative examinations have shown that, although the expansion of the rhithral and potamal is dependent on the altitude and geographical latitude of an area, it is directly controlled by the temperature. The surface area of the rhithral decreases and the potamal increases at the same altitude from the polar regions to the tropics. The same shift in zones is recognizable in a mountain region from the high plateau surface down to the plains. By damming the water, artificial rhithral and potamal conditions can be created in a lowland river.

Similarly, with other animal species, such as for example the flatworms and molluscs, corresponding distributions can be observed in the Central European rivers and streams. There are also bird species which prefer certain types of flowing water. The water ouzels *Cinclus cinclus* and *C. mexicanus* live along the trout waters of Europe and North America. A close relative, *C. leucocephallus,* occurs in and along the Andean stream and river courses from Venezuela to Bolivia. In the thundering mountain creeks of the Andes (from Columbia to Tierra del Fuego), there are duck species (*Merganetta armata*)

TABLE 19 TYPES OF FLOWING WATER (according to Weber-Oldecop 1977)

		Lime-deficient water	Lime-rich water
Salmonidae region (water cold during the summer)	Mountains	Type I Lemaneetum-fluviatilis Hildenbrandietum-rivularis Chiloscypho-Scapanietum	Type II Vaucherio-Cladophoretum
	Lowlands	Type III Callitricho-Myriophylletum	Type IV Ranunculo-Sietum
Cyprinidae region (water warm during the summer)	Lowlands	Type V Sparganio-Elodeetum	Type VI Sparganio-Potamogeton pectinati

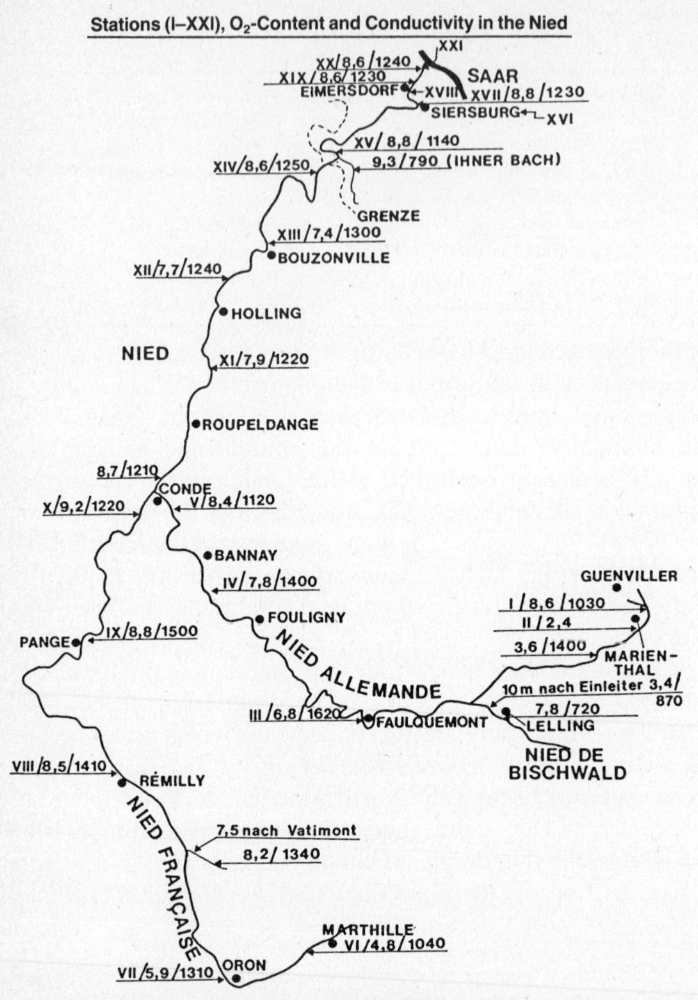

Figure 50 (above and opposite) Fish species in a German/French border river (Nied) in 1977. Electric fishing techniques were used to record fish species at 21 stations (I–XXI). The oxygen content and the conductivity at the time of collection are listed after the location number.

which survive as torrent ducks or "Pato-corta-corientes" by means of a special adaptation to the fast-flowing rivers.

The vegetation at high altitude shows a distribution pattern which is dependent upon the river regions. The water moss *Fontinalis antipyretica* is characteristic of fast-flowing mountain creeks. This species also occurs in well oxygenated lakes; individuals, however, distinguish themselves by their location with respect to the creek's water velocity. Due to an increase in the strength of the epidermis, the rupture limit of specimens from a river amounts to 535 g/mm^2, whereas for those from a lake it is 350 g/mm^2. *Potamogeton* species dominate in the potamal. In the tropical mountains, the Podostemonaceae (America, Africa, Asia), and the Hydrostachyaceae (the Cape, Madagascar) are characteristic, and exhibit striking convergences as a means of

Fish Species confirmed in Stations I–XXI

Station / Species	GERMAN NIED					FRENCH NIED					NIED										
	I	II	III	IV	V	VI	VII	VIII	IX	X	XI	XII	XIII	XIV	XV	XVI	XVII	XVIII	XIX	XX	XXI
Cottus gobio	/		/	/	/		/	/	/	/			/	/	/	/	/	/	/	/	/
Gobio gobio			/	/	/			/	/	/	/	/	/	/	/	/	/	/	/	/	/
Noemacheilus barbatulus	/		/	/	/			/	/	/	/	/	/	/	/	/	/	/	/	/	/
Rutilus rutilus	/		/	/	/			/	/	/	/	/	/	/	/	/	/	/	/	/	/
Scardinius erythropht.			/	/	/			/	/	/	/	/	/	/	/	/	/	/	/	/	/
Perca fluviatilis			/	/	/			/	/	/	/	/	/	/	/	/	/	/	/	/	/
Leuciscus cephalus			/	/	/			/	/	/	/	/	/	/	/	/	/	/	/	/	/
Chondrostoma nasus					/			/	/	/	/	/	/	/		/	/		/	/	
Abramis brama											/	/			/	/		/			/
Acerina cernua												/	/		/	/		/			
Alburnus alburnus												/			/	/		/	/	/	
Carassius carassius															/			/			
Cyprinus carpio																		/			/
Lepomis gibbosus																/					
Leucaspius delineatus																/					
Leuciscus idus																			/	/	
Lucioperca lucioperca																			/		
Alburnoides bipunctatus								/	/				/		/	/			/	/	/
Anguilla anguilla									/				/	/	/	/	/	/	/	/	/
Salmo gairdneri								/	/					/			/				
Barbus barbus				/				/	/					/			/		/	/	/
Esox lucius					/			/	/									/		/	
Gasterosteus aculeatus	/		/	/			/														/
Phoxinus phoxinus	/							/						/							
Tinca tinca			/	/									/		/	/		/			
Salmo trutta f. fario								/	/												

adaptation to the fast velocity flow. Neither of these groups of plants possess any intercellular spaces. The seeds of the Podostemonaceae quickly swell in water and resist the current by attaching themselves to the substrate (Tables 19 and 20).

In 1971, using submerged macrophytes, Kohler classified the Moosach in the plains of Munich, which has a high lime content and is characterized by steep gradients. He was able to determine a close relationship between the distribution of plant species and physico-chemical water components. Numerous animal species are characterized by a change from river to land ecosystems, such changes being determined by the period of reproduction.

The spotted salamander *Salamandra salamandra* is found in the distribution area of the European beech; it deposits its larvae in the oxygen-rich rhithral.

TABLE 20 CLASSIFICATION OF CANADIAN RIVERS INTO SIX HYDROPHYTIC GROUPS (according to Ricker 1934, Whitton 1976)

		Weak flow	Strong flow
Salmonidae region	Ca-deficient water	*Brasenia, Castalia*	*Zygnema, Mougeotia*
	Ca-rich water	*Chara*	*Cladophora, Fissidens*
Cyprinidae region		*Potamogeton pectinatus* etc.	*Cladophora*

4.2.4 Estuaries

Flowing water ecosystems are not only closely associated with terrestrial ecosystems, but in the estuaries, are constantly exchanging materials and energy with marine ecosystems. In the mouths of rivers, specific relationships are formed, whose rhythm is dependent on the tides of the ocean and the numerous water properties of the rivers (Burton and Liss 1976, McCluskey 1971). Fresh- and saltwater animals exist alongside species which have optimally adapted to the salinity of this zone. Apart from salinity, various transition phases can be distinguished (Table 21).

Close to the bottom, the oxygen supply is greatly reduced. The fauna of estuaries is built upon food chains whose existence is generally based on detritus eaters. Among the Turbellaria, *Procerodes ulvae;* among the Coelenterata, *Cordylophora caspia;* and among the Annelida, *Nereis diversicolor* and *N. virens* are typical European estuary species; and numerous molluscs (*Sphaeroma,* etc.) and fish (*Platichthyes flesus, Gasterosteus acileatus*) are also found. Estuaries are young formations. They are the direct result of the eustatic fluctuations in sea level during the Pleistocene.

4.2.5 Related Marine Groups

Initially marine species may migrate into the fresh water via the estuaries. The question of whether the numerous marine elements among the water fauna in the Amazon chose this path or whether they are relicts from the marine past of the river system can only be answered via fossil history. In the Amazon waters, the fresh water dolphins and the stingray are particularly out of place since these species usually belong to the marine environment. This

TABLE 21 BREAKDOWN OF ESTUARY ZONES ACCORDING TO SALINITY

Zone	Salinity (% NaCl)
Hyperhaline zone	>40
Euhaline zone	40 –30
Mixohaline zone	(40) 30 – 0.5
Mixoeuhaline	>30 but < at the estuary/ocean interface
Mixopolyhaline	30 –18
Mixomesohaline	18 – 5
Mixooligohaline	5 – 0.5
Limnic zone	< 0.5

also applies to the Amazonian freshwater sardines and soles. The shape of the small, slender, elliptic Amazon sole *Achiropsis nattereri* reminds one of its considerably larger marine relative, *Achirus achirus;* a rounder sole of the Amazon and Ucayali can grow to hand size. Also, the Amazonian needlefish (*Potamorhaphis guianensis, Strongylura amazonica*) have very close relatives in the Pacific. The original river dolphins (Platanistoidea) occupy a special position among the toothed whales. The Iniidae family, of which the Amazon dolphin *Inia geoffrensis* is a member, is represented by another species, the Chinese river dolphin *Lipotes vexillifer,* uniquely occurring in Lake Tung-Ting in Central China (Hunan Province). The La Plata river dolphins living within the La Plata system, however, belong to yet another family (Stenoldelphidae, represented by the species *Stenodelphis blainvillei*) which, also occurs in brackish and salt waters. The Ganges dolphins (Platanistoidae) represent a third family of river dolphins in the form of *Platanista gangetica* found in the Ganges and the Indus. In addition to *Inia geoffrensis,* another dolphin species occurs in the Amazon, i.e., *Sotalia fluvatilis,* whose closest relative (*Sotalia guianensis*), however, lives in the littoral zone of the ocean.

Several freshwater crabs have marine relatives to this day. *Atya lanipes* (cf. Holthuis 1963) so far has only been known to live in the freshwater bodies on St. Thomas (Virgin Islands). Its closest relative occurs along the West African coast (*A. africana*). *Troglocubanus jamaicensis,* among others, lives in the cave waters of Jamaica. *Palaemonias ganteri* is known to exist in caves in Kentucky in the eastern U.S.A., and *P. alabamae* in caves in northern Alabama (family Atyidae). The numerous neotropical freshwater crabs and prawns are also descended from marine ancestors. Some species of the freshwater prawn family Palaemonidae, particularly the subfamily Palaemoninae still exhibit features manifesting the transition between fresh- and saltwater. Within the group as revised by Holthuis (1952), there exists nine genera in North and South America. Like *Leander tenuicornis, Brachycarpus* (monotypic, *B. biunguiculatus*) is a widely dispersed marine species which also lives in the Mediterranean. All the intermediate stages occur with the 28 known *Macrobrachium* species. The eggs of some of the freshwater species are laid in saltwater. Such transitions also exist in the case of the genera *Palaemon* and *Palaemonetes.*

Many species in Central and South America which are closely associated with freshwater, show a clear distinction between Atlantic and Pacific species (Chile to Peru). Cave forms, too, occur with the Palaemoninae. *Creaseria morleyi* inhabits the caves of Yucatan (Mexico), and *Troglocubanos* which is obviously associated with *Palaemonetes,* has four closely-related species in the caves of Cuba.

Among the anomures, only the Aeglidae family containing the genus *Aegla,* penetrates freshwater (Schmitt 1942, Buckup and Rossi 1977). This genus is associated with flowing water. It occurs in numerous species in southern Bolivia and southern Minas Gerais (Brazil), as far as Chiloe and the Rio Negro in Argentina.

4.2.6 Pollution in Flowing Water

Advantages of location and the utilization of flowing water bodies for cheap transportation and as a safe means of access has meant that since the end of the 18th century, river valleys have become the preferred energy centers of the industrial nations. The roads and railways follow the naturally outlined routes of the terrain. The possibility of using the rivers' fresh water and of discharging the waste water into them decided the question of the

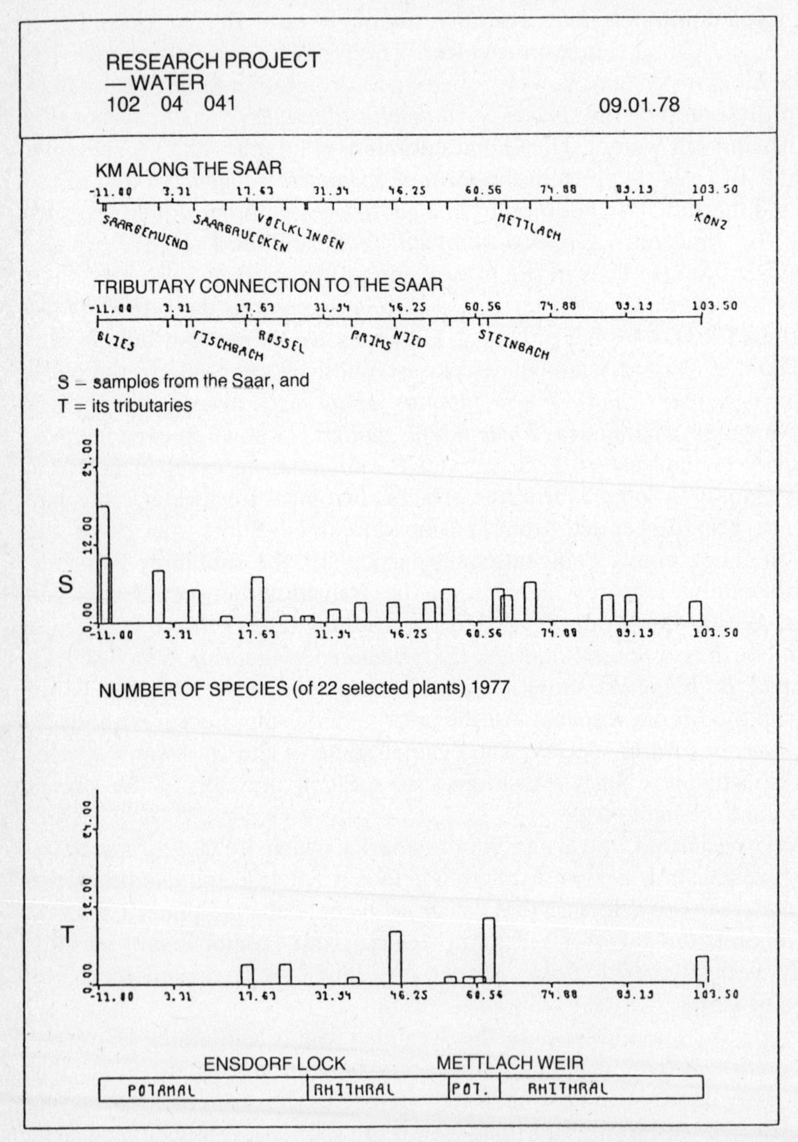

Figure 51 Distribution of 22 submerged water plants in the Saar (Saargemünd to Konz; = S) and the mouths of its tributaries (N). Numbers indicate length in kilometers along the Saar. Along the edge of the diagram, the hydrographic conditions (potamal, rhithral) are shown.

location of large industries, including today's nuclear power plants. The pollution in flowing water systems thus caused, led to the extinction of the original biotic communities, to a considerable change in the composition of species, and/or to the occurrence of completely new species and relationships (Fig. 51). This development can be followed particularly well in flowing water bodies which have been regularly examined since the turn of the century.

Many species occurring today in the Saar, for instance, owe their existence to man. This is true for the migration mussel *Dreissena polymorpha* and the snail *Physa acuta* (Fig. 52) and for the tubificids *Branchiura sowerbyi* and the prawn *Atyaephyra desmaresti.* Some of the introductions were along the 311 km Rhine–Marne canal, built between 1838 and 1853 and the Saar–Kohle-Canal, constructed between 1862 and 1866. Others (for example, *Branchiura*) were probably introduced directly into the river where they preferred to live in the heated sections. A number of highly toxic substances, mud loads, and thermal outflows (for instance, from power plants) produce a mosaic which, due to the strictest selection conditions, favors special wastewater biotic communities. Increased danger exists in the vicinity of thermal outflows, the danger being greater due to faster chemical reactions. As long as this pressure can be tolerated, heat-loving species may appear which cannot survive in rivers at normal temperatures. This applies, for instance, to the tropical Tubificidae *Branchiura sowerbyi.* Many fish living in the heated water of the rivers exhibit faster growth. It is difficult, however, to determine their age, since due to the often limited or absent hibernation period in winter, the annual rings on their scales melt into each other. The shift in time of spawning maturity and the favored development of eggs at high water

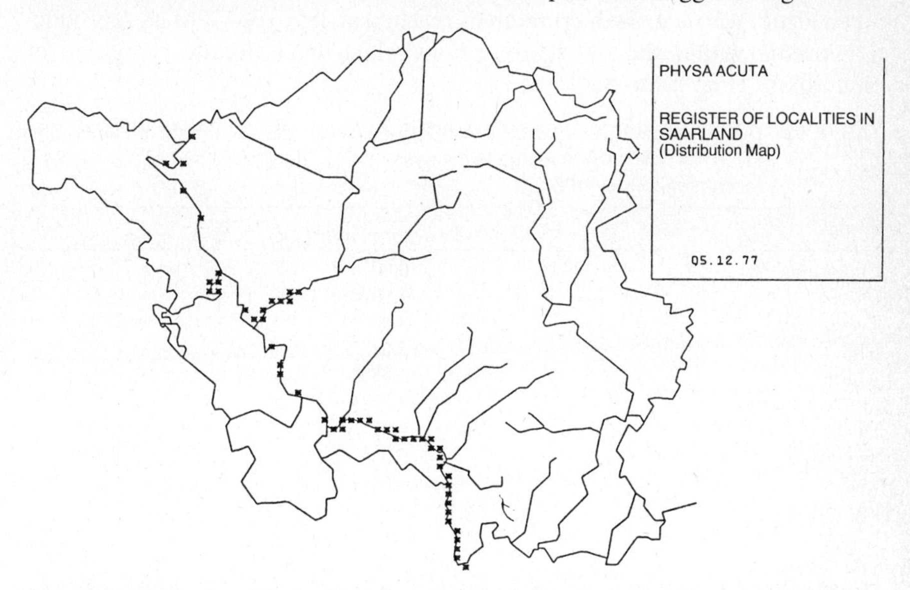

Figure 52 Distribution of the introduced water snail *Physa acuta* in the Saar and its tributaries. This distribution pattern, however, shows characteristic seasonal fluctuations which are related to the development of individual populations (cf. Fig. 53 a, b, and c).

TABLE 22 HEAVY METAL CONCENTRATION (ppm) IN THE SEDIMENTS OF GERMAN RIVERS (according to Förstner and Müller 1974)

River	Pb	Zn	Cd	Cu	Co	Ni	Cr	Hg
Rhine	251	903	9	192	26	164	330	6.3
Main	218	810	12	208	51	128	211	5.0
Neckar	211	999	37	203	55	190	382	1.1
Danube	156	699	14	232	47	125	187	1.5
Ems	112	642	10	55	54	104	134	4.4
Weser	241	1572	14	115	57	98	281	2.3
Elbe	430	1425	21	161	51	126	175	7.6

temperatures lead us to conclude that the time of hatching of the fry may fall into the period of reduced food supply (winter).

The effects of increased temperatures on the development of the fish spawn are more serious, however, since the spawn and young fry are more sensitive than mature fish. Although the eggs may not die due to excess temperatures, simultaneous deficiencies in oxygen can lead to disturbances in their development, deformities, or an early death. Warm water outflows also affect the chemical balance of the water. They change the conversion of substances which occurs on the levels of production, consumption and decomposition. The heating of the water leads to an increase in the number of heterotrophic organisms and thus to a predominance of heterotrophic processes. Because of the warming up, the microbial mineralization further reduces the already diminished O_2-content. This effect is particularly pronounced in water bodies having a strong organic contamination. The increase in water temperatures leads to a shift from diatom populations to green and blue-green algae, whose growth optimum is reached at 28° to 30°C. Mass reproduction occurs within the flat bank regions which leads to the formation of phytotoxins (Fig. 53 to 53c).

TABLE 23 POISONOUS EFFECTS OF VARIOUS PESTICIDES ON FRESHWATER FISH (*TILAPIA MARIAE*). Average lethal doses (LD_{50}) in mg/l in a 48-hour test at 24°C (according to Klee 1969)

Pesticide	LD_{50} in mg/l
Endosulfan	0.0015
Endrin	0.0021
Toxaphene	0.0078
Dieldrin	0.037
Aldrin	0.059
Methoxychlor	0.072
DDT	0.075
Heptachlor	0.080
Lindane	0.083
Chlordane	0.10
Parathion	0.24
Chlorthion	0.52
Diazinon	0.60
Systox	7.5
Malathion	15.0

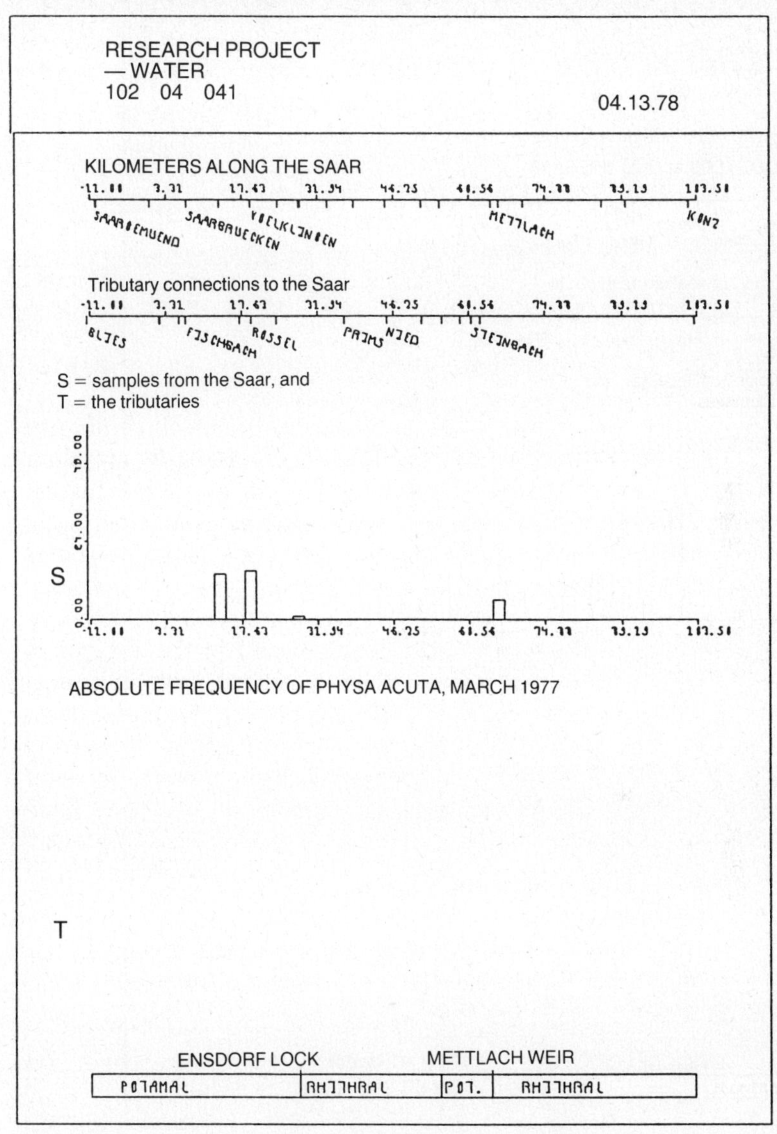

Figure 53a Occurrence of *Physa acuta* based on a controlled sample taken in March 1977 in the Saar (S) (river kilometers plotted above) and its tributaries (T).

As a result of the higher temperatures the effects of the pollutants dissolved in the water are usually magnified. In this connection, surface reactive agents, organic compounds, and heavy metals deserve particular mention (Burton and Liss 1976) (Tables 22 and 23).

When examining their biological effect, however, care must be taken to consider not only the dissolved metals but also their transformation into sediment and nutritional chains. Examinations of Virginian oysters (*Crassostrea virginica*) from the Rappahannok River in Virginia which contained up to

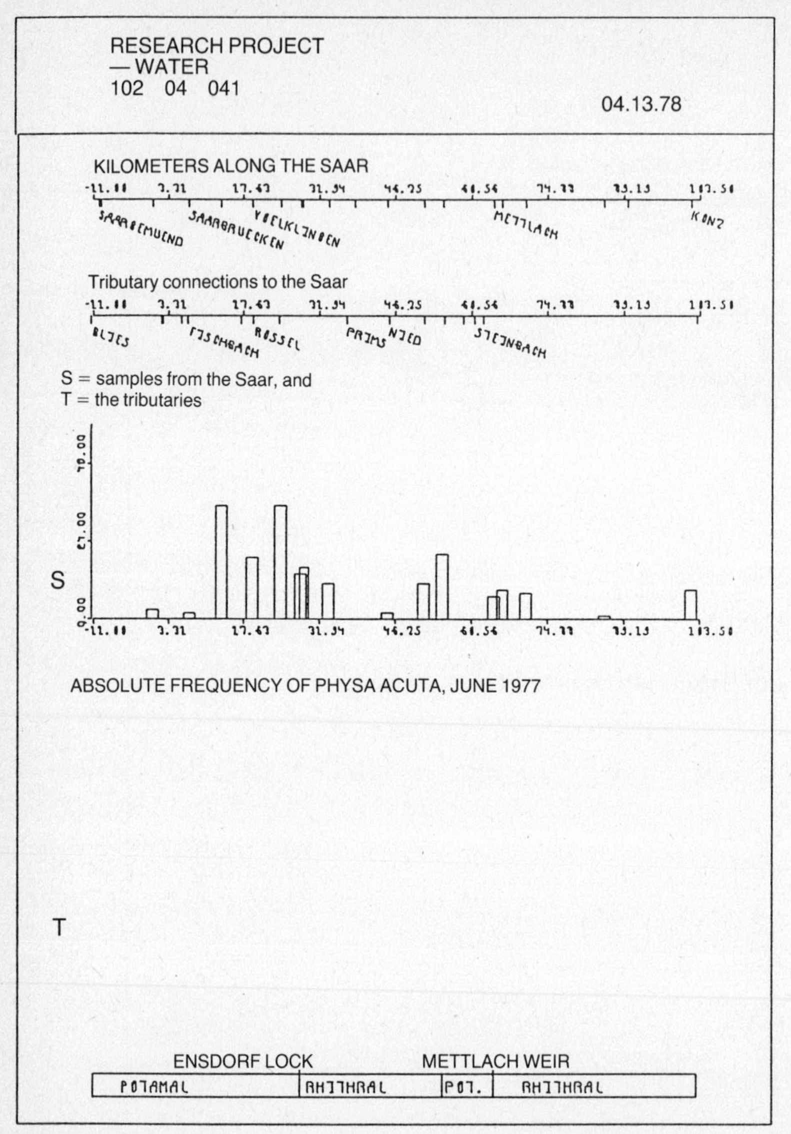

Figure 53b Occurrence of *Physa acuta* in June 1977 in the Saar (S) and its tributaries (T).

600 ppm of zinc, show clearly that the salinity of the environment also controls the absorption of heavy metals (Huggett, Cross, and Bender 1975). An animal weighing 20 g and having a through flow rate of about 575 l/day, with a particle solution of 30 mg/liter and 2 ppm of zinc, absorbs and retains 600 ppm of this zinc contamination per year.

Tests in the estuary region show distinctly that — with an increase in salinity, irrespective of the sediment concentration — oysters have a lower zinc and copper absorption rate than individuals in a stronger freshwater environment.

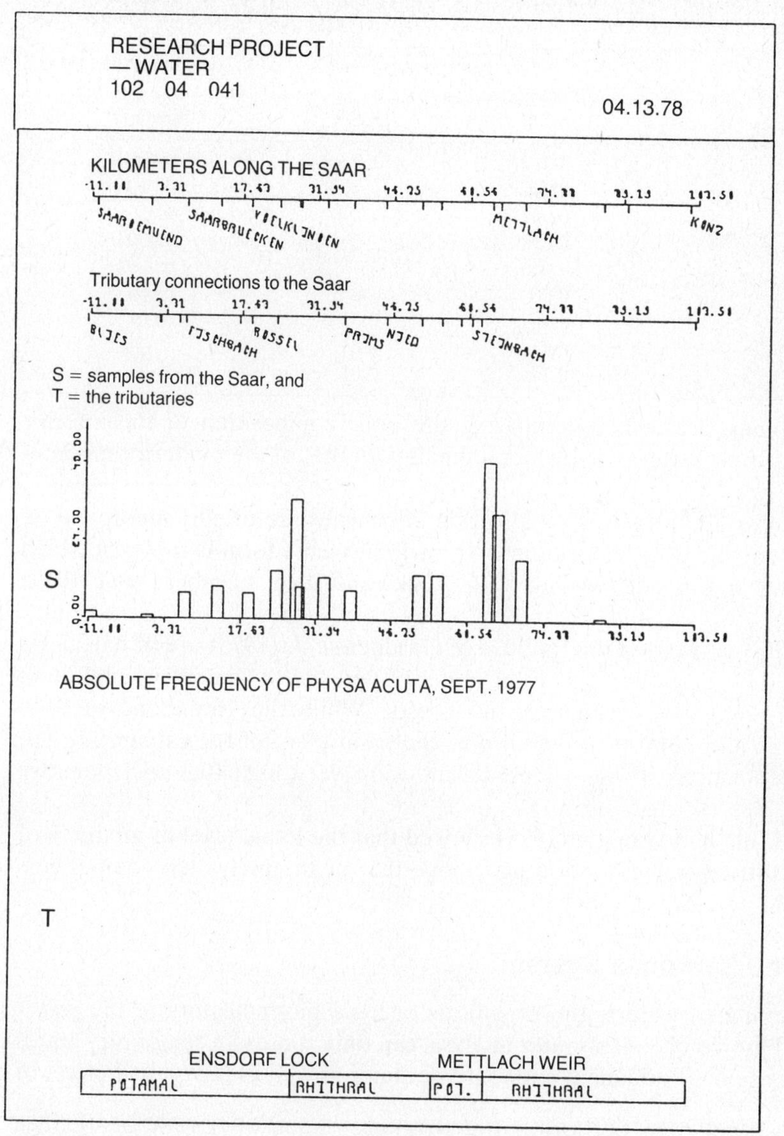

Figure 53c Occurrence of *Physa acuta* in September 1977 in the Saar (S) and its tributaries (T). While in the Saar — due to the inflow of warm water — most populations can be recognized in May, gatherings in tributaries are successful only during the summer months.

The tolerances given by Quentin (1970) for pesticides dissolved in water in the U.S.A. and U.S.S.R. show an uncertainty in their evaluation (Table 24).

Since limnic organisms can serve as "catching substrate" organisms for pollutants, species from differently contaminated waters are generally characterized by a concentration of the pollutants in the corresponding habitat. There are, however, some significant species-related differences.

TABLE 24 TOLERANCES OF PESTICIDES IN WATER (according to Quentin 1970; in mg/l)

Agent	Tolerances		Toxic to freshwater fauna (not yet lethal, however)
	USA	USSR	
Aldrin	0.017	1.0	0.00004
DDT	0.042	–	0.0006
Endrin	0.001	–	0.0001
Heptachlor	0.018	–	0.0002
Lindan	0.056	–	0.0002
Toxaphene	0.05	–	0.003
Parathion	0.1	0.03	0.001
Malathion	0.1	0.05	–
Systox	0.1	0.01	–

This becomes clear when comparing the ionic composition of the interior medium of freshwater or saltwater animals with that of their exterior medium (Table 25).

Heavy metals and pesticides in the organisms are usually multiplied by the environmental concentration and, in food webs, form complexes which are difficult to degrade (Archer 1976, Lockwood 1976, Vernberg *et al.* 1977) (Table 26).

In this connection the findings of Gardner *et al.* (1975) are of particular interest; they examined the total mercury and methyl-mercury content in organisms of various estuaries and coasts. While they found no methyl-mercury at all in *Spartina alterniflora,* the main plant of the estuary region, the concentrations in animal organisms were very high (0.1 to 1 ppm/dry weight).

Marking and Dawson (1975) showed that the lethal level of mixtures of harmful substances very often lies above that of the individual components (Tables 27 and 28).

4.2.7 The Saprobes System

In contaminated waters, the organisms become bio-indicators of the water quality. Physico-chemical water analysis can only represent temporary states of a water body. Permanent and extreme states as well as the overall effect of

TABLE 25 IONIC COMPOSITION OF THE INTERNAL MEDIUM OF ASTEROIDEANS, THE MUSSELS, AND SHORE CRABS OF THE BALTIC SEA IN mMol/l AND OF THE BRACKISH WATER (16‰ NaCl) SURROUNDING THEM

	Na^+	K^+	Ca^{2+}	Mg^{2+}	Cl^-	SO_4^{2-}
Brackish water	215	5.0	5.6	24.1	253	13.1
Asterias rubens (Coelom liquid)	216	5.4	5.6	24.2	255	13.1
Mytilus edulis (blood)	213	7.5	5.8	24.5	253	13.2
Carcinus maenas (blood)	358	10.1	9.7	12.6	384	8.1

TABLE 26 DDT AND METABOLIC RESIDUES IN FOOD CHAINS OF THE EVERGLADES NATIONAL PARK (1966–1968; according to Bevenue 1976)

Substrate	Residues (ppb)
Surface water	0.01
Freshwater	0.46
Flood plain	0 – 49
Algae banks	0.2 – 34
Swamp snail eggs	14
Oysters	0 – 27
Snails	0
Crabs	0
Freshwater prawns	0 –133
Crustaceans	0 – 37
Mosquito fish (Gambusia)	16 –848

the harmful substances can only really be confirmed by living systems (Kohler 1975, Hildenbrand 1974, Müller 1974, 1977, Müller and Schäfer 1976). The saprobes system originally established for flowing waters by Kolkwitz and Marsson (1902, 1908) in order to evaluate quality was, after some vehement criticisms, continuously improved and expanded to include lakes (among others by Bick and Kunze 1971, Fjerdingstadt 1971). Liebmannn (1969) distinguishes four saprobe phases:

oligosaprobe = quality class I
β-mesosaprobe = quality class II
α-mesosaprobe = quality class III
polysaprobe = quality class IV

Other systems (for example, Fjerdingstadt 1971) make a further differentiation and emphasize, in particular, the change from heterotrophic to autotrophic in the self-purification process of a water body. The *German Standard Process for the Examination of Water* distinguishes four biologically characteristic areas as saprobe phases depending on the degree of pollution or the phase of self-purification reached in a flowing water body.

Polysaprobe Zone Section of a water body which is very heavily polluted. Micro-organisms dominate and bacteria exhibit mass reproduction. The density of individuals is generally quite high. Decomposers predominate while producers are almost completely absent. Organisms with high oxygen content are also absent.

TABLE 27 Hg-CONTENT (in ppm) IN FISH OF VARYING ORIGINS (according to Reichenbach-Klinke 1974)

River	Trout	Eel	Bream	Perch	Pike	Sander	Carp
Amper	0.089	0.066	–	–	0.551	–	0.162
Danube (Dam)	0.8	0.5	0.5	0.79	0.78	0.9	0.5
Lech	0.1		–	0.65	0.25	–	–
Rhine	–	0.27	0.10	0.15	0.51	0.10	0.29

TABLE 28 CONCENTRATION (ppm/dry weight) OF TOTAL MERCURY AND METHYL MERCURY IN THE BODIES OF COASTAL ORGANISMS IN THE SOUTHEASTERN PART OF THE UNITED STATES

Species	Number of specimens examined	Total mercury	Methyl-mercury
Anguilla rostrata (Anguillidae)	1	1.2	0.85
Brevoortia tyrannus (Clupeidae)	2	0.30	0.25
Arius felis (Ariida)	3	1.8	1.41
Opsanus tau (Batrachoididae)	1	2.9	1.4
Fundulus heteroclitus (Cyprinodontidae)	1	0.47	0.14
Centropristis striata (Serranidae)	2	0.84	0.47
Rachycentron canadus (Rachydentridae)	1	0.93	0.70
Archosargus probatocephalus (Sparidae)	1	0.75	0.46
Negaprion brevirostris (Carcharhinidae)			
Muscles	4	3.8	3.1
Carcharhinus leucas (Carcharhinidae)			
Muscles	1	10.2	3.9
Squalus acanthias (Carcharhinidae)	1	4.0	3.0
Callinectes sapidus (Crustacea, Portlinidae)	8	0.45	0.31
Spartina alterniflora (Gramineae)			
Roots	8	0.78	–
Stems	9	0.23	–
Leaves	8	0.23	–

α-Mesosaprobe Zone Section of a water body which is heavily polluted. Numerous species of micro-organisms; macro-organisms are also heavily represented. The decomposers still dominate; however, the producers, as well as the animal consumers, are increasing. The total number of species is higher than in the polysaprobe-zone.

β-Mesosaprobe Zone Sections of a water body with moderate organic pollution and optimum living conditions for most organisms. Large increase in the producers and consumers and decrease in the decomposers. Biotic communities show great constancy and presence of species.

Oligosaprobe Zone Section of a water body containing pure water with practically no organic pollution. Macro-organisms dominate. Producers are in the majority. The number of species is great, the individuals' density per species is generally low.

The *German Standard Process for the Examination of Water* emphasizes the fact that the classification of organisms is undertaken according to their ecological potential. However, any biological evaluation must include the biotic community's total structure consisting of the animal and plant biomes.

"If the type of organic pollution (oxygen balance, degrading pollution such as H_2S, NH_4-ions, etc.) can be derived from the qualitative composition of the population, then the number of species and the density of individuals allow valuable conclusions with respect to the degree of contamination, the

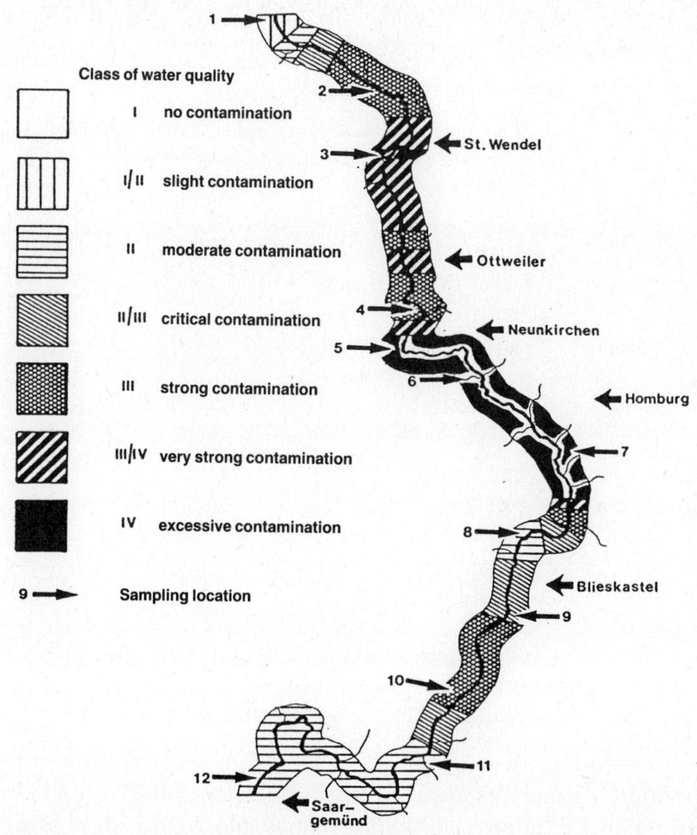

Figure 54 Water quality in the Blies (tributary of the Saar) in August 1978.

intensity of self-purification and the influences of waste waters having a toxic or mechanical effect."

Dead zoological communities, which, however, are still traceable through their relicts (thanatocenoses), often allow conclusions about the causes of harmful effects (Fig. 54).

An examination of permanent biological conditions and their seasonal changes must be undertaken in considerations of the fluctuation of abiotic factors of the flowing water system. Hydrographic relationships are of direct importance in this connection. A prerequisite for as complete an analysis as possible of the colonization are frequent examinations such as, for example, samples on a monthly basis. At present, 16 quality requirements are stipulated for surface waters in the Federal Republic of Germany (Table 29).

The flowing water bodies of most industrial states, and increasingly those of the so-called Third World, however, no longer possess completely natural biotic communities. Various contaminations destroyed and changed them. The marine ecosystems are large collection basins for all pollutants (Fig. 54a). The contaminants remain in the biosphere and in turn affect us all. Starting from the ideal assumption that the waste waters (about 140 km^3/year;

TABLE 29 QUALITY REQUIREMENTS FOR SURFACE WATERS IN THE FEDERAL REPUBLIC OF GERMANY (Environmental Report 1974)

Parameter	Standard value
1. Temperature	<28°C; generally, however, not higher than 3°C above the natural equilibrium temperature of water bodies
2. pH	6.5–8.5
3. Oxygen content	Day/night mean > 70% saturation: if discharge < MNQ : at least 60%
4. BSB_5	<5 mg/l
5. $KMnO_4$-consumption	<20 mg/l
6. Dissolved organic carbon	<5mg/l
7. Biological state	β-mesosaprobe or better
8. Chlorides	<200 mg/l
9. Sulfates	<150 mg/l
10. Ammonium	<0.5 mg/l
11. Total iron	<1 mg/l
12. Manganese	<0.25 mg/l
13. Total phosphates (P)	<0.2 mg/l
14. Phenoles	<0.005 mg/l
15. Radioactive substances	<100 pCi/l
16. Toxic substances	No concentrations which are in excess of the tolerance doses for drinking water, which would impair the self-purification of the water or which would be harmful to the fish

Müller 1976) distribute themselves evenly among the oceans, they would be diluted $1350 \times 10^6 : 140$, i.e. approx. 10 million times and would, therefore, not cause any grave pollution. The fact is, however, that the inflowing concentrations are, in many cases, far above the necessary dilution gradients (at least 500 times) (Table 30).

4.3 URBAN ECOSYSTEMS

A city is not an independent, closed system. Its social, economic, and ecological interdependence extends far beyond the city limits. If, therefore,we attempt to define a city as an ecosystem, we have to keep these supraregional exchange processes in mind (input and output relationships).

4.3.1 The Ecological Structure of the Cities

Urban ecosystems possess a specific structure which is the result of interactions between human planning and natural, spatial factors (Müller 1977).

Figure 54a The Pasig River in the north of Manila (Philippines) is polluted by domestic waste water. The water hyacinth (*Eichhornia*) indicates by its growth the high content of organic floating matter (8/24/1980; near Fort Santiago).

Figure 54b i New York 1972.

Figure 54b ii Village near Korhogo (Ivory Coast, November 1979).

Figure 54b iii Slums in north Manila (8/27/80).

Figure 54b iv Covered shopping street in Osaka (Japan, August 1980).

Figure 54b v Traditional and modern Japan (above: Tokyo; below: Osaka; August 1980).

Figure 54b vi Street scenes in Shanghai (September 1980).

Figure 54b vii Street scenes in Peking (9/9/1980).

Figure 54b viii Settlements along the periphery of Caracas (Venezuela, 2/12/1979).

Figure 54b ix Bogota (Colombia, 1979) with a view of Montserrate.

Figure 54b x Downtown Porto Alegre (Brazil, 10/2/1980).

Figure 54b xi Sao Paulo (downtown, 10/22/1964).

Historical development and geographical location influence this structure. The load capacity of urban ecosystems, therefore, depends on the specific behaviour of their inhabitants (production goals, consumption habits) (Fig. 55), on the synergetic interplay between spatial factors and on the ability of the systems close to nature which are interspersed — mosaic-like — in the developed area (green belts, lakes, etc.), and organisms to tolerate the existing load and, if necessary, to reduce it or to break it down. In cities with river terraces and in basin locations, local contamination will normally have a

Figure 54b xii Sao Paulo (Brazil, 10/4/1980).

different effect from that in coastal cities or in cities located at the foot of mountains.

4.3.2 The City Climate

Each city has a specific mesoclimate characterized by a macro-climatic situa-

TABLE 30 LIMITING VALUES OF SUBSTANCES CONTAINED IN WATER (according to Müller 1976; in mg/l)

Substance	For fish	For self-purification of water	For the biological purification of waste waters	For the purpose of land irrigation	For utilization as raw water for processing into drinking water
Arsenic	15 – 23	–	0.7	0.3	0.01
Boron	–	–	–	0.3	1.0
Lead	0.2 – 10	0.1	5	–	0.03
Cadmium	3 – 20	0.1	2–5	–	0.005
Chromium	15 – 80	0.3	2–10	–	0.03
Cyanide	0.03– 0.25	0.1	0.3–2	–	0.01
Cobalt	30 –100	5	–	0.5–1.0	0.05
Copper	0.08– 0.8	0.01	1–5	5	0.03
Nickel	25 – 55	0.1	2–10	0.5–1.0	0.03
Mercury	0.1 – 0.9	0.018	–	–	0.0005
Zinc	0.1 – 2	0.1	5–20	2	0.5

tion and micro-climatic influences. In industrial cities, during calm weather conditions, a layer of smog forms, due to increased concentration of aerosols, whose grain or drop size is generally smaller than 10^{-3} mm (floating capacity), in the city air; this layer of smog, by weakening the effective incoming and outgoing radiation has a direct effect on the radiation balance and, by affecting the formation of condensation droplets, also affects the formation of clouds or precipitation over the city. The duration of sunshine, intensity of the rays and thus local brightness are diminished in the urban area. Due to the reduction in outgoing radiation (glasshouse or greenhouse effect) and the

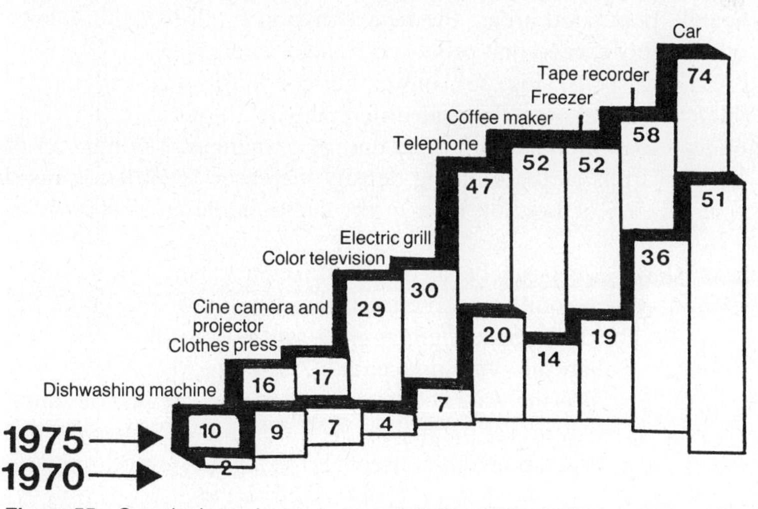

Figure 55 Growth always has consequential effects. The desire for more comfort for the entire family leads to an additional load on our environment. This diagram represents 100 households (4 persons per household, working families with average income), and compares 1970 and 1975 (Federal Republic of Germany).

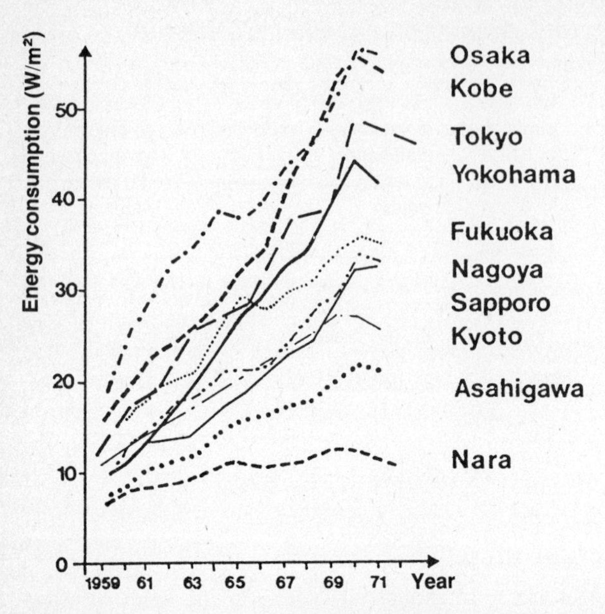

Figure 56 While the world-wide anthropogenically-produced energy amounts to approx. 1‰ of the irradiated solar energy, local energy production and consumption can often exceed solar energies. As an example of the consumption, 10 cities in Japan are listed (according to Bach 1976).

anthropogenic energy produced, the temperature in the cities outside the tropical regions rises distinctly (by about 0.5 to 1.7°C) when compared with the surrounding areas. Viewed globally, anthropogenic energy production is below 1‰ of the solar energy reaching the earth; however, concentrating on cities and heavily populated areas, the heat emission locally attributable to man may considerably exceed that produced by solar energy (Bach, Beck and Goettling 1973, Bach 1976) (Fig. 56). In weather conditions with little energy exchange (high pressure), the thermal differences in the cities are usually more pronounced than in those with great energy exchanges. The number of frost days usually drops as the building density increases. 25% less ground frost occurs in the city of London than in the surrounding areas (Chandler 1970).

The heat island effect can be increased through location in a valley, through the road system, building structures, and density of the buildings, and can be modified or decreased by green areas and water surfaces. Differences in exposure are therefore very noticeable. Asphalt-covered roads normally have the highest, flowing water bodies — unless they are precontaminated by warm inflows (from power plants, etc.) (cf. flowing water ecosystems) — and surfaces bare of vegetation and houseroofs the lowest surface temperatures.

During high radiation weather, due to the rising warm air, low pressure can form over the city which is fed from the surrounding areas. This process can become effective as a transportation mechanism for emissions toward the center of a built-up area. Unless the inversion layer is penetrated by warm air rising from the center of the city, a temperature rise with a reduction in the normal day amplitude will then result during this type of weather situation. The high temperatures at all hours of the day and relatively high night temperatures have a distinctive ecological effect. Earlier blooming times and the occurrence of warmth-loving, introduced animal and plant species absent from the surrounding countryside, are just a few indications of this.

The city is generally drier than its surroundings. With the increase in temperature, the relative air humidity, on the one hand, falls, which, during hot days can lead to the formation of an artificial desert climate (≈40% relative humidity); the capacity for the evaporation of water, on the other hand, increases which in turn elevates the relative humidity, especially in areas where water surfaces suitable for evaporation are located in the vicinity of heat islands. The sensation of muggy air results from this increase in relative humidity (Eriksen 1975). In spite of low air humidity, the larger number of condensation nuclei can lead to an increase in the formation of fog and clouds.

A significant increase (5 to 16%) in precipitation must be expected in the cities (Chicago 5%, St. Louis 7%, Kiel 10%, Moscow 11%, Bremen 16%; for Hamburg, cf. Reidat 1971). The heaviest rain usually falls leeward of the center of a city. Cities attract, particularly in summer, short and violent thunderstorms with heavy rains which pose problems for the drainage systems. From 1893 to 1907, Munich recorded 756 thunderstorms, while Maisach, located 25 km to the west, only recorded 303 thunderstorms. Chicago reported 13%, St. Louis 21% more days of thunderstorms during the summer than the surrounding areas. An analysis of the travelling speed of rainfronts with a tendency to thunderstorms showed in a considerable retardation effect through the dense areas. In La Porte, leeward of the heavy industrial complex of Chicago-Gary, precipitation during the summer is about 30%, and the number of days with thunderstorms about 63% greater than in the surrounding areas (Changon 1970). The effect of the heavy industry is as great here as it is in Paris where, on weekdays, up to 45% more precipitation is recorded than at weekends.

Stiehl (1970) was also able to determine a distinct orographic component for the distribution of precipitation in the metropolitan area of Marburg. Similar findings were made for the city of Saarbrücken. It must be added that rain may also carry harmful substances with it (cf. Seekamp and Fassbender 1974, among others). Södergren (1973) detected up to 35 ng PCB/liter per month in the rainwater. The effect of built-up areas on the water drainage can be observed in all cities and towns.

Wind current conditions in a city are also determined by the structure of buildings, orography and general weather conditions at a given moment. Cities in basin locations are often characterized by a large percentage of calm days. The direction of the wind is greatly influenced by orography (the valley structure), the openness of the roads, and the shape, size, and location of individual building complexes. The daily dynamics of cold air drainage is also of great significance in the aeration of metropolitan areas.

4.3.3 Types of Emission

Apart from the climatic peculiarities found in the city biotope, a wealth of substances foreign to nature and resulting from manufacturing processes are present, particularly in industrial cities.

Emissions in the form of gas and dust, radioactive substances, and noise affect human beings, animals and plants living in the city. Together with the factors relating specifically to the area, they constitute the type of pollution peculiar to a city (Friedländer 1977, Hartkamp 1975, Yordanov 1976).

How closely types of climate and pollution are related to each other is best illustrated by the formation of smog (smoke + fog). Two basic types have been distinguished: the Los Angeles-type and the London-type (Table 31). The most reactive components of the Los Angeles smog are especially dangerous to plants (ozone = O_3, PAN = Peroxacylnitrate) (Fig. 57). With stronger PAN-reactions, plants develop intercostal necroses which assemble into transverse bands. In older, dicotyledon plants the base, and in younger plants the tip of the leaf, is affected.

In monocotyledon plants, PAN produces light-brown, necrotic stripes on the leaves. Depending on the plant species, the stripes run either cross- or lengthwise on the leaves.

Phytogenic indicators were used to prove the presence of smog in Germany (Knabe, Brandt, van Haut and Brandt 1973). The effect of ozone on plants and animals was thoroughly examined. Forage plants showed varying degrees of susceptibility. Among the legumes, resistance to ozone increases — from *Trifolium pratense* (very susceptible), *Melilotus albus, Trifolium repens,* and *Trifolium hybridum,* via *Medicago sativa* to *Coronilla varia,* whereas in the case of grasses, it increases from *Agrostis palustris, Poa annua, Lolium perenne,* via *Festuca rubra* to *Cynodon dactylon.*

TABLE 31 CHARACTERISTICS OF THE LOS ANGELES AND LONDON SMOGS (according to Claussen 1975)

Characteristics	Los Angeles	London
Air temperature	24–32°C	1–4°C
Relative air humidity	Below 70%	85% (+fog)
Inversion type	Falling inversion	Radiation inversion
Wind velocity	Below 2 m/s	Calm
Visibility	600–1600 m	90 m
Most frequent occurrence	August – September	December – January
Most used fuel	Petroleum products	Coal and petroleum products
Important components	NO_x, O_3, CH, CO, PAN	SO_2, Soot, CO
Type of development	Photochemical/thermal	Thermal
Effect on reactant	Oxidation	Reduction
Maximum concentration	Noon	Morning and evening
Irritation	Eye irritation	Irritation of the respiratory organs

Gases and Dusts The origin of air pollution is subject to significant regional fluctuations (point, line and surface sources) (Table 32).

Such overviews, however, can only give an initial reference point for the receipt load to be expected from the surface of a given city. Today, the emission records established in most industrial cities of Europe, the U.S.S.R., and U.S.A. provide more precise figures. They show, however, that great care must be taken when using such statistics in reference to the toxicity of individual harmful substances (Ashley *et al.* 1976).

Just looking at the organic components to be expected in the Cologne city air, based on the emission records of 1972 for that city, it becomes clear that one cannot draw conclusions about the "load" in an area from the measurements of SO_2 or CO alone. The domination and — technically speaking — the relatively unproblematic measurement of CO, dusts, and sulfur dioxide, have led to these substances being viewed as conducting

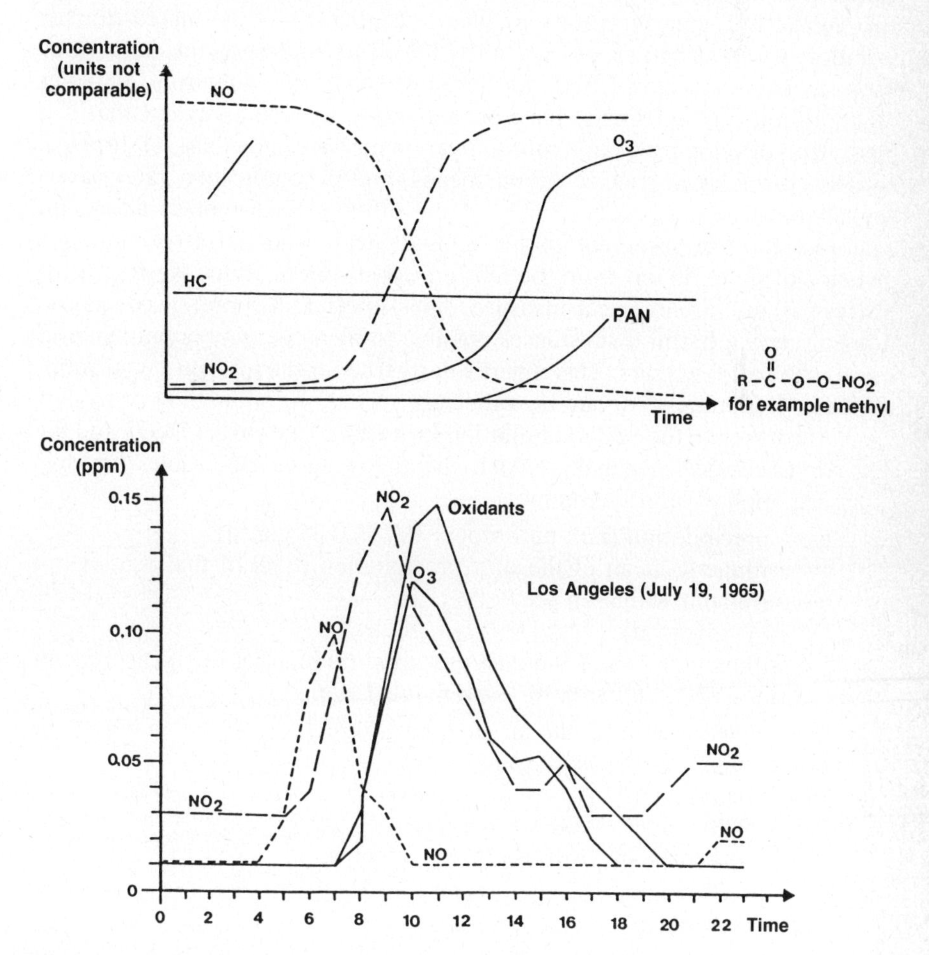

Figure 57 Theoretical process (top) of the photochemical smog formation and proven gas concentrations (above) in the air during a smog occurrence in Los Angeles.

TABLE 32 SOURCES OF POLLUTION IN THE FEDERAL REPUBLIC OF GERMANY AND IN THE USA IN 1970 (in 10^6 t/year; according to Leithe 1974)

Origin	NO_x		CO		Dust		SO_2		Hydrocarbons	
	FRG	USA	FRG	USA	FRG	USA	FRG	USA	FRG	USA
Traffic	1.1	8.1	9	63.8	0.2	1.2	0.1	0.8	2.4	16.6
Energy production and domestic	1.5	10.0	0.3	1.9	1.0	8.9	3.7	24.4	0.1	0.7
Industry and trade	0.04	0.2	1.9	9.7	1.0	7.5	1.5	7.3	0.9	4.6
Waste disposal	0.10	0.6	1.5	7.8	0.2	1.1	0.02	0.1	0.3	1.6
Total	2.7	18.9	12.7	83.2	2.4	18.7	5.3	32.6	3.7	23.5

substances for the receipt load of localities and smog-alarm plans. Monitoring of other substances, however, is generally being neglected (Hettche 1975, Burlotte 1976, Perry and Young 1977).

While SO_2-concentrations are clearly controlled by the macro-climatic situation, CO-concentrations show a rhythm clearly reflecting the daily traffic pattern. The connections between SO_2-concentrations, quantities of fumes emitted, and type of climate have been clarified to such an extent that it is possible to develop prediction safe simulation models on the basis of dispersal models, variables in transportation speeds, and decomposition rates (average: about 12%/hr) (Giebel 1977, among others). Such models allow, for example, the establishment of the annual mean value of 10 $\mu g/m^3$ at a distance of 60 to 80 km from the SO_2 emissions in the Ruhr. Restructuring processes and technological measures have made reductions of concentrations of certain harmful substances possible in many densely-populated ares during the past few years. This is particularly true for the total dust concentrations (cf. Benjamin 1976, among others).

According to the Federal Pollution Protection Law, of 3/15/1974 and the TM-Air (Technical Manual — Air), the following receipt values for dust types are applicable in Germany:

(a) Dust precipitation (non-hazardous) $IV_1 = 0.35$ g/m^{2}d)
(= arithmetic mean of the dust precipitation in all of the locations of measurement within an area)
$IV_2 = 0.65$ g/m^{2}d)
(= arithmetic mean of the dust precipitation in all of the measurement locations within an area for each month)

(b) Dust concentration in the air (non-hazardous)
Size of particles below 10 μm
$IV_1 = 0.10$ mg/m^3
$IV^2 = 0.20$ mg/m^3
Size of particles above 10 μm
$IV_1 = 0.20$ mg/m^3
$IV_2 = 0.40$ mg/m^3

TM-Air gives dust precipitation figures in the form of annual and monthly mean values, and daily mean values for concentrations of dusts with a particular size over 10 μm. The dust precipitation for each location is calculated from the following formula:

$$\text{Dust precipitation} = m/AT \ (\text{g/m}^2\text{d})$$

where:

m = measurements of dust precipitation
A = collection surface in m^2
t = sampling time in days

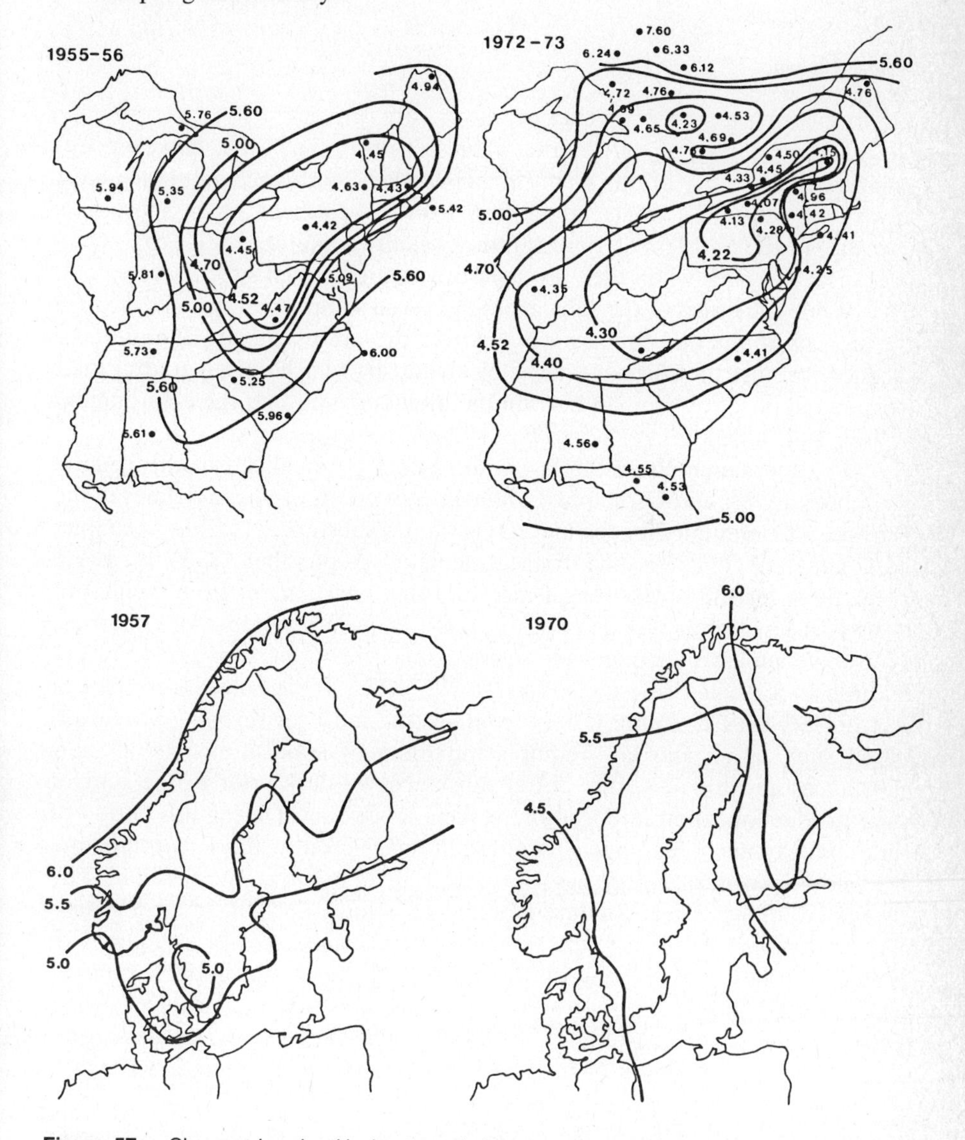

Figure 57a Changes in rain pH, due to variations in sulfur ion concentrations from 1955 to 1973 in the USA and Scandinavia as a result of sulfur ion concentrations (according to Likens, 1976).

TABLE 33 LIMITING VALUES ACCORDING TO THE TECHNICAL MANUAL FOR THE PREVENTION OF AIR POLLUTION (TM-Air), ISSUED IN 1977

	mg/m^3	mg/m^3
Gas	I_1	I_2
Chlorine	0.10	0.30
Hydrochloric acid	0.10	0.20
Hydrofluoric acid	0.0020	0.0040
Carbon monoxide	10.0	30.0
SO_2	0.140	0.40
H_2S	0.0050	0.010
Nitrogen dioxide	0.10	0.30
Nitrogen monoxide	0.20	0.60

The reliability of the limiting values given in Table 33, however, depends on the quality of our scientific knowledge. Since each substance must always be evaluated in its interaction with a wealth of other substances (because of the problem of synergy), they usually do not do justice to the real biological pressure when they are measured as individual substances.

While the effect of the substances discussed so far generally acts through the air, environmental chemicals may arrive by various means. Environmental chemicals are products of the chemical industry which during or after their intended application, spread beyond the intended area into the environment (water, soil, air).

The fundamentals of the law with respect to chemicals of the Federal Republic of Germany correspond to the requirements of the guidelines of the European Community for the marketing of new substances (see, for example, EEC guideline 79/831 issued by the Council on September 18, 1979). Legal regulations for the protection of man and the environment from dangerous chemicals have already existed for several years in a number of countries. Thus, Sweden and Japan enacted such laws in 1973, Canada in 1975, Norway and the U.S.A. in 1976, and France in 1977. The Toxic Substances Control Act (TOSCA 1976) in the U.S.A. contains the most far-reaching provisions of all such laws. The law requires the filing of applications for all new substances and for new uses of old substances 90 days prior to the start of their production, their importation or their processing. During this period of time, which can be extended by an additional 90 days, the Environmental Protection Agency (EPA) has the power of restrictive intervention. The stipulated application procedure is practically the equivalent of a procedure for admission. In contrast, the fundamental principle of German law with respect to chemicals is a registration procedure with the implicit right to proceed without specific authorization. This means that the producer or the importer may start marketing a new substance at the latest 45 days after receipt of the registration of the new substance by the Registration Bureau, provided the documentation is complete and correct. Japanese Law provides an admission procedure for the marketing of new substances. On the basis of a pre-examination, the substances are classified into: harmless (white list), harmful (black list), or into substances where uncertainties with respect to

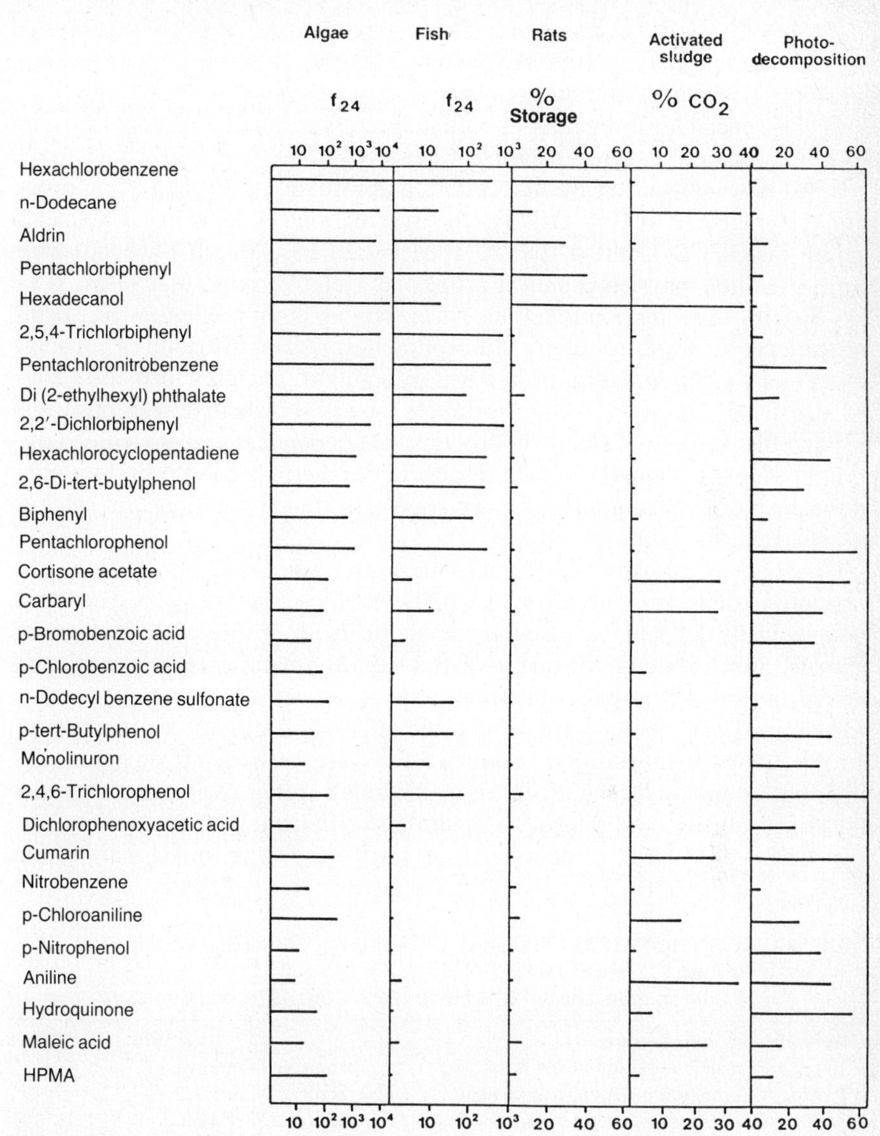

Figure 57b Comparison of the reaction of chemicals in an ecotoxicological profile analysis. The substances were listed in accordance with their increasing water solubility. The diagram shows that, for instance, accumulation in algae can be correlated with water solubility; i.e., low water solubility means high concentration (f_{24} = concentration factor in relation to surrounding medium; percent CO_2 within the activated sludge and the photo-decomposition test are a measure of total decomposition in relation to the substances used; according to Korte 1980).

their harmlessness or toxicity exist (grey list). The latter are subject to extensive testing.

The Law for Protection against Harmful Substances (Law concerning chemicals) in the Federal Republic of Germany became effective on January 1, 1982. Its purpose is "to protect man and the environment from the harmful effects of hazardous substances by means of an obligation on the part of the marketer to test and register substances and to classify, mark and package hazardous substances and preparations, and by mean of bans and limitations as well as special legal regulations with respect to protection from poison and from occupational hazards" (para. 1 ChemG).

All new substances are first tested in accordance with the testing program for chemicals of the OECD (the International Control of Chemicals within the OECD Context ENV/CHEM/79.22/Paris, March 14, 1980) with respect to their physico-chemical properties (relative molecular mass, UV-VIS-Spectra, melting point/melting range, boiling point/boiling range, steam pressure curve, water solubility, adsorption/desorption, distribution constant (*n*-octanol/water), volatility from a watery solution, complex formation ability, density of liquid and solid substances, size and distribution of particles/size and distribution of fibers, hydrolyses, pH-dependency, dissociation constant in water, screening test with respect to the thermal stability and stability of the air, viscosity of liquids, surface tension of watery solutions, fat solubility, permeability and corrosiveness).

Acute toxicity (toxicity of a substance after a single administration) and subacute toxicity are determined from 'Lethal Dose-50' (LD_{50}) tests on rats — statistically determined doses leading to the death of 50% of the laboratory animals. Subchronic or chronic toxicity is toxicity manifested after repeated administration over a longer period of time. The tests usually last 28 days for subacute toxicity, 90 days for subchronic effects and at least six months for chronic toxicity. In addition, extensive tests are carried out on numerous other environmentally significant risks, including mutagenicity, carcinogenicity, teratogenicity, and effects on organisms as indicators of ecotoxicity.

So far, standardized open land tests with respect to an estimate of the

TABLE 34 MEAN RADIATION EXPOSURE OF A HUMAN BEING IN THE FEDERAL REPUBLIC OF GERMANY
(according to Aurand 1976, Schwibach and Gans 1976, and others)

1. Natural radiation exposure		approx. 110 mrem/a
1.1 due to cosmic radiation at sea level	approx. 30 mrem/a	
1.2 due to terrestrial radiation from outside	approx. 60 mrem/a	
1.3 due to incorporated radioactive substances	approx. 20 mrem/a	
2. Artificial radiation exposure		approx. 60 mrem/a
2.1 Medicine (for example, X-ray diagnosis, nuclear medicine, radiation therapy)	approx. 51 mrem/a	
2.2 Nuclear bomb tests	approx. 8 mrem/a	
2.3 Technology	approx. 2 mrem/a	
2.3.1 Radiation sources	approx. 1 mrem/a	
2.3.2 Industrial products	approx. 1 mrem/a	
2.3.3 Interference (e.g. television)	approx. 0.7 mrem/a	
2.3.4 for persons professionally exposed to radiation	approx. 0.7 mrem/a	
2.3.5 due to peaceful utilization of nuclear energy	approx. 1.0 mrem/a	

TABLE 35 RADIATION EXPOSURE WITHIN THE FEDERAL REPUBLIC OF GERMANY (according to Aurand 1976)

Land (State)	In the open air (mrem/a)	Inside apartments (mrem/a)
Baden-Wuerttemberg	54	69
Bavaria	60	74
Berlin	51	61
Bremen	37	46
Hamburg	49	49
Hessen	53	79
Lower Saxony	42	57
North Rhine-Westphalia	52	67
Rhineland-Palatinate	60	88
Saarland	69	106
Schleswig-Holstein	50	53
Mean value	52	70

ecosystem effect of environmental chemicals are missing. The test procedures undertaken so far, however, permit risk estimates.

Radiation Contamination and Radionuclides Radiation exposure and absence of radionuclides in the food chains of built-up areas and their populations are gaining increased attention. Depending on the type of radiation, the biological effect can differ, even though the rad-number may be the same; this is why it is necessary to determine the relative biological effectiveness of radiation (RBE). The radiation load depends on a number of factors (natural and artificial radiation exposure) (Tables 34, 35 and 36).

Units of measure for radioactivity (absolute measures):

1 Curie = 1 Ci = 3.7×10^{10} disintegrations/s
1 Millicurie = 1 mCi = 1/1000 Ci = 10^{-3}Ci;
1 Microcurie = 1 μCi = 1/1000 mCi = 10^{-6}Ci;
1 Nanocurie = 1 nCi = 10^{-9}Ci;
1 Picocurie = 1 pCi = 10^{-12}Ci.

On the basis of international agreements, the new unit Becquerel (Bq) was introduced in 1976.

1 Bq = 1 disintegration/s

thus, 1 Ci = 3.7×10^{10} Bq = 37 GBq (Giga-Becquerel).

TABLE 36 NATURAL RADIATION EXPOSURE DUE TO EXTERNAL RADIATION SOURCES (according to Aurand 1976)

Region	Population (in millions)	Mean dose (mrem/a)	Highest dose (mrem/a)
Regions with normal radiation	2500	120	–
Granite regions in France	7	300	400
Monazite regions in India (Kerala)	0.1	1300	4000
Monazite region in Brazil	0.05	500	12000

Measuring units for ionizing reactions (for example, gamma rays):
1 Röntgen = 1 R = 2.58×10^7 Ci/g air.

Generally speaking, it is not the number of disintegrations or ionizations that are of interest, but rather the energy transfer which is defined in rad:
1 rad = 100 erg/g
1 mrad = 10^{-3} rad.

This is the energy which is absorbed per gram of a substance from the radioactive processes (surface dose).
1 rem = roentgen equivalent man per time unit = dose rate.

In addition to this exogenous radiation, that absorbed through food, for example, is very significant. If mixtures of radioactive substances are present, the MAC (maximum admissible concentration) is calculated by means of a formula. Food chain effects may cause phenomena which cannot always be ascertained via such limiting values. Miettinen (1976) examined the residues of the plutonium isotopes ^{239,240}Pu and ^{238}Pu in a terrestrial food chain (lichens → reindeer → man) and an aquatic food chain (sediment → benthal fauna → fish). It was shown that in 1963/64, up to 220 pCi/kg fresh weight of ^{239,240}Pu were found in *Cladonia alpestris,* whereas in 1972/74, it decreased to 20 pCi/kg. ^{238}Pu, on the other hand, has increased continuously since 1960. In 1964, the liver of a reindeer from the same area contained 20 pCi/kg (fresh weight); in 1974, the content had decreased to 2 pCi/kg. Brisbin and Smith (1975) performed similar analyses with the North American *Odocoileus virginianus* and Rossley, Duke and Waide (1975) with the food chain of ground arthropods, in which ^{137}Cs was predominant. ^{137}Cs was analyzed in herbivorous and carnivorous arthropods in the area of the Piedmont region of Georgia. The highest concentrations were found in the lichen (*Parmelia*), although the lichen does not directly get into the food chain. The concentrations of herbivorous arthropods vary between 1.9±0.71 and 10.7±4.74, those of predatory species between 5.4–2.54 (*Collops*) and 11.7–0.30 (*Trimerotropis*). Between 50 and 70 pCi radioactive caesium/g dry weight were found in the muscles, kidney and faeces of 17 individual *O. virgianus,* whereas concentrations in the liver and bones were considerably lower. Holleman and Luick (1975) in their study of *Rangifer tarandus* investigated the influence of calcium on the position of radioactive caesium in controlled food intake conditions.

Noise The noise emission from a town is influenced by the number and speed of vehicles, the width and depth of roads, the road gradient and the road's state of repair. Also of importance are crossings and traffic lights which regulate traffic. The average noise of a city is in general 70 dB(A).

In industrial towns this level, however, is usually surpassed. Despite different subjective evaluations of the problem (independent of social group, genetically determined modes of normal reaction, age, profession) certain areas are subject to legally binding limits, which derive from the effect that noise has on a person's health. At levels of 65 to 70 dB(A) physiologically detectable effects occur in a person who is awake (peristalsis in the stomach and saliva secretion are hindered, metabolism is increased, pupils dilate, blood pressure increases, diastolic pressure increases, peripheral vessels narrow resulting in a decrease of skin temperature and blood circulation, the volume of the heartbeat decreases slightly, and hormones are secreted by the adrenal gland.

"Noise charts" already exist for numerous large cities showing the noise pollution in various districts. They serve as the basis for concerted efforts toward noise reduction (cf. Feldhaus and Hansel 1975).

For the determination of noise emissions, the permanent background level must be established. This corresponds to the constant sound which supplies the same noise energy within the test period at the location of observation as would be supplied by the actual sound. Special sound characteristics (individual tones or sounds having noticeable level changes) are taken into consideration by adding up to 5 dB (A) to the permanent sound level.

Waste Economy It is not, however, just the emissions enumerated above which act as a load on the system, but the waste resulting from the production and consumption processes is also important.

According to the Waste Removal Law in the Federal Republic of Germany (AbfG), version dated January 5th, 1977, wastes are "movable objects of which the owner wants to dispose, or whose regulated removal is advisable for the sake of the public" (para. 1). The composition of waste naturally lets us draw our own conclusions about the technological abilities and consumption habits of the population having produced it. Domestic waste shows great qualitative and quantitative differences between rural communities and large cities. In the Federal Republic, about 19 Mt/year of domestic waste are produced (in 1975, about 95 M m^3/year, see Kumpf *et al.* 1975). By way of comparison, the inhabitants of a rural community produced about 0.15 t (0.7 m^3); those of a large city about 0.32 t (2 m^3) per year. The specific weight of the domestic trash, too, shows remarkable differences between city and countryside, and the qualitative and quantitative composition of the trash has undergone drastic changes in recent years.

Although the basic principles of trash processing have been regulated by the Waste Removal Law (AbfG), dated June 7, 1972, not only are supplements on a regional level urgently needed, but, moreover, a complete reconsideration of the whole problem is called for. "Trash may no longer remain trash" (Keller 1976), if we want to preserve the self-regulation of our cities.

Keller (1977) showed clearly the ways to achieve this goal:

(a) Reduction of waste generation

(b) Application of ecologically beneficial manufacturing processes
(c) An examination of the new material utilized in context from the point of view of the intended purpose of the product
(d) Increased durability of products
(e) Utilization of waste as raw material in the production process
(f) Utilization of the energy contained in the waste
(g) Return of waste to biological cycles

From an ecological as well as an economic point of view, waste substances should always be evaluated according to their harmfulness and their usefulness. Harmfulness is described in ecological terms. An estimate of usefulness, however, is only possible by means of economic analyses. In contrast to ecology, economics deals with the flow of goods between people. Macroeconomics used to be based on the assumption that enough raw materials would be supplied by the environment and waste generated by manufacturing and consumption could be returned arbitrarily to nature. These assumptions changed most rapidly in industrial nations which are poor in raw materials and are export-oriented. The need for ecological recycling is putting ecologists and economists under pressure to find quick solutions (Keller 1977, Müller 1977, Schenkel 1977).

4.3.4 City Biota

Cities with their different load factors present challenges to life. The adaptive capabilities of living systems and their limits manifest themselves within these challenges. Certain populations exclude cities, which they treat as island forms, whereas others penetrate deeply into them to form special habitats; others again live exclusively or predominantly in cities and their phylogeny is closely related to the development of human settlements. The reaction of organisms to the most diverse forms of utilization by human life can be most distinctly analyzed in the cities. Although this fact is uncontested, the city became a research subject for ecologists and biogeographers only relatively late.

4.3.4.1 City Flora Species composition, phenology, and vitality of the flora are shaped by the city and pollution type in question (Förstner and Hübel 1971, Froebe and Oesau 1969, Kreh 1951, 1960, Kunick 1974, Lanphear 1971, Sukopp 1972, 1973, 1976, Sukopp, Kunick, Runge and Zacharias 1974, Saarisalo-Taubert 1963, Schmid 1975, etc.). The most prominent characteristics can be the extinction of a whole taxon. The vegetation reacts with its available phenotypes to the patchily distributed and seasonally fluctuating abiotic complexes of factors. The various blooming phases of plants in different habitats have been known for a long time. Inner city habitats are usually conspicuous for an earlier start of the blooming season. Franken (1955) mapped Hamburg and showed there were at least three zones of forsythia blooming. In zone 1 (northeast of St. Pauli and Altona), the forsythia were in bloom before April 21, 1955; in zone 2 (Wohlsdorf, Volksdorf), after April

28, 1955, and in zone 3 (the Hamburger mountains s. Hamburg), not until May.

Flowering Plants Correlated with various load phases, species and associations occur which are usually absent in the surrounding areas.

Bornkamm (1974) described 50 ruderal populations for the city of Cologne. Special habitats (for example, roofs containing pebbles, roadsides) have their own populations and show habitat-specific successions (Almquist 1967, Bonte 1930, Bornkamm 1961, Goebbels 1946, Heine 1952, Hupka 1938, Scheuermann 1934, 1941, Schroeder 1969, Stieglitz 1977, etc.). Sukopp (1973) was able to demonstrate characteristic zonations of plant populations for the city area of Berlin (Table 37).

These zones are distinguished by different percentages of hemerochores (= species which owe their present distribution to man) (Kunich 1974, among others). In zone 1 of Berlin (closed, built-up area), the proportion of hemerochores amounts to 49.8%, in zone 2 (less densely built-up area) 46.9%, and in the inner peripheral area (zone 3) 43.4%; in the outer peripheral zone (zone 4), it is only 28.5%. Increasing urbanization is marked by an increase in the number of hemerochores.

The extent of anthropogenic influence on habitats can be defined by the hemerobic system of Jalas (1955) and Sukopp (1972, 1973, 1976). This distinguished the following habitat qualities:

Ahemeroby	= not affected by civilization
Oliogohemeroby	= slightly affected by civilization
Mesohemeroby	= moderately affected by civilization
Euhemerophy	= strongly affected by civilization
Polyhemeroby	= very strongly affected by civilization
Metahemeroby	= excessively and unilaterally affected by civilization. Organisms destroyed according to vulnerability.

TABLE 37 VEGETATION DEVELOPMENT IN THE INNER CITY HABITATS OF BERLIN (according to Sukopp 1973)

Sand	Ruderal soil (mortar soil)	Piled-up soil rich in nutrients and humus
Bromo-Corispermetum	*Chenopodium botrys* populations	Chenopodietum stricti
↓		↓
Berberoetum incanae	*Oehnothera* stage	Lactuco-Sisymbrietum altissimi
↓	↓ ↘ Mililotetum ↙	↓
Festuca trachyphylla-stage	*Poa*-Tussilaginetum	Artemisietum
↓	↓	↓
Robinia (and *Lycium*) bushes, respectively	Chelidonio-Robinietum	*Sambucus nigra* associations
↓	↓	↓
		Acer pseudoplatanus-*Acer platanoides-stages*

Certain plant and animal species are characteristic of the various hemerobic habitats (hemeroby indicators); their developmental stages, however, are known for only a few cities.

Stieglitz (1977) examined waste deposits, sewage treatment plants, and other locations influenced by man within the areas of Mettmann and Wuppertal-Elberfeld and confirmed the existence of a number of remarkable plant species introduced by man. As is well known, during the Second World War, and in the post-war years, many plants showed a relatively wide dispersal on the rubble areas of the cities, although, previously to that time, these plants had practically never or only very seldom settled in urban areas. Some of these species were able to conquer large areas (for example, *Epilobium angustifolium* or *Buddleia davidii*).

In Central European cities, a load gradient can often be observed, starting with delicate tree species (lime, for example) to more resistant species (among others, *Robinia pseudacacia*). Numerous city plants were only introduced into Central Europe by man. Mahonia (*Mahonia aquifolia*), phlox (*Phlox drummondii,* among others), sumach (*Rhus* species) and evening primrose (*Oenothera* species) came from North America, the dahlia was introduced from Central America into Europe for the first time in 1784; the ground cherry (*Physalis olkekengi*) from South America, gladiolus from South Africa and the Orient, the lilac (*Syringa*), the butterfly bush (*Buddleia davidii*), the magnolia (*Magnolia precia*) and the chrysanthemum from East Asia and Japan.

This is why city parks usually contain a large number of alien plant species. The beautiful city parks of Porto Alegre, Farroupilha, and the Paulo Gama park alone contain 171 tree and shrub species of which only about one-third are endemic to Rio Grande do Sul.

4.3.4.2 Lichens as Biotic Indicators The bryophyte and lichen flora in particular, whose indicator value has been used for a long time to measure air quality criteria, exhibits striking distribution patterns. The first observations on lichens as indicators for air pollution were made in 1859 by Grindon in southern Lancashire and in 1866 by Nylander in the Luxembourg Gardens in Paris. Since that time, the lichen vegetation of numerous large cities has been mapped (among others in Augburg, Munich, Stockholm, Oslo, Helsinki, Zurich, Danzig, Debreczin, Krakow, Vienna, Caracas, Lublin, Bonn, Hanover, Quebec, Goteborg, New York, London, Linz, Montreal, Stuttgart, Hamburg, Saarbrucken). In all studies which have been published so far on the dispersal of lichens in cities, the authors arrive at clear zonings which are closely related to air-climatic factors and to the reception load of the sites.

In 1976, Sernander introduced the terms normal zone, combat zone, and lichen desert into the literature. In accordance with most authors, five zones are generally distinguished in cities (Fig. 58).

Zone 1 = lichen desert — lichens appear only in rare cases (*Lepraria*

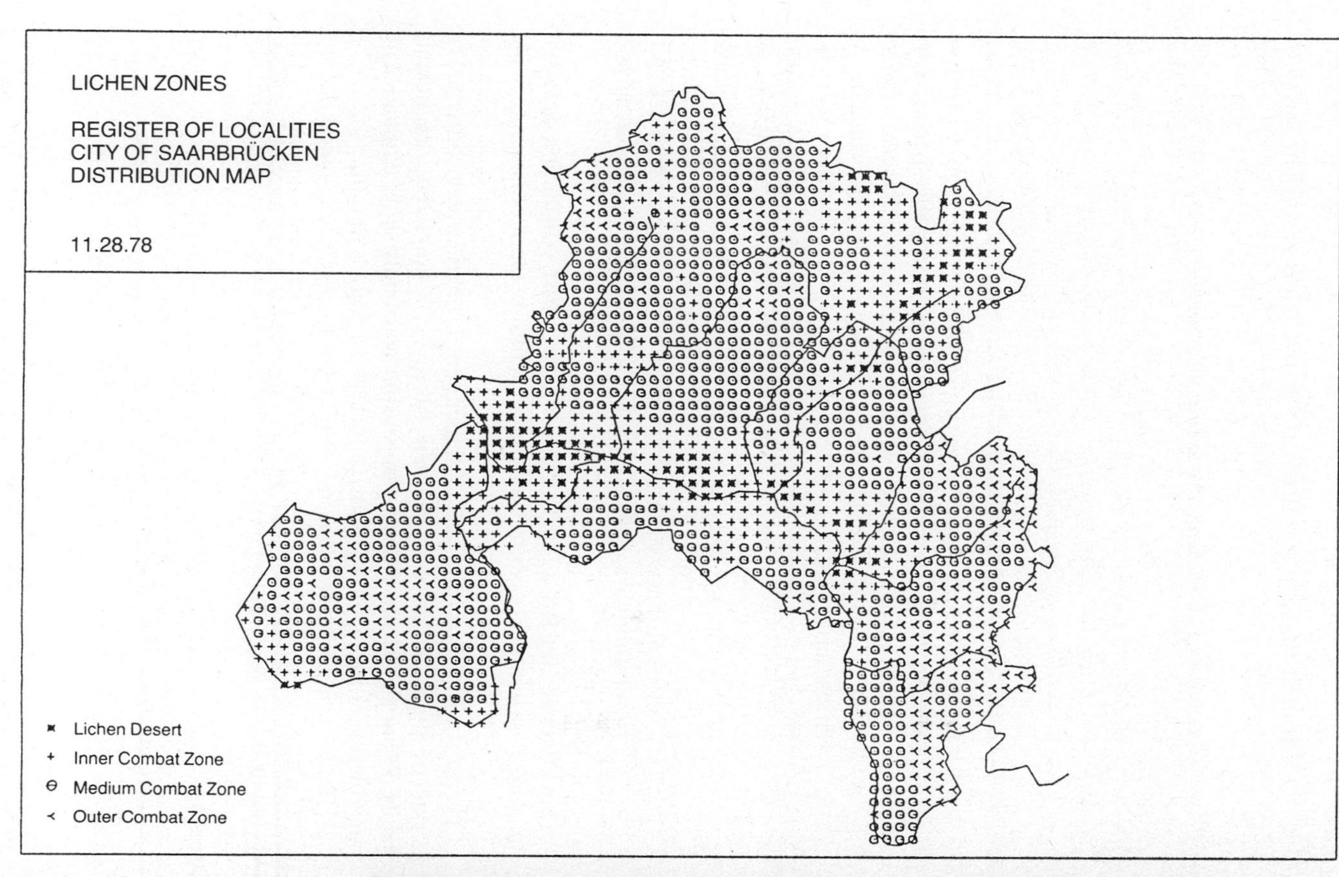

Figure 58 Lichen zones within the city limits of the city of Saarbrücken.

aeruginosa, Lecanora hageni). Mostly, however, only airborne algae (*Pleurococcus viridis*) are present.

Zone 2 = Inner Combat Zone — contains only impoverished, subneutrophile societies on the bark of deciduous trees. The barks of coniferous trees are not inhabited because of the low pH-value. Only individual lichen species occur (*Physcia orbicularis, P. ascendens, Lecidea, Parasema, Lecanora varia,* and *L. conizaeoides*).

Zone 3 = Medium Combat Zone = neutrophiles dominate. Main societies are: *Physcietum orbicularis,* including *Physcia orbicularis, P. sciastra,* and *P. nigricans,* as well as *Lecanoretum subfuscae,* including *Lecanora subfusca, L. coelocarpa, Caloplaca cerina,* and *Rhinodina exigua.*

Zone 4 = Outer Combat Zone — oxyphile societies are dominant, they are, however, already accompanied by nitrophile species. Parmelietum furfuraceae, a society encompassing *Parmelia fuliginosa, P. exasperatula, P. sulcata,* and *Evernia prunastri,* is characteristic of this zone.

Zone 5 = Fresh Air Zone — settlement influences no longer have a lethal efffect on lichens. Oxyphile societies on the barks of deciduous and coniferous trees as well as on wood and silicate (Beschel 1958) are dominant. The following are typical societies: Usneetum dasypogae, an oxyphile beard lichen society, Laborion pulmonariae, moss-leaf lichen societies, Parmelion physodis, etc.

Hawksworth and Rose (1976) point out the close correlation between the average SO_2 winter concentration and the sulfur content of lichens (Table 38).

The SO_2 resistance of lichens is dependent on their water content, the pH-value of their thallus, and of their substrate. Completely dry lichens probably do not absorb any SO_2 and thus suffer no or only slight damage. The damage increases proportionately with increasing water content.

Hypogymnia physodes, among others, was closely examined for its reaction with gases and dusts. It was not only SO_2 that led to damage in this case, but also HF, HCl, NO_x, NH_3, formaldehyde, 2,4-dichlorophenoxyacetic

TABLE 38 SULFUR CONTENT (PPM/DRY WEIGHT) OF LICHENS AND SO_2 CONCENTRATES (IN $\mu g/m^3$ IN WINTER) IN WESTERN CENTRAL SCOTLAND (according to Hawksworth and Rose 1976)

Species	Mean winter SO_2 concentrations in $\mu g/m^3$				
	30	35	40–50	55	60–70
Evernia prunastri	382	589	794	1129	–
Hypogymnia physodes	537	–	545	–	1509
Usnea subfloridana	254	676	1101	–	–

acid derivatives, gasoline, xylene, ethyl alcohol, acetone, carbon tetrachloride, parathion derivatives, Zn, Pb, Cd, Cu, cement, and potassium dusts.

In fumigation tests with HCl, HF, and SO_2, *Hypogymnia physodes* reacted with necrosis, at the following low concentrations:

mg/m^3	Hours	% Dead lichen thallus
HCl 0.09	598	10
HF 0.004	72	27
SO_2 0.46	216	10

The acidity and buffer qualities of the substrate on which the lichens grow in certain areas play an important role, particularly in the case of crustaceous lichens. A parasitic growth of bacteria, viruses, or fungi is significant (similar to higher plants) with respect to their resistance to harmful substances. In heavily pressurized areas, lichens can survive on limestone, asbestos roofs, or in wound areas of trees even at a point where other habitats have already become free of lichens. The bark of individual tree species contains specific pH values which may change under the influence of pollution (Lötschert and Köhm 1973). On a single tree, physical as well as the chemical properties may change from one small area to another.

Within the Federal Republic of Germany in particular, Guderian (1977) and Schönbeck (1962, 1972, 1974) studied the role of sulfur dioxide in relation to other noxious gases in numerous tests. Thus, bush beans, red clover, radishes, tobacco (BEL W_3), garden cress, petunias, and African violets, for example, were tested not only with respect to their reactions with SO_2, but also both organic air components (dimethylformamide, toluene, ethylene, propylene, acetone, trichloroethylene, and methylene chloride, for example). The effect of ethylene (dispersed primarily via car exhausts in city areas) on petunias and African violets was a discoloration of the leaves and a dropping of the blooms. At a gassing concentration of 0.6 mg ethylene/m^3 air, a yield reduction of 25% occurred within five days in bush beans, as compared with a control sample. Strong growth reductions similar to those in the bush beans also occurred in tobacco plants which are relatively resistant to acid exhaust fumes (SO_2, HF, HCl). In the case of radishes, formation of thickened epidermis as well as cracks occurred in the body of the plants as a result of the effect of ethylene. In the test plants examined, i.e., bush bean, red clover, garden cress, tobacco and radish, the reaction criterion of growth performance showed that ethylene is at least four times more toxic than SO_2.

The joint effect of ethylene and SO_2 is additive. After 5 days of gassing with 0.6 mg ethylene/m^3 air and 1.8 mg SO_2/m^3 air, yield reduction in the bush bean of 40% occurred, while under the effect of ethylene alone, the reduced yield amounted to 25% and of SO^2 alone, 12%.

In the case of the garden cress, the combined effect of ethylene and SO_2 also presented a good visual picture of the damage caused. With ethylene, the leaves showed yellowing, with SO_2, leaf necroses occurred, and with the effect of ethylene and SO_2 combined, both yellowing and necroses occurred, with leaves more severely damaged than in the individual reactions. Plants react differently to dimethyl formamide which is the most widely present organic component in the Cologne pollution records (3,600 t/year).

While radish, bush bean, and cress show only slight reactions, red clover died at concentrations of 20 mg/m^3 air within five days. Krause (1975) and Duderian *et al.* (1976) arrived at similar conclusions with sulfur dioxide concentrations (0.2 mg SO_2/m^3 air) on bush beans which had been dusted with cadmium, zinc, and a cadmium-zinc mixture. The reaction increased for individual dusts as well as for the mixture. In particular, plants treated with cadmium were permanently damaged.

4.3.4.3 Mosses Gilbert (1970, 1971) attempted to define moss biomes specifically for cities. He showed that the sensitivity of mosses is heavily dependent on the substrate: it varies from terrestrial mosses on limestone, through mosses on rocks free of lime and those on nutrient-rich bark, to the most sensitive of all, mosses on nutrient-deficient bark.

Gilbert attempted to bring out clearly the relationships between growth form and SO_2-resistance. The faster growing a moss species, the more resistant it is, generally speaking, against SO_2 (because of the problem of the duration of exposure of the thalli). Winkler (1976), in experiments involving the addition of lead nitrate, noted that toxicity is dependent on the water storage capacity of the mosses. The higher the water storage capacity of the mosses, the worse the damage. While in SO_2 gassing a revival of photosynthesis may occur, a treatment with lead will result in progressive, irreversible damage. In 1974, Dull defined moss zones for the metropolitan area of Duisburg according to air purity standards.

$$\text{Index of Atmospheric Purity} = \text{IAP} = \sum_{n}^{1} (Q \cdot f)/10$$

n = Number of species per area
f = Frequency of species
Q = Average number of species growing in the area together with species X.

IAP:O: Species resistant to SO_2, for example: *Tortula muralis, Bryum caespiticium, Bryum argenteum, Ceratodon purpureus.*

IAP:1: Species which are rare within the range above 0.55mg/SO_2/m^3 and thus occur only in protected locations, for example: *Dicranella heteromalla, Unium affine, Unium cuspidatum, Plagiothecium curvifolium.*

IAP:2: *Polytrichum formosum, Unium punctatum, Fissidens bryoides, Hypnum cuppressiforme, Blasia pusilla, Plagiothecium laetum, Lophozia bicrenata.*

IAP:3: Fresh air indicators

For example, *Dicranoweisia cirrata, Lepidozia reptans, Nardia scalaris, Tortula subulata.*

IAP:4: Sensitive clean air indicators
Encalypta streptocarpa, Campylopus flexuosus, Homalia trichomanoides, Dicranum scoparium, Metzgeria furcata, Oxyrrhynchium pumilum, Porella platyphylla.

4.3.5 City Fauna

Cities are marked by a characteristic fauna which can be distinguished from the populations in surrounding areas by a different history and slightly deviating population-genetic structure.

Particularly striking are the species which, by nature, only occur in cities or which, conversely, are always absent from them.

4.3.5.1 Invertebrate Fauna Numerous invertebrate species, particularly insects, have become city fauna; others avoid cities or only penetrate them in the form of individual, pre-adapted allele types. Numerous abiotic factors and trophic conditions provide the reasons for the permanent presence of individual species. In the case of others, a connection can be established, at least temporarily, to the city via individual factors.

Many animal species are attracted by illuminated shop windows, etc. Lepidoptera, Coleoptera, and Diptera appear during the night at such locations. The Diptera fauna of a large city possess a characteristic composition (Greenberg 1971, Nourteva 1971, Povolny 1971); in Central Europe, *Fannia, Musca,* and *Drosophila* species dominate. There is no doubt that they are of great ecological significance to the city locations, since not only are individual species disease carriers, they are also important consumers of organic residue matter.

Annual periodicity is very pronounced in city insects. In 1946/47, Peters (1949) examined the annual, periodical occurrence of *Periplaneta americana* in 61 bakeries through Stuttgart. Significant fluctuations in the number of individuals were noted, depending on the month. Green surfaces, garbage dumps, ruderal places, children's playgrounds, etc., interspersed throughout the city in a mosaic-like pattern possessed a location-specific fauna (Berhausen 1973, Iglisch 1975, Müller 1977, Tischler 1952, Topp 1971, etc.).

Among the locusts in Central Europe, only *Stenobothrus bicolor* settles in the inner city. Aerial plankton on the outer limits of the city exhibits a totally different structure from that found within the city limits. Tardigrades seem to be frequently absent from the inner cities, even though their shells tolerate extreme environmental conditions. Their world-wide dispersal is mainly due to the drifting of their shells in the wind. In places, snails penetrate into the densely built-up areas via isolated habitat islands. It may be assumed that during this immigration, a genetic reconstitution of the populations takes place. As an example, mention should be made of *Cepaea* populations in the vicinity of Saarbrücken. The inner city locations of Saarbrücken are dominated by the red-unbanded species and yellow-unbanded

Figure 58a The goshawk (*Accipiter gentilis*) (in the illustration a young female to whose tail a transmitter has been attached) has also become a city bird in some cities. By means of a remotely monitored specimen it was ascertained that individual animals prefer to hunt their preferred prey (pigeons, among others) within the city.

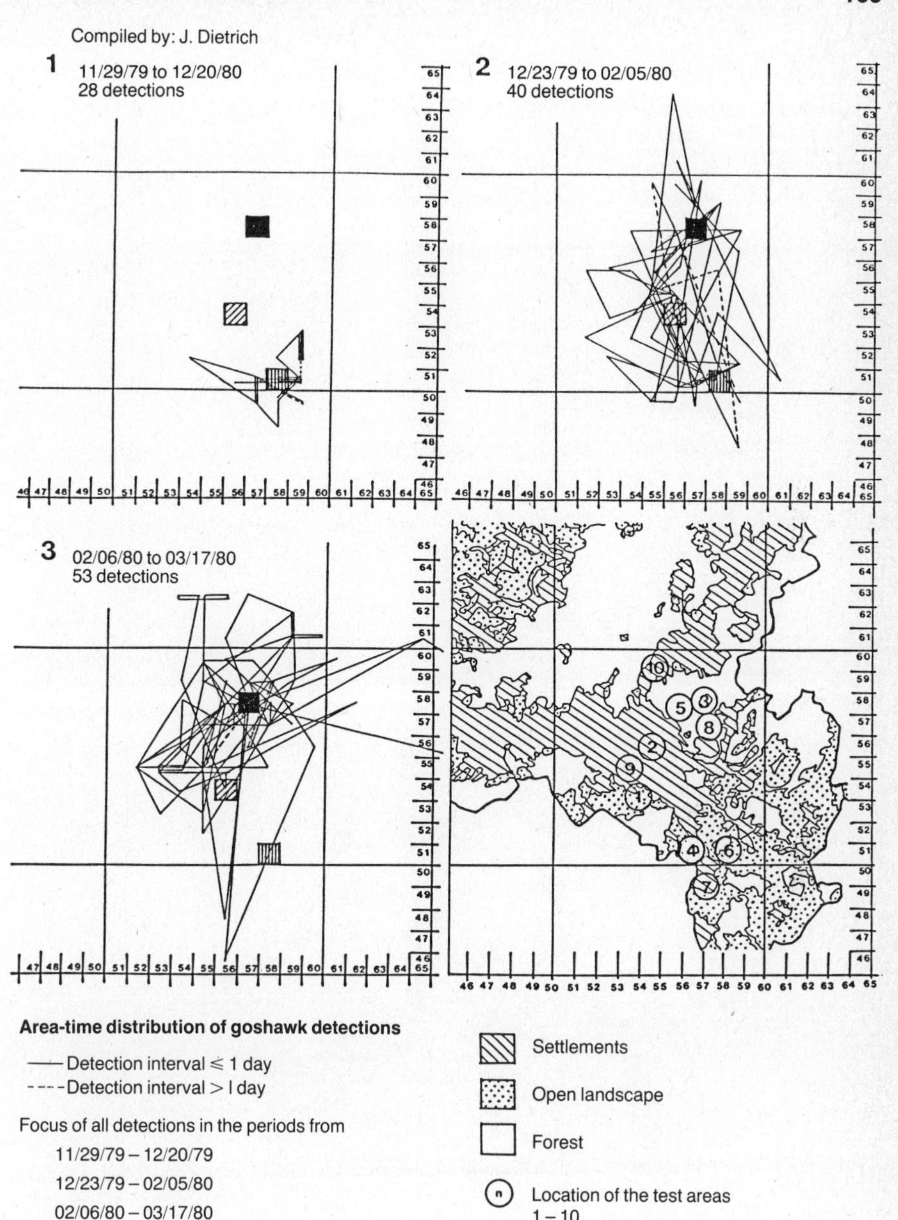

Figure 58b Sites visited by a remotely monitored goshawk female (Fig. 58a) in the city region of Saarbrücken.

species (Müller 1977). The distribution of the band characteristic among the individual city sites illustrates the fact that city conditions (increase in isolated habitats, climate type, pollution type, etc.) can probably also accelerate the evolutionary processes.

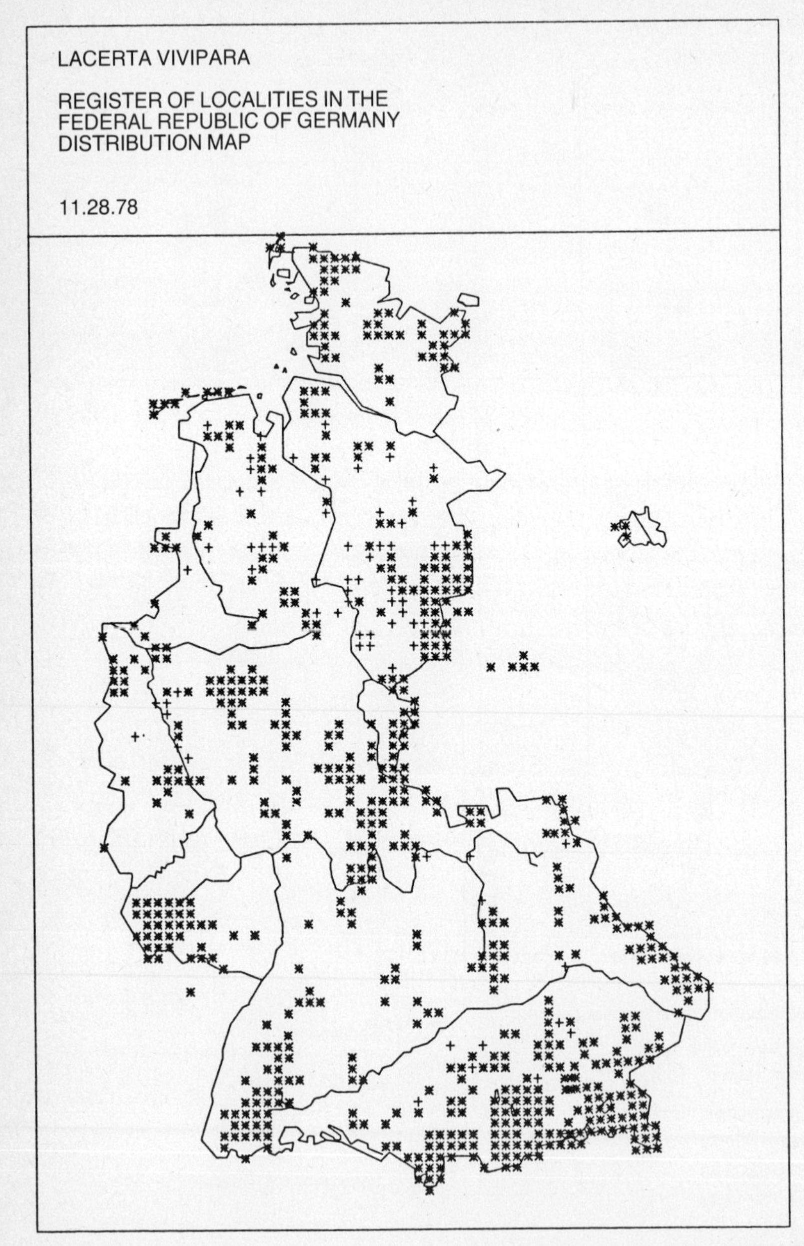

Figure 59 Distribution of the common lizard (*Lacerta vivipara*) within the Federal Republic of Germany (each character corresponds to an area of 10×10 km^2; ⋆ = findings after 1960; + = findings prior to 1960).

4.3.5.2 Vertebrate Fauna Deer penetrate — via habitat islands — into city areas. Foxes and, in Manitoba, polar bears, regularly visit the garbage dumps on the outskirts of the cities. Mass reproduction in rabbits is avoided through the use of ferrets, thus preventing damage to flower beds in the inner city.

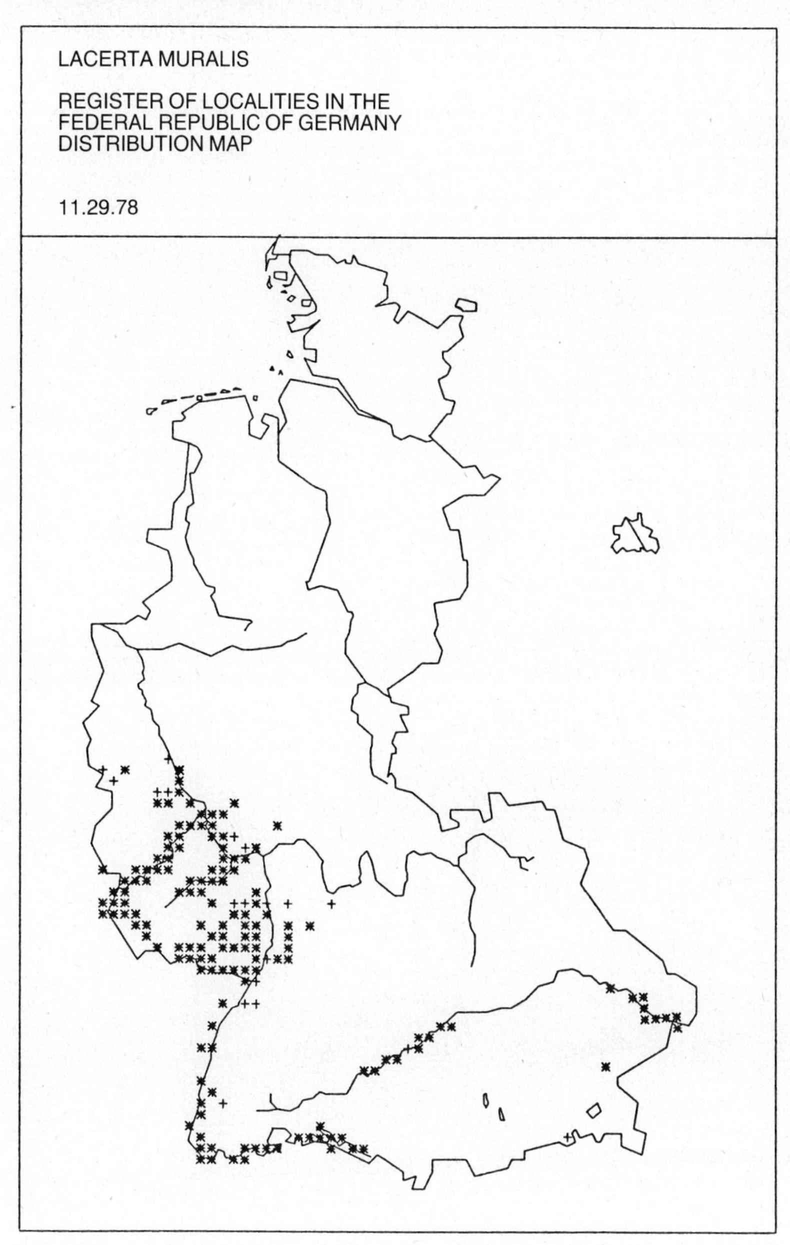

Figure 60 Distribution of the warmth-loving wall lizard (*Lacerta muralis*) within the Federal Republic of Germany.

The sand lizard (*Lacerta agilis*), and its warmth-loving relative, the wall lizard (*L. muralis*) (Figs. 59, 60, 61, and 62) and *Bufo viridis* penetrate into the center of European cities (Müller 1976, Wendland 1971). In tropical cities, reptiles are usually represented by geckos. *Hemidactylus mabouia*, intro-

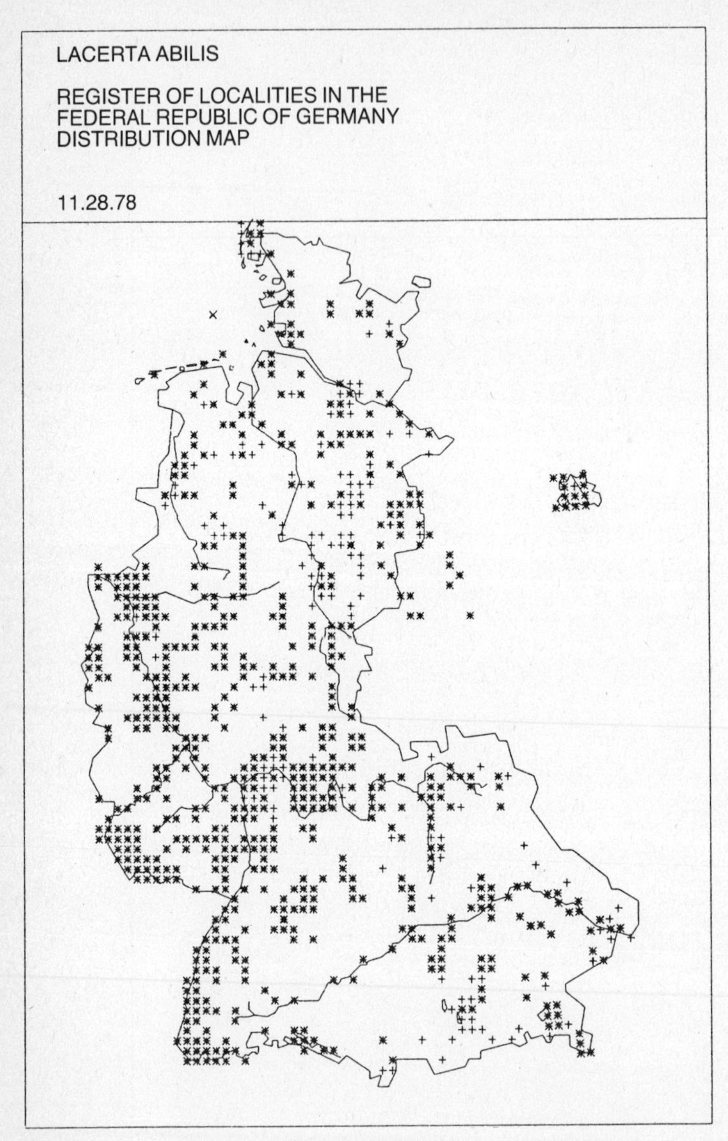

Figure 61 Distribution of the sand lizard (*Lacerta agilis*) within the Federal Republic of Germany.

duced from Africa into South America, is a typical house gecko in Belém, Bahia, Porto Alegre, and Santos. A close relative, *H. triedus,* lives in houses in Karachi, Madras, and Colombo, where it exists together with *H. brooki, H. depressus,* and *H. frenatus,* the latter of which was introduced by man into Australia, Africa and St. Helena.

The avifauna has been particularly well examined in recent years, and some bird species reach high population densities in the inner cities. Among the birds of Central Europe, this in summer includes the redstart (*Phoenicur-*

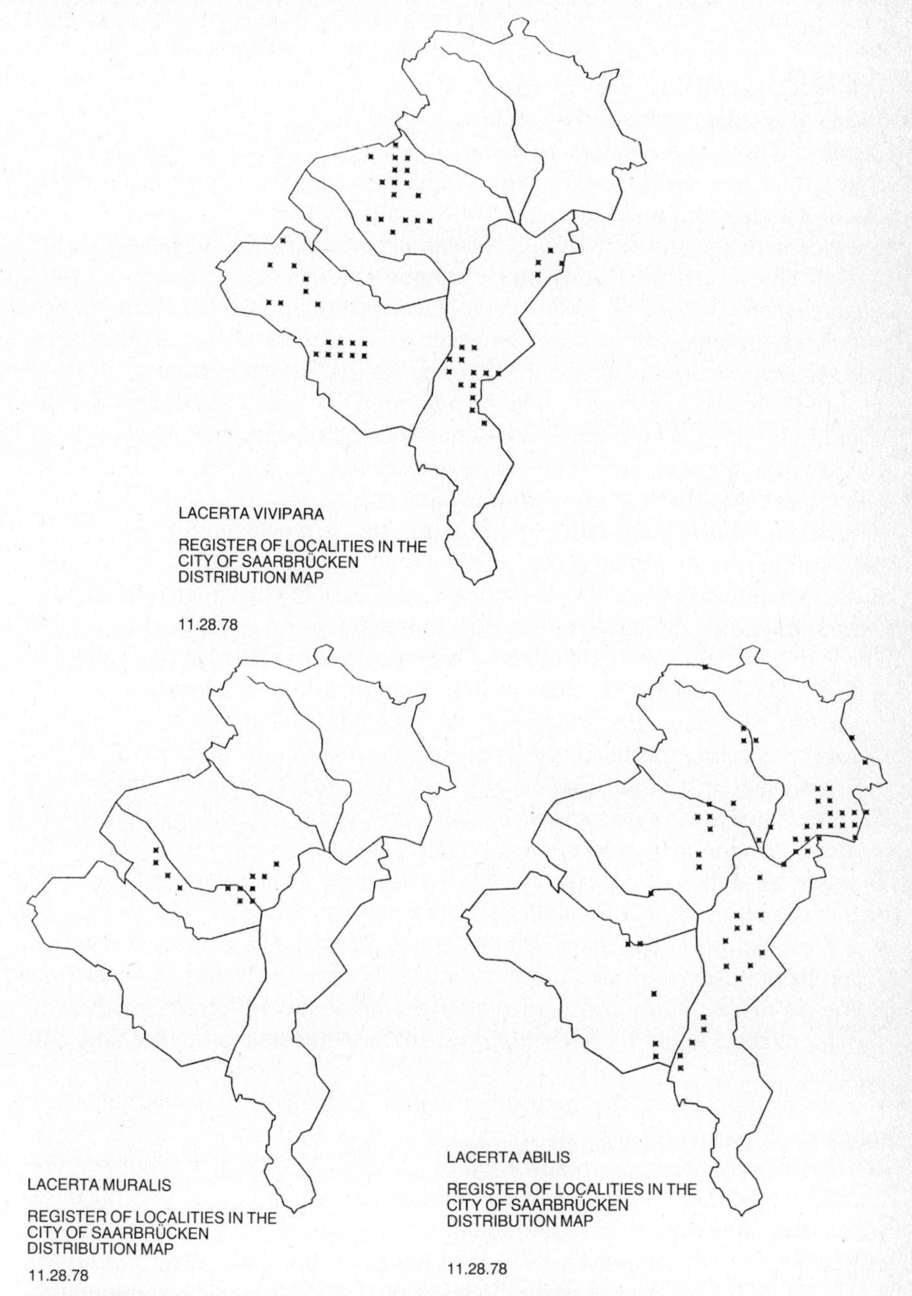

Figure 62 Distribution of the lizards (*Lacerta vivipara, Lacerta muralis, Lacerta agilis*) within the Federal Republic of Germany (each character ⋆ corresponds to an area of 250 m × 250 m).

us ochruros) and swift (*Apus apus*). They are replaced in winter by the starling (*Sturnus vulgaris* which remains throughout the winter). Non-migratory birds include: the house sparrow (*Passer domesticus*), and the pigeon

(*Streptopelia decaocto*). *P. domesticus* was transmitted world-wide by man. Today, it is present in the center of North American cities as well as in cities such as Rio de Janeiro, São Paulo, and Porto Alegre. Where *P. domesticus* is absent, it is usually replaced by a close relative. In Japanese cities, this is the introduced tree sparrow (*P. montanus*), which in Central Europe usually avoids the inner cities (it is a typical species along the city limits); in the Canary Islands, and on Madeira, it is the Spanish sparrow (*P. hispaniolensis*), together with the thrush (*Turdus merula,* species *cabrerae, agnetae*). In Sri Lankan cities, such as Kandy and Colombo, the two crow species *Corvus splendens* and *C. macrorhynchos* have become city birds, while vultures are completely absent; vultures are, however, characteristic of South American cities (*Coragyps atratus* in Rio de Janeiro, Santos, Manaus, Belém, etc.).

In Paris, the ring dove (*Columba palumbus*) — which is generally absent from the interior of German cities — has become the city pigeon whereas in Moscow, the hooded crow (*Corvus corone cornix*) has become the city crow. It is remarkable that the hooded crow also prefers to be near human settlements in the European North Cape. Their nests are mostly built in birch trees close to houses. A similar occurrence is noted still further south, in Mojøen on the Vesfnfjord (Norway). In African cities, it is in particular the milvines (*Milvus migrans*), alongside old world vultures (*Gyps rueppeli,* among others) which give the cities their character. This is particularly true for the capital of Senegal, Dakar, where *M. migrans* is actually breeding in the trees along the roads. In Surinam, the following birds penetrate into the inner cities: *Coragyps citratus,* the falcons *Milvago chimachima* and *Polyborus plancus,* the small pigeon *Columbigallina talpacoti,* the cuckoo relative *Crotophaga ani,* the barn owl *Tyto alba hellmayri*, the tyrant flycatchers *Muscivora tyrannus, Tyrannus melancholicus,* and *Pitangus sulphuratus,* the wren *Troglodytes aedon,* the thrush *Turdus leucomelas,* the brood-parasitic cowbird *Molothrus bonariensis,* and the crown sparrow *Zonotrichia capensis.*

In Australia, the Senegalese turtledove *Streptopelia senegalensis,* introduced from Africa and India, prefers cities and their environs. The cockatoo *Eolophus roseicapillus,* east Australian Rosella parakeet *Platycercus eximius,* and parakeet *Psephotus haematonotus* are also present in parks and city gardens.

In harbor cities, the pattern of avifauna changes. Seagulls and their relatives in particular take the place of the species described above. Birds clearly demonstrate how urbanization of animals can be a process of adaptation (Erz 1964, Lenz 1971, Tenovuo 1967). Classic examples are the ring dove (*Columba palumbus*) in Paris and the thrush (*Turdus merula*). The thrush was first described as a city bird in 1795 in volume 5 of the *European Fauna* (Part I, p. 121) by the pastor and natural scientist J.A.E. Goeze of Quendlingburg, but he described it as a shy forest dweller constantly taking cover when in cities. Since then, it has been fully integrated into the Central European city fauna (Heyder 1955, 1969/70).

Transformation phenomena have also been confirmed for other thrush species. The fieldfare (*Turdus pilaris*) is a characteristic bird of the Wilhemshöhe Park near Kassel and penetrates from there into the inner city districts (cf. Saemann 1968, 1974). In Mo (Rana) (Norway), 20 km south of the Arctic Circle, it builds its nests in the highest trees of the city. Other bird species, such as the magpie (*Pica pica*) also exhibit tendencies towards urbanization.

In the city areas of Oldenburg (103 km^2), a settlement density of 1 breeding pair/km^2 was recorded in 1976. Higher settlement densities were reported in Wilhelmshafen (1.6 pairs/km^2), Bonn (1.8 pairs/km^2) and Oberhausen (2 pairs/km^2). The nest locations in the city are always in less densely built-up areas, usually in close vicinity to single family houses (20.7% in Oldenburg). The city birds in Central Europe can be classified into at least two groups according to their breeding period. Species which start breeding in March/April (thrush, greenfinch, house sparrow, pigeon) are without exception non-migratory birds, while many species of the May/June breeding period must be grouped into the migratory birds (common swift, warbler, swallow). At the same time, these species show characteristic dispersal movements. Mulsow (1968) subdivided Hamburg into a "house sparrow-common swift-inner city zone" and a "thrush-greenfinch-city edge zone." Eggers (1975, 1976) made a further differentiation: in 1976, he analyzed the settlement densities of the Hamburg avifauna within the harbor-industrial zone, in the inner city, in the apartment housing zone and garden city, the parks, the hedgerow landscapes, the heathlands, the meadows, the swamps, and the forests. The avifauna of the industrial and harbor zone as well as that of the inner city are of particular interest.

Numerous problems, however, are created by birds in densely populated areas. Betweren 1966 and 1970 alone, 711 collisions occurred between birds and aeroplanes in Canada. Furthermore, birds can act as carriers of numerous diseases. Salmonella (*Salmonella gallinarum,* for instance), pseudotuberculosis pathogens (*Pasteurella pseudotuberculosis*) and arboviruses are transmitted by various species from their wintering locations to the summer breeding sites. The damage caused by birds in cultivated areas can be great (the African weaverbird *Quelea quelea* is an example).

4.3.5.3 Domestic Fauna and Domestic Animals The domestic fauna occupies a special position among the city fauna, which is composed to a large extent of warmth-loving species. The cockroach *Periplaneta americana* has made its home especially in university cities. The Indian pharaoh ant (*Monomorium pharaonis*) is present in various European capitals, and the North American termite *Reticulitermes flavipes* was introduced into Hamburg in 1937 whereupon it settled down in the heating system of the city. Insects and mites are regularly shipped in with spices and seeds. Among the 72 insect and mite species found in the drugs and spices sold in the drug stores of Hamburg, the tropical insects *Plodia interpunctella, Ortyzaephilus mercator, Lasioderma serricorne,* and *Tribolium castaneum* have been imported and are now part of the city fauna. According to Weidner (1963, 1964), the butterfly *Ephestia elutella,* a relative of the Mediterranean flour moth, and *Stegobium paniceum,* are often found in German drug stores.

Domestic animals also belong to the city fauna; they often present considerable hygienic problems in large cities (e.g., dogs, pigeons, etc; cf. Beck 1973, Schmidt 1975). In cities, the most frequently kept domestic animals are dogs, cats, pigeons, ornamental birds (parrots, budgerigars, canaries, goldfinches, among others), ornamental fish, guinea pigs, golden hamsters, and various reptile species. In 1975, for instance, 7,560 dogs were registered with the Tax Bureau of the City of Saarbrücken. Their distribution within the city area shows clear correlations with the living standards of the various regions of the city. While small animals do not diminish toward the center of a city (for example, in Saarbrücken), cats and parrots decrease noticeably. When classifying dogs according to large, small, and pedigree breeds, significant differences can be noted between the city's edge and the center. To determine the of the amount of pollution, it is necessary to ascertain the absolute number of domestic animals per area (the density of domestic animals).

Low dog densities (1 to 59 dogs/km^2) characterize communities with open, less dense developments along the edges of the city. Larger dog densities (120 and more dogs/km^2) are concentrated in the inner city of Saarbrücken and in a few residential areas along the city's edge. If we assume that a large dog produces one liter of urine and ½ kg of excrement per day, the dogs of Saarbrücken will produce at least 3,715 liters of urine and 1,858 kg of excrement each day.

This means a considerable "load" for densely-populated inner cities. At the same time, the hazard of transmission, particularly of pathogenic nematodes, to both children and adults, is increased.

Most domestic animals have a long history of domestication (Herre and Röhrs 1973, Simoons 1974), which, in the case of the dog, reaches back to the Mesolithic period (16,000 to 6,000 B.C.), in the case of the pigeon to 3,000 B.C. (cf. Nicolai 1975) and the cat started to be kept as a pet about 2,000 B.C.

Reptiles, on the other hand, only became house animals in more recent times. Land tortoises, especially *Testudo hermanni, Testudo graeca,* and *Testudo horsfieldi* are imported into Central Europe in great numbers every year. Yugoslavia alone exported 124,236 specimens to the Federal Republic of Germany in 1971 (= 30.98% of its total export of tortoises). In 1974, a poll among 10,319 students in 30 schools of the Saarland was conducted (Müller and Blatt 1975). 1,925 of these (= 18.65%) owned a tortoise. Most tortoise keepers (over 30%) live in densely-populated areas. The mortality and loss rate of the animals imported into the Saarland alone amounts to 82.81% during the first two years of their life as a pet.

4.3.5.4 Species Diversity In urban ecosystems, various factors not only cause a diminution and selection of animal species and biotic communities, but also bring about numerous changes in the diversity of their tightly-knit interrelations. The reaction of individual organisms revealed by their ecological potential, as well as changes in the variety of relationships can be used as an evaluation criterion (Müller, Klomann, Nagel, Reis, and Schäfer 1975,

Cyr 1977). In disturbed populations consisting of many species, one species often becomes dominant, whereas in undisturbed populations, the ratio between predator and prey animals is balanced. It therefore seems reasonable to use "H" as a measure of the pollution of a system. During examinations of contaminated and uncontaminated sites, it has been found that diversity may be applied as a measure of the degree of contamination in a city system. A prerequisite, however, is the comparability of localities and historical values (diversity values specifically applicable to the site). It is not therefore possible to transfer diversity values of — for example — airport sites (cf. Becker 1977) to road sites. On 16 long-term sample sites within the densely-populated areas of Saarbrücken, 45,039 ground arthropods were caught from May to December 1972 by means of the Barber trap method; the qualitative distribution of these ground anthropods, however, varies greatly between sites. Ten of the sample sites were located in new red sandstone, six on shelly limestone. In correlation with ground arthropods, a plant-sociological study of the vegetation was undertaken as well as the recording of the daily rhythm of temperature and evaporation, pH-value, and dust and SO_2 pollution. Because of the absence or presence of certain species, the individual sites sometimes differed greatly from each other. *Abax ater,* a euryeco forest species dominates most meadows, but is absent from some which, by their structure, appear to be comparable. Similar conditions also exist with other species. Only the establishment of the surface-specific diversity values results in a direct relationship between the carabid populations of the individual sites and their specific "loads". Nagel (1975) made similar observations in connection with polluted and unpolluted semi-dry lawns within the Saar-Moselle region, as did Joos (1975) on turnpikes. Diversity analyses, however, only prove valid for systems with a reasonably large number of elements (Scherner 1977).

4.3.5.5 Evolutionary Processes The dispersal structures characteristic of cities are closely related to the isolation barriers and selection gradients, which were first built by man. Thus, it is not surprising that evolutionary processes occur in cities. Original isolation barriers can be removed and species which are isolated from each other in nature, may become introgressive (Jalas 1961, Sukopp 1973, 1976, among others). Totally new species complexes may develop as is now assumed for the Lecanora-conizaeoide lichen group whose evolution apparently ran parallel to the industrial revolution.

As in the case of plants, evolutionary processes in city regions can be confirmed for animals (Askew *et al.* 1971, Bishop 1972, Cleve 1970, Kettlewell 1955, 1973, Lees *et al.* 1973, Steiniger 1978, Weidner 1958).

The occurrence of melanistic specimens among the Lepidoptera (particularly in the case of the geometrides) has been noted in many industrial countries. The peppered moth (*Biston betularia*), in its nominate form, is a white moth with black lines, cell spots and rough, blackish dust. The form *carbonaria,* on the other hand, is totally black with a white dot at the foot of the forewings and a few small white scales in the costal field of the secondary wings. Apart from this form, there are numerous additional varieties which may all be viewed as intermediate between the nominate form and *f. carbonaria.* All of these transition forms are usually summarized by the term *insulanica.* In 1955, Kettlewell discovered a correlation between the industrial areas in Great Britain and the occurrence of a higher percentage rate of the melanistic *Biston betularia* and other industrial melanistic butterflies. Together with 170 other entomologists, he established an observation network throughout Great Britain in order to determine these connections.

Junk (1975) conducted a similar experiment in the Saarland. His results show that within the densely-populated area of Saarbrücken, the *carbonaria* form is dominant.

For Coleoptera within city sites, Steiniger (1978) was able to ascertain clear changes in the polymorphism of the allele.

4.3.6 Urban Soils

The soil — as the product of the physio-chemical weathering of rock and of biogenetic reactions — represents an information system which occupies a central position in the evaluation of landscapes (Brauns 1968, Bridges 1970, Bunting 1965, Franz 1975, Gansen 1965, Gerasimov and Glazovskaja 1960, Kubiena 1948, Mückenhausen 1962).

The soils of the urban system are extremely heterogenous, and the content of the fine soil and humus often fluctuates considerably. All transitions exist from rubble grounds, soils poor in fine grades and humus, to humic garden and sandy soils with a negligible portion of rubble (Gray 1972, Runge 1973, Sukopp 1973). The rate of profile differentiation and soil formation in ruderal soils is surprising (5 to 25 years). It manifests itself in the fractioning of the fine soil by an increase in the calcium carbonate content at depth from about 6 to 10%, an increase in the pH-value from 6.7 to 7.5 and a decrease in the content of organic carbon from *c.* 3 to 0.5%. The fine ruderal soils often contain greater available nutrients than natural soils. The high stone content (25 to 50%), however, causes the water supply to be unfavourable (i.e. there is a rapid runoff of water and less storage capacity).

4.3.7 Exposure Tests with Organisms in Rural Areas (active monitoring)

Due to the genetic variability and daily and seasonal population fluctuations, conclusions with respect to spatial quality can only be drawn if the information from the existing genetic structure (plants, animals) is as complete as possible. Therefore, the recording of effects and trends for standardized organisms as an additional source of information is necessary (Fig. 62a).

Records showing the effects of pollution on exposed animals have only been developed in recent years. Limnic organisms (Müller and Schäfer 1976) and agricultural animals (Hahn *et al.* 1972, Moore 1966, 1974, Stöfen 1975, Vetter 1974) were predominant in this research. Plants and materials, on the other hand, have been used for a much longer time as effect criteria in the

Figure 62a Exposed panels for lichens within the city area of Saarbrücken (1978). The panels were exposed in all four wind directions.

monitoring of surface effects (Sorauer 1911, Schönbeck and Van Haut 1974, Prinz 1975, Müller 1975) (Fig. 63).

One of the guidelines of the "Association of German Engineers" (VDI 3792) defines the exposure of the lichen *Hypogymnia physodes.* The simple exposure methods used are based on the utilization of *Lepidium* (cf. among others, Thomé 1976), *Lolium multiflorum* (Prinz 1975, among others) and wood samples (cf. Arndt 1974, among others).

Records of limnic organisms were also set up for flowing water bodies in cities. To evaluate the quality of the individual river sections, organisms exposed in boxes are used. By analogy with the records showing the pollution effects in North Rhine-Westphalia (Schönbeck and Van Haut 1974, Prinz 1975, Prinz and Scholl 1975, Müller 1975), the exposed organisms are tested for two effects:

(a) What is the reaction of the exposed animals in river sections with varying pollution?

(b) Which substances are absorbed by the organisms (and perhaps even concentrated after absorption), leading, for example, to effects that can be diagnosed?

In addition to its significance as an indicator, this exposure system has the natural function of storing the evidence. In the case of micro-organisms, growth panels have been used for some time for exposure in water. Their population density, population succession, and species composition is checked at regular intervals (cf. among others, Friedrich 1973).

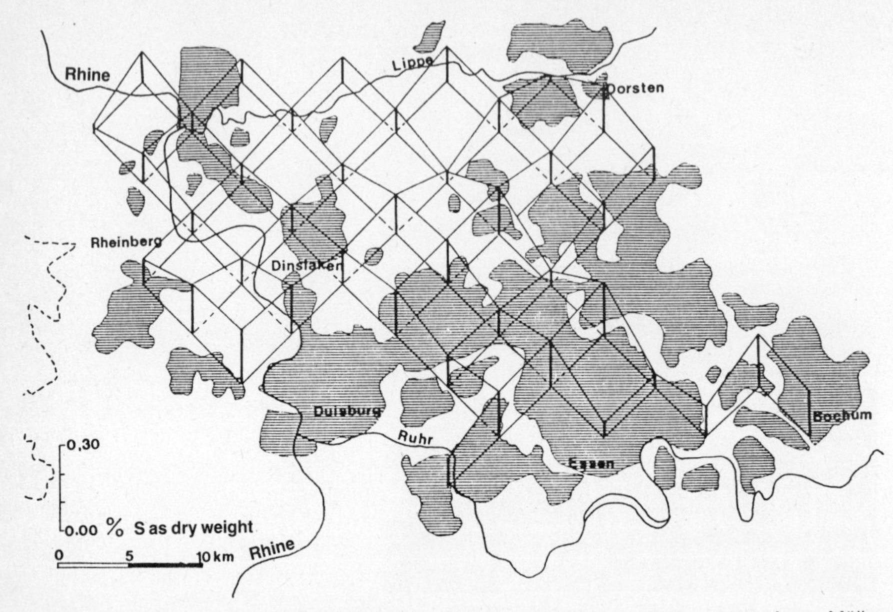

Figure 63 Record covering North Rhine-Westphalia (according to Prince 1975, from: Müller 1975). The record is based on exposed grass cultures (*Lolium multiflorum*). These are cut at regular intervals and the content of harmful substances (heavy metals or sulfur, for example) is analyzed.

4.3.8 Residue Analysis in Exposed Organisms and Rural Populations

The duration of stay of exposed organisms is often closely related to contamination with harmful substances at a site. The analysis of the substances contained in the exposed subjects will, therefore, often allow one to draw conclusions with respect to the harmful effects of the substances involved. The transfer of this information to human populations is thus possible from the site. A relationship to the "load" on rural populations can be established. This offers a realistic possibility of appropriately discussing the adaptive capabilities of living systems.

Numerous data are available concerning contamination with harmful substances of populations in the rural landscape, particularly in England (Moore 1966, 1974, Prestt and Ratcliff 1972, among others), the Netherlands (Koeman 1975, among others), and Belgium (Joiris 1974, among others). The studies concerning the accumulation of poisons are naturally significant since these poisons show an increasing concentration in food pyramids (predatory birds, sea birds). However, there are still many unsolved problems in this area. The toxicological evaluation of mercury concentration in marine animals and birds is still problematic. The highest Hg-concentrations were found in seals, porpoises, and dolphins (Gerlach 1975, Koeman 1975). On the other hand, considerably lower concentrations exist particularly in the tissues of guillemots and auks. Data is also available concerning agricultural animals living outside urban areas and cultivated plants (Hahn and Aehnelt 1972,

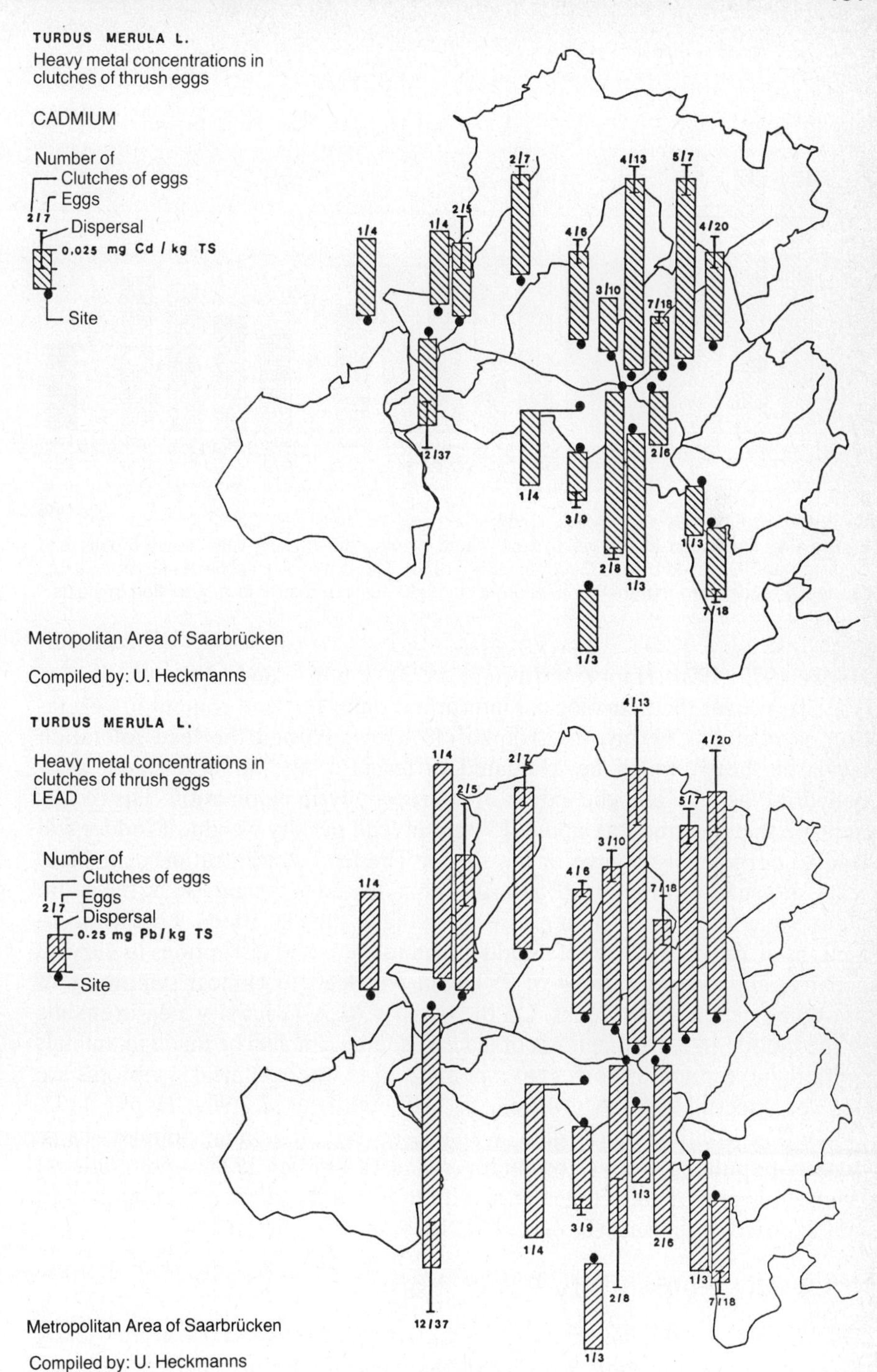

Figures 63a and **63b** Cadmium (above) and lead (below) contents in the eggs of thrushes (*Turdus merula*) in the city area of Saarbrücken in 1979.

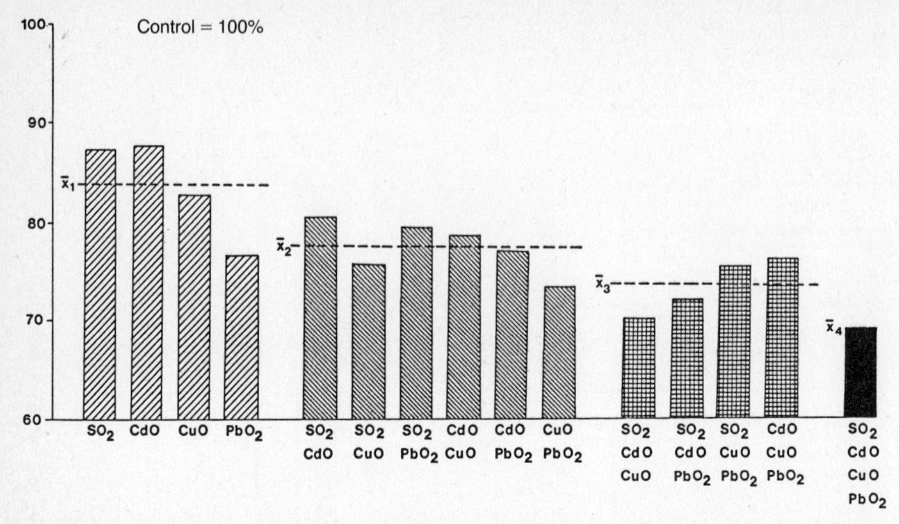

Figure 64 Reduction in the yield of cultivated plants after dusting with heavy metals and gassing with SO_2 (according to Guderian *et al.*, 1976). The control is established at 100%. It can be clearly recognized that the combined gassing and dusting, results in a reduction in yields.

Hapke 1972, 1974, Hapke and Prigge 1973) (Fig. 63a and 63b, and 64).

To permit the toxicological interpretation of the lead content in vegetation as it affects herbivores, Hapke (1972) established the lead toleration levels in sheep, which he evaluated in terms of the aminolevulinic acid-dehydrase activity and the extent of coproporphyrin elimination. His results indicate that any quantity above 15 ppm of lead per dry weight of fodder will lead to permanent damage in the sheep. The lead concentration near highways is usually between 50 and 100 ppm lead/dry weight; it is therefore inadvisable to feed sheep on grasses growing in such locations. Lead concentrations of 100 ppm in the total fodder lead to increased disruptions in enzyme activity, and concentrations of 250 ppm will lead to clinical symptoms of poisoning within a few weeks. On the basis of ALA-D activity measurements in the blood, an early diagnosis of lead poisoning can also be made in animals (with the exception of horses) at a point in time where clinical symptoms are not yet discernible (Haeger-Aronson 1960, Wada *et al.* 1969, Hapke 1972, 1974). The contamination data available so far for human populations in densely-populated areas (Rosmanith *et al.* 1975, Stöfen 1975, among others) often show remarkable coincidence with those obtained for rural populations and exposed organisms (Hower *et al.* 1974, among others).

4.3.9 The Human City Population

The rapidly developing science of ecopsychology teaches us more clearly every day how the daily routine of man constitutes his "action biotope" (Boesch 1976) in which much — but as we know, not everything — is functional.

The home, an obvious starting point, is governed only in part by practical considerations and exhibits biome-specific adaptations. In a much wider sense, it is the prerequisite for the development of the ego. Personality and home are so closely interwoven that often an analysis of one of them can provide clues about the other. From the crowding effects among animals, many behavioral scientists have concluded that a certain increase in population may in some instances lead to catastrophic consequences for human beings, since "the one true danger to human beings is human beings, that is, too many human beings" (Lorenz and Leyhausen 1968). The results of experimental analyses by psychologists (Hutt and Vaizey 1967, Kälin 1972, among others) show that this assumption is justified, but that, the densities of individuals most advantageous to the population need not be biologically determined in the same manner as is the case with animals. The relationship between psychological stress and population density can be interpreted as the possibility of satisfying individual desires which, generally speaking, is becoming more difficult as the population density increases.

Human populations and individuals often have different reactions to external physico-chemical interferences (for instance, alcohol, poisons), depending on their genetic structure and variable psychological condition. If the interference remains below the tolerance threshold, often no reaction occurs. It may result in the phenomenon known in psychology as "creeping". Small external doses (below the reaction threshold) displace the tolerance limits. A later increase in the dose also remains without defence reaction which, in extreme cases, leads to the physiological adaptation limit (for instance, alcoholism, drug abuse, excessive nicotine consumption, adaptation of the city residents to big-city noise). This adaptation capability does not necessarily exist with all noxious agents in the environment. Their presence must not obscure the fact that individual-specific physiological and psychic adaptation limits exist which, once exceeded, no longer permit any corrective action. Physiological alarm devices exist to detect the numerous external interference factors; these alarm devices have been analyzed by Levi (1972), among others, and can be characterized by an increased production of adrenaline and noradrenaline.

Stress is physiologically definable via the production of these two adrenal gland hormones; however, additional stress indicators have been established for some animal species, for instance SST-value = tail-bristling value in the case of Tupaiidae; cf. von Holst (1975). Lazarus (1966) and Sells (1970) consider stress a threat to the well-being which the organisms cannot counteract with innate or well-established compensation measures. Boesch's definition (1976) is more general in that he would like to "define as stress any complication of action" (p. 335).

Changes in health (epidemiology, medical geography) and physiological processes or urban populations are of major biogeographical interest (Fig. 64a).

The effects of city life or of the working hours in the daily biological rhythm of man on the frequency of work-related accidents, suicide attempts,

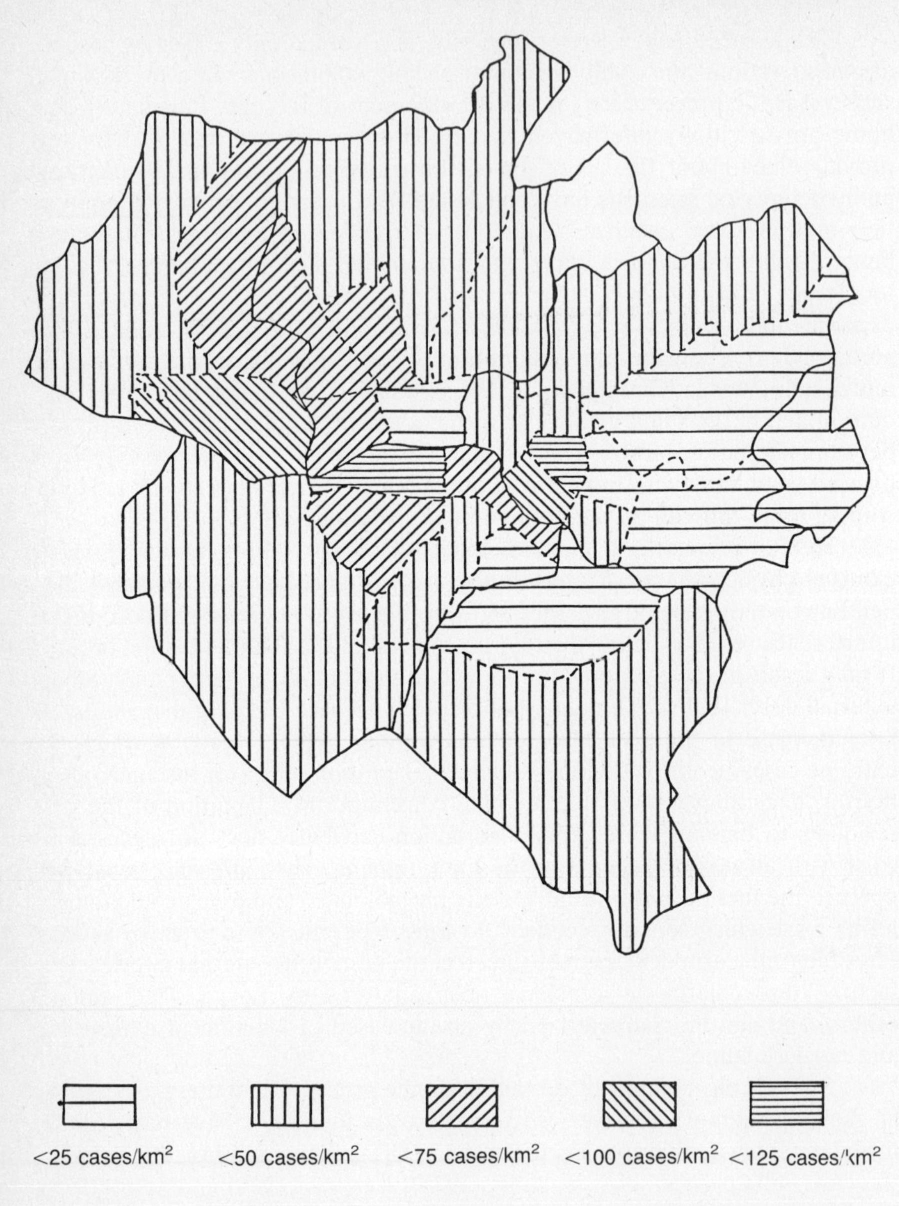

Compiled by: Helga Meier

Fig. 64a Broncho-pulmonary malignancies per residential area in downtown Saarbrücken during the years 1970–1974. The apparently causal link between the frequency of malignant cases occurring and the nearness of smelting works may be purely coincidental.

and mortality rate have by no means been completely clarified as yet (Hildebrandt 1976), although voluminous literature exists in this area. The distribution centers of certain diseases (chronic bronchitis, cancer, etc.) can be clearly delineated, without, however, it being possible to define clearly the causality of their relationship to the urban site — a few exceptions notwithstanding (Fig. 64a). The genetic and social variability of urban populations as well as their high mobility make a causal interpretation more difficult. Mobility, on the other hand, provides the basis for continental distribution of certain diseases (Jusatz 1966). In years past, because of the often reduced sunlight in densely-populated areas, frequent cases of rachitis occurred in industrial areas (cf. Umschau 1971, p. 249). Today, however, the negative effect of the lack of sunlight is practically nullified by increased oral doses of vitamin D to infants.

The quality of drinking water, too, often deeply affects the health of urban populations. Although it is well-known that strict national and international standards ensure that no toxic substances enter the water, factors which often seem to be secondary — such as the hardness of the drinking water — do play a certain role. In some regions, it has been ascertained that the mortality rates from cardiovascular diseases are higher in areas with soft water than in those with hard water. "It has not yet been satisfactorily clarified as to whether soft drinking water itself plays a causal role or whether it correlates with still unknown causal factors" (Keil *et al.* 1975).

4.3.10 Food Chains in Urban Ecosystems as Pollution Integrators

It is becoming increasingly difficult to classify all of the multi-dimensional processes occurring within a city as an evaluation matrix for human beings (Tables 39, 40, and 41). Numerous unsolved problems in the area of carcinogenic, mutagenic, and teratogenic substances require environmental compatibility tests. These, however, are only usable as relative reflections of the real risk of urban living if the metabolic process of the substances in the food chains (Table 42) is analyzed within the corresponding city environment

TABLE 39 EXAMPLE OF THE LOCATION OF DDT AFTER TREATMENT OF ELM DISEASE (according to Bevenue 1976)

	DDT and metabolite residues (ppm)
Elm	–
Ground surface	–
Soil (depth ≈ 10 cm)	6.6– 14.9
Leaves (directly after spraying)	183 –293
Leaves (after several months)	21 – 32
Earthworms (directly after spraying)	36 –237
Earthworms (after several months)	119
Robins brain	39 –342
liver	763
heart	247
Pigeon (brain)	179
House sparrow (brain)	196
Starling (brain)	29
Grey squirrel	4

TABLE 40 LEAD CONCENTRATIONS IN AN ARTIFICIAL ECOSYSTEM (according to Po-Young *et al.* 1975, in ppm)

	Sand	Loamy sand	Clay-mud sand	Control sample
Sand	0.013	0.002	0.002	0.001
Algae	275	153	114	0.02
Daphnia	187	154	85	0.02
Snails	334	88	56	0.05
Mosquito larvae	403	247	80	0.02
Fish	13	2	1	0.02
Sorghum leaves	497	1	1	1.5
Sorghum roots	695	15	5	1.6

TABLE 41 ENRICHMENT OF CADMIUM IN SOILS AND IN EARTHWORMS ALONG ROADSIDES IN THE USA (according to Gish-Christensen 1973; in ppm dry substance)

Distance from road	3 m	6.1 m	12.2 m	24.4 m	48.8 m	Control
Soil	1.23	0.72	0.72	0.68	0.78	0.66
Earthworms	12.6	8.8	8.3	6.9	7.1	3.0

TABLE 42 CADMIUM CONCENTRATIONS IN VARIOUS SAMPLES OF A FOOD CHAIN ON THE LEEWARD SIDE OF A ZINC WORKS (in ppm dry weight, according to Martin and Coughtrey)

Sample			3 km distance from works	28 km distance from works
Plants: leaves of	*Quercus*		6	0.8
	Brachypodium		2	–
	Glechoma		8	–
	Urtica		7–9	2.1
	Allium		–	0.9
	Mercurialis		–	2.2
	Sambucus		15	–
	Hedera		18	2.1
	Bryophytes		25	1.3
Residues from the forest stand:	Wood		3	–
	Bark		14	2.6
	Oak leaves		19	0.8
	Other leaves		29	1.9
Soil:	5 cm below litter		42	2.0
	10 cm below litter		14	–
Animals:	*Arion ater*	Snails	9.9	3.7
	Arion fasciatus		29.7	4.8
	Arion hortensis		56.5	3.5
	Agriolimax		43.0	9.2
	Clausilia		76.1	–
	Oniscus	Isopod	171.0	28.2
	Lumbricus	Earthworm	122.8	25.2

TABLE 43 ANNUAL PRODUCTION OF BIOMASS IN CONJUNCTION WITH FORESTRY AND AGRICULTURAL UTILIZATION ON THE STREITBERG NEAR ETTENHEIM (in t/ha, from Mitscherlich 1975)

	Forestry				
	Spruce	Larch Beech	Larch with Beech	Larch Fir	Meadow
1967	8.9	8.9	7.1	6.2	5.0
1968	10.9	9.7	8.8	6.8	5.5
1969	10.3	8.7	7.8	6.2	6.7
	Agriculture				
	Wheat		Rye		Potato
	Grain	Straw	Grain	Straw	
1967	3.9	4.5	2.7	3.7	11.9
1968	3.1	3.9	2.6	4.6	7.9
1969	3.2	3.5	2.9	3.7	11.3

(Beavington 1975, Blau and Neely 1976, Jeffries and French 1976, Lagerwerff and Specht 1970, Williams and David 1976). At the same time, a spatial evaluation must be performed by means of standardized organisms within effective parameters, whose reactions and residues of harmful substances are rendered transferable through epidemiological examinations to human populations.

4.3.11 City Environs and Reduction in Pollution

No city is an island unto itself. Exchange proccesses with its environs are constantly occurring. The structures of these environs, on the other hand, have a manifold effect on the city configuration. The biomes (Müller 1974, 1977) in which the cities are located require biome-specific adaptations. The city configuration and architecture of cities located in rainforest, desert, or taiga regions show unmistakable characteristics which are determined by natural factors. Often, these adaptations are connected to social structures.

TABLE 44 FILTER EFFECT OF A BROWN FOREST SOIL AFFORESTED WITH BEECH (in kg/ha (rounded off), according to Mayer 1972)

	Na Na	K K	Ca Ca	Mg Mg	Fe Fe	Mn Mn	N N	Cl Cl	S S
Element content in the precipitation in rural areas	7.6	3.8	13.1	3.5	1.0	0.3	21.5	17.2	22.2
Element content in the precipitation in the forest stand	13.5	44.7	41.4	5.1	2.7	9.7	75.7	34.9	49.1
Element content in the seeping ground waters at a measuring depth of 100 cm	6.7	1.2	10.1	2.1	0.1	3.4	4.7	19.1	10.4
Filter effect of the soil as %	50	97	76	59	96	65	94	45	79

The immediately adjacent city surroundings, on the other hand, with their forests, agrarian ecosystems, fallow fields, rivers, and lakes are in constant interaction. They are polluted by the city; however, they contribute to the reduction in pollution, since the pressures originating in the city do not remain limited to it (Buchwald 1974, Mayer *et al.* 1975, Müller 1975). Man controls the annual production of biomass in his city surroundings (Table 43).

In the U.S.A. alone, 15×10^6 t of ethylene per year are sprayed into the air. Two mechanisms exist for the degradation of ethylene: oxidation with ozone or reaction with nitrogen oxide in light. The ground functions as an important means of degradation, i.e., through microbial degradation. Calculation shows that annually about 7×10^6 t of ethylene which reached the air was removed by ground microbes.

The filter effect is dependent on the element and on physical, chemical, and biological soil factors and on the rate of degradation of a harmful substance in living systems (Tables 44 and 45).

Trees and park areas may contribute to a sometimes significant reduction in pollution by harmful substances, dust, and noise, even to a reduction or stabilization of radioactive substances (Table 46 and Fig. 64b).

The ability to absorb gaseous, harmful substances differs considerably depending on the species. These differences are, in part, correlated with the sensitivity of a species to a particular element. Seeds for example, store sulfur

TABLE 45 AVERAGE RATE OF DEGRADATION OF SOME OF THE PLANT-PROTECTIVE AGENTS USED IN FORESTS (according to Mitscherlich 1975)

	Quantities of active substance used (kg/ha)	Rate of degradation
Insecticides		
DDT	1 - 30	Stable (very slow)
HCH, Lindan	0.2– 1	Stable (very slow)
Toxaphene	1.5– 5	4– 6 months
Aldrin	0.5– 5	Very slow
Parathion	0.5– 1.5	2– 4 months
Fungicides		
TMTD	10 - 50	2– 3 months
Herbicides		
2,4-D	1 - 2	2– 4 months
MCPA	0.5– 1.5	2– 6 months
2,4,5-T	0.5– 1	3– 6 weeks and longer
Pentachlorophenol	2	5– 7 months
Simazine	0.5– 2	6–10 months
Dalapon	4 - 25	4– 6 weeks
Sodium chlorate	100 - 250	up to 2 years
Decontamination agents		
Methyl bromide	150 -1000	1– 2 weeks
Chloropicrin	300 - 500	1– 2 weeks
Allyl alcohol	150 - 200	4– 8 days

Figure 64b (below) Protective Plantings ("native forest", according to Miyawaki) around the plant compound of the car manufacturer Honda in Japan, designed in accordance with plant-sociological findings. The protective plantings are composed of the evergreen species which have been raised in the Japanese temple groves (above) for centuries.

TABLE 46 FILTER EFFECT OF FOREST AND PARK TREES AGAINST DUST (according to Keller 1971)

Tree species	Dust quantity mg/g of dry leaf weight	Crown weight (leaves) (kg/ha)	Amount of dust caught (kg/ha)
Beech	70	4000	280
Oak	90	6000	540
Spruce	30	14000	420
Mountain Pine	200	5000	1000

in quantities of up to 1.31% of their dry weight; whereas some tomato species store 0.5, and beech trees up to 2.0% S/dry weight (leaf). Much higher concentration rates are known in the case of sulfur-binding micro-organisms. Park areas are of great importance (cf: Urban Climate) in that they cause the temperature to fall (Baumgartner 1971, Müller 1975). Bernatzky (1972) was able to prove in the case of Frankfurt that a park area measuring 50 to 100 m in width causes a decrease in temperature of up to 3.5°C. It is possible that in the immediate vicinity of leaves, the micro-climate may change (Burrage 1976). The storage capacities of animals and plants in non-toxic areas also contribute to a reduction in pollution. Examinations during the past few years have led to the quantification of the noise-reducing qualities of tree species aand park areas. It has been proven that the depth of a protective plantation is much less responsible for its noise reduction capabilities than are its structure and species composition. Plantations containing large-leafed species, dead needles or twigs in the interior, and closed leaf screens facing the sound source have the greatest sound-reducing effect.

These few short examples will suffice to show that park areas around the city assume ecological functions for the city which go above and beyond the superficial recreational function for the city dweller. The pollution-reducing effect of living systems is at present being incorporated into many phases of planning (border planting, forests providing protection against pollution).

Chapter 5

Propagation Areas and Biomes

Biomes, or physiognomically definable symbioses, are an integral part of the major climate and vegetation belts of our earth. Plant formations represent the outer belt of these symbioses. Plant formations include, among others, tropical rain forests, mountain forests, savannas, tundras, taigas, and deserts. "The basis for the preservation of all life in the biosphere is the primary production in green plants which store energy, having transformed the light energy of solar energy into chemical energy by means of photosynthesis. This chemical energy serves to maintain all other living processes, those of man included" (Walter 1971). Each plant formation possesses a specific structure. Plant formations together with the animal species living within them are defined as biomes. This designation originated with Clements and Shelford (1939) and Carpenter (1939) who defined a biome as a "plant matrix with the total number of included animals."

5.1 BIOMES OF THE LAND

5.1.1 The Tropical Rainforest Biomes

The evergreen tropical rainforest of the lowlands, whose core areas follow the equatorial zone between 10°N and 10°S is distinguished — at nearly double the productivity of forest types outside the tropics — by an unfavorable agrarian-economic production potential which is strongly determined by the soil and which is an "ecological handicap" for the cultivation of plants in the tropics (Weischet 1977). Daily and annual temperature variations (6 to 11°C and 25 to 27°C maximum amplitudes respectively) and high precipitation

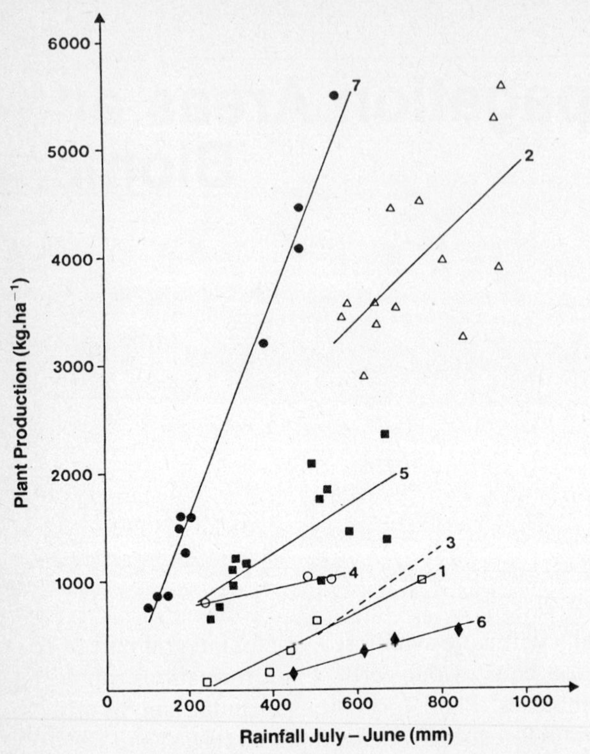

Figure 65 Correlation between production and precipitation in various Steppes and Savannas of South Africa. 1 = East Cape Acacia-Savanna. 2 = Natal Grasslands. 3 = South African Hochveld (western part) Savanna. 4 = North Transvaal Savanna. 5 = Zimbabwe Savanna. 6 = Zimbabwe Brachystegia Forest. 7 = Southwest African Dry Savanna.

(usually in excess of 2,000 mm) with frequent daily maxima of over 100 mm characterize the climatic conditions in this biome (Fig. 65).

Frost is absent from the daily climate of the tropical rainforest. Usually, two rain periods occur correlating with the sun's position at the zenith. The natural limit of typical tropical rainforest biota in the tropical mountains is generally located at between 1400 and 1500 m. Here, the mean annual temperature can fall to 15 or 20°C. At altitudes of 3,000 m, frosts and precipitation modifications occur (Lauer 1976, Weischet 1969) (Table 46a).

Semi-evergreen forests and savannas group themselves around the tropical rainforest biomes in accordance with the climatic gradient which includes a reduction in precipitation and a lengthening of the dry period. As far as the interaction between the rate of precipitation and arid periods is concerned, it must be emphasized that the length of the dry period is of the greatest importance to the more humid types of vegetation, whereas the amount of rain is the decisive factor for dry species (Walter 1964) (Fig. 65). The structure of the tropical rainforest must be understood as a functional adaptation to the humid, tropical climatic conditions and soils. At the same time, it is the stage on which the richest fauna in the world was able to

TABLE 46a ALTITUDINAL DISPERSAL OF INSECTIVORE BIRD FAMILIES IN THE CORDILLERA DE VILCABAMBA, PERU (according to Terbogh 1977)

Family	Altitude (m)									
	585	685	930	1350	1520	1730	1835	2215	2640	3510
Dendrocolaptidae	4	4	4	3	2	1	3	3	0	0
Furnariidae	4	4	5	10	9	8	7	4	5	3
Formicariidae	15	12	9	8	9	8	6	4	1	0
Tyrannidae	4	9	11	10	11	8	9	6	10	6
Troglodytidae	1	1	1	3	2	3	2	3	1	0
Parulidae	0	2	2	1	3	3	2	3	3	1
Total	28	32	32	35	36	31	29	23	20	10

develop and survive. Given the high (and increasing) entropy rates of the nutrient-deficient soils (for instance, in Central Amazonia), the relatively stable state of this system is dependent on strong, negentropic processes within the biotic area in order to be able to maintain the total structure. Stratification into three or four stories is a remarkable structural characteristic of the rainforests in the tropical lowlands. In addition, the mixing quotient (indicating how many trunks of the average timberstand are of a particular tree species) is very strong. The mixture in the uppermost stratum is especially extreme, where on average a species is represented by only two individuals. The mean number of tree species increases in close proximity to the evergreen tropical rainforest and again decreases with increasing altitude. As a consequence of the lack of a distinct seasonal climate, annual rings can usually not be found in the cross-sections of the tree trunks. The buttress roots protect seed trees and roots closest to the surface display a tendency to grow towards the rainwater running down the tree trunks. The leaf edges are usually unbroken and have a drip apex, as with many tertiary tree species in Central Europe. Examinations by Stark (1971), as well by Went and Stark (1968) show that the mycorrhizal fungi, which also take part in the mineralization, lead the nutrients liberated from the litter directly to the roots without their entering into the soil solution. Thus, the shallow root network in the tropical rainforest soil is not only dependent on the nutrients supplied by the litter, but also on the tendency of the vegetation to filter the nutrients directly out of the water dripping off the leaves and running down the trunks. The life cycle of leaves in the tropical forest trees fluctuates between 7 and 20 months. Leaf development often takes such a short time, that it is appropriate to use the term bursting into leaf. Cauliflory often occurs and, in contrast to desert plants, the seeds live only a limited time. The tropical rainforest is rich in fruit trees and thus provides a large food reservoir to numerous frugivores (among the birds, the Neotropic Cotingidae, Rhamphastidae, for example). Numerous protective devices (for example, huge husks) have developed to shield the fruit from being eaten. About 90% of all Recent liana species are endemic to the tropical rainforest.

The plants of the herb layer, where high CO_2-concentrations are no

TABLE 47 NUTRIENT CONCENTRATIONS IN LITTER, DEADWOOD AND MINERAL SOILS (0 – 10 cm) OF TROPICAL FOREST STANDS (according to Klinge 1976)

Forest type	%							
	N	P	K	Ca	Mg	Na	Total without Na	Total without Na;N
Litter								
Mountain rainforest, New Guinea	1.38	0.08	0.22	1.49	0.23	–	3.40	2.02
Rainforest, Manaus	1.28	0.015	0.05	0.34	0.13	0.02	1.82	0.54
Moist forest, Ghana	1.51	0.05	0.44	1.95	0.24	–	4.19	2.68
Rainforest, Colombia	1.55	0.06	0.10	0.71	0.15	–	2.57	1.02
Dead Wood								
Mountain rainforest, New Guinea	0.32	0.03	0.09	0.29	0.08	0.01	0.81	0.48
Rainforest, Manaus	0.86	0.01	0.02	0.002	0.04	0.006	0.94	0.08
Moist forest, Ghana	0.32	0.03	0.05	0.87	0.07	–	1.34	1.02
Rainforest, Colombia	0.46	0.015	0.05	0.44	0.08	–	1.05	0.59
Mineral Soil								
Mountain rainforest, New Guinea	1.12	0.02	0.037	0.25	0.05	0.01	1.48	0.36
Rainforest, Manaus	0.12	0.002	0.002	0	0.005	0.001	0.126	0.06
Moist forest, Ghana	0.10	0.0003	0.015	0.06	0.008	–	0.183	0.08
Rainforest, Colombia	0.09	0.01	0.002	0.0006	0.0007	–	0.103	0.01

rarity and where the air humidity is generally above 90% exhibit numerous shade adaptations to the darkness of the forest floor.

The rainforests of the tropical lowlands often stand on red-colored, humus- and nutrient-deficient lateritic soils. The soil surface consists mainly of yellow and reddish-yellow latosols. These include the ferrisol and kaolisol. Andosols and podsols can occur in isolated instances; the latter occupy huge areas in the Guyana-massif and are characterized by an atypical vegetation (Table 47).

The humic substances of tropical podsols and blackwater tributaries make a developmental connection plausible. The rates of erosion and transport of the solution in Amazonian rivers fluctuate greatly depending on the different types of soil (Gibbs 1967).

In the soils of the rainforests cleared by burning (shifting cultivation), superficially present organic material is being degraded almost completely by washout. The cosmopolitan bracken species (*Pteridium aquilinum*) often invades these destroyed surfaces.

The calcium concentrations are of particular interest since Ca has proven to exist only in the minutest quantities in the Central American mineral soils.

The great profusion of plant species in the tropical rainforest is dependent on an enormous phytomass. According to Klinge and Rodrigues (1968),

TABLE 48 NUTRIENT EXCHANGE AND NUTRIENT INFLUX VIA PRECIPITATION, AND NUTRIENT CONTENT OF THE LITTER IN CENTRAL AMAZONIA (according to Klinge 1976)

	Nutrient washout ($kg.ha^{-1}.a^{-1}$)	Nutrient influx ($kg.ha^{-1}.a^{-1}$)	Nutrient content of the litter ($kg.ha^{-1}.a^{-1}$)
Total N	5.8	10	105.6
Organic N	4.9	3	negligible
Total P	0.1	0.3	2.1
P (PO_4^{3-})	0.09	0.07	negligible
Total Fe	6.5	2.1	1.2
Ca^{2+}	4.9	3.7	18.3
Mg^{2+}	2.6	3	12.6
K^+	6.5	negligible	12.7
Na^+	11.0	negligible	5.0
Mn	0.1	negligible	0.7

7.3 tons of dry mass alone (of which 76.6% is leaves) per ha are scattered yearly in the Central Amazonian rainforest, through which about 2.2 kg P, 12.7 kg K, 5 kg Na, 18.4 kg Ca and 11.6 mg Mg per unit area are returned annually to the soil (Table 48).

Nutrient concentrations in the leaves of tropical forests, however, vary depending on the species and site. To this litter quantity, 2 tons of branches and 1 ton of trunk wood per year must be added. The surface phytomass was estimated at 1,000 tons of fresh mass and 44 tons of dead wood, and the root mass at 200 tons. Brünig (1968) arrived at higher figures in the forests of Borneo; these, however, can be explained by the fact that the Central Amazonian soils contain less nutrients and that the forests of Sarawak have a different species composition (high quantity of Dipterocarpaceae). The extensive lack of big game (exceptions: okapi and giant forest hog in the rainforest of the Congo; tapirs in the rainforests of South Africa and Southeast Asia) is remarkable, and this lack of big game results in the fact that the zoomass, when compared with the phytomass, is considerably lower. However, it is still of great importance to the functioning of the total system. The animal communities of the tropical rainforest are related to a large extent to the processed, dead plant materials populated by fungi (Table 49).

TABLE 49 ANIMAL BIOMES IN VARIOUS TROPICAL ECOSYSTEMS (from Farnworth and Golley 1974)

Ecosystem	Animal biomass (g/m^2)	Ratio of the plant mass to the zoomass	Author
Tropical rainforest (Panama)	7.3	4383	Golley *et al.* 1962
Tropical rainforest (Puerto Rico)	12.0	2264	Udum 1970
Mangrove swamps (Puerto Rico)	6.4	1762	Golley *et al.* 1962
Coral reefs	143.0	5	Odum and Odum 1955
Savanna (East Africa)	45.0	15	Dasmann 1964

The basic structure of the rainforest ecosystems is supported by plant organisms and requires the supply of inorganic substances which in the long term generally cannot be processed by the soil (Fittkau 1971, 1973). In Malaysia alone, there are about 30,000 flowering plants (of which 5,000 are orchids and 5,000 tree species) with 2,400 genera. In Sarawak, the Diptero-carpaceae represent about 10% of all tree species and between 25% and 80% of the phytomass of the forests. Although the Angiosperms dominate the rainforests of the tropical lowlands, there are noteworthy occurrences of isolated conifers in some regions. In the Western Malayan archipelago, these are represented by the families Podocarpaceae, Araucariaceae, and Pinaceae in more than 20 species (Fig. 65a). The poikilothermic vertebrates, such as the amphibians and reptiles, have a particularly rich diversity.

Whereas there have been numerous diversity investigations into the avifauna of different types of rainforest (MacArthur 1969, MacArthur, Recher and Cody 1966; Orions 1969) the quantitative evaluation of the reptiles in tropical forests is limited in general to small systems (amphibian fauna of individual trees — amongst others; cf. Heyer and Berven 1973) and are therefore mostly not representative of the typical fauna of a forest. Investigations on a 10 ha plot of forest in Reserva Ducke in the Central Amazon rainforest near Manaus (Müller 1976) show that there is a considerably greater number of species in this small area than for example in a similar sized area of forest in Central Europe. The tropical rainforest with its diverse habitats, determined by the vegetation structure, and its favourable climatic conditions provides the ecological background for a unique animal kingdom. As with the varying structure of the individual types of forest a stratification can be detected in the fauna of the rainforest. For example, there are many kinds of butterfly that never touch the forest floor. Near the ground is where animals live whose need for light and humidity is minimal. In flood areas (Amazonia and Congo, amongst others) these species undertake mass emigration (as in the case of insects, mites, certain spiders and molluscs amongst others). Other species specialize in living at mid-tree level and show typical signs of having adapted to trees. In this category we find tree prickers, capucin apes, Palaeotropic scaly ant-eaters and the papuan *Chiruromys* species with their gripping tails.

Tree snakes, tree agamids, tree iguanas and tree varanids have special scaly tails which ensure a firm grip. Obligate symbionts have made tree trunks their home and have adapted with specific forms and colors. Phytophagous species have in common with their hostplants mosaic-like distribution patterns. Species capable of flying are found in the amphibia and reptilia (e.g. the dragon lizard *Draco fimbriatus* and the flying frog *Rhacophorus reinwartii*). As well as groups poor in species there are groups rich in species (e.g. ants, termites, pollinating insects, anurae, cf. Irmler 1977, Schubart and Beck 1968). Amongst the Coleoptera of the tropical rainforest soils Staphilinidae, Pselaphidae, Ptiliidae and Carabidae dominate. Also noteworthy is the proportion of phylogenetically old taxa found alongside the very young taxa in tropical rainforests. Cycadaceae, found in the New Red Sandstone, is one of

Figure 65a Sites of *Araucaria angustifolia* near Itaimbezinko (Rio Grande do Sul, Brazil; 1967).

the old groups. Amongst the animals the *Peripatus* species deserve special mention. The fact that numerous phylogenetically old animal groupings are to be found in the tropical rainforest caused many writers to assume that the rainforest even in its present day expanse is 'ancient' and has survived the ice ages without any great change. The results of more recent biogeographical and palynological investigations demonstrate, however (Müller 1973 summarizes them), that biome shifts took place during the last glacial and post-glacial epochs during alternating wet and dry phases, which led to a strong dissipation of rainforest biomes. The rainforests of Southeast Asia, Australia,

Africa and South America were much more isolated than today by savanna zones, and savanna species were able to penetrate forests which are still isolated today (Ashton 1972, Crocker 1959, Fränzit 1976, Keast 1961, Müller 1965, 1972, 1973, 1975, 1976, Morcali 1966, Van Der Hammen 1974).

Limiting factors for many propagation areas in tropical rainforests are at present open country, high mountains, wide lowland rivers and also paucity of nutrient and chemical composition of many of the soils. Fittkau (1971) has repeatedly suggested that the latter applies to the Central Amazon. He attributes the absence of rubber trees, the brazil nut, *Ceiba,* lime-loving water and land molluscs and a water plant flora rich in species to the chemical nature of the soil. On the basis of the geochemical conditions he divided the Amazon rainforest into two ecological areas: the central region and its periphery. Calcium, magnesium and calium are under-represented in Central Amazonian locations. Furch, however, showed that traces of other substances were present in sufficient quantities. They are stored selectively in the plants (vanadium in the brazil nut, amongst others). The elements K, Ca, Na and Mg clearly decrease from Panama to Manaus. Such locations which are geochemically impoverished, may act as separation barriers in the same way as savanna zones or rivers (cf. distribution centers in the Neotropic).

Throughout the world present day rainforest biomes are in the process of being destroyed by anthropogenic factors. If these developments continue then "it looks as though tropical rainforests by the turn of the century will have been stripped clean or completely destroyed" (Lamprecht 1971, Goodland and Irwin 1975). The recognition that the former rainforest ecosystems should be permanently protected by a sufficient number of national parks has come too late in many cases. Amazonia, formerly the largest continuous forest ecosystem on our planet, reacts with extreme sensitivity to outside intervention and is being covered with roads. Since research into the Amazon fauna is still in its infancy, we are unable even to foresee what will be lost forever with the changes and destruction of the forests (Table 50).

Some of the almost natural ecosystems of Central Amazonia are characterized by significant rates of erosion, caused either by the climate, the relief, or the physical properties of the soil associations. Thus, the latter have a distinct effect on the specific limit of the pollution capacity of the individual phases of cultivation (Table 50 and Fig. 65B).

5.1.2 The Savanna Biomes

The term was derived from the Spanish *sabana* (= grassy plain) and is used by us exclusively to indicate the grassy fields of the periodically humid tropics, independent of the manner in which they developed and to what extent they are interspersed with trees or shrubs. According to Schmithüsen (1968), three belts can be distinguished within the periodically dry tropical vegetation belts (monsoon forests, dry forests, scrubs and savannas):

1. The most humid of the three belts can be characterized by the tropical

Figure 65b Mountain forest in Rancho Grande (Venezuela 2/14/1979).

TABLE 50 CHANGES IN THE WATER BALANCE IN AMAZONIA DUE TO CLEARING OF FORESTS (according to Fränzle 1976)

Water balance	Rainforest biome	Cultivation formation 1	Cultivation formation 2
Precipitation	2100 mm/a	2100	2100
Evaporation	1100 mm/a	950	830
Discharge	1000 mm/a	1150	1270
Discharge rate	31.71 $km^{-1}.s^{-1}$	36.4	40.2

monsoon forests, the humid rain-green and semi-evergreen forests which are most widely dispersed in the Asiatic tropics, as well as by the campos cerrados and tall grass savannas with their evergreen gallery forests along rivers (moist savannas; three to five arid months).

2. The middle belt consists of deciduous, dry forests, succulent-deficient scrubs and savannas with medium-tall to low grass without evergreen gallery forests (dry savannas; six to seven arid months).
3. The third belt, Waibel's "scrub steppe", consists of scrub and succulent formations and forests (Caatinga) and dry grass fields devoid of any scrub (scrub savanna; 8 to 9.5 arid months).

Within the tropics, the savanna is the total opposite of the rainforest. The number of fire-proof trees (pyrophytes) is great. On thick laterites, they grow to a height of 6 to 12 m, and are usually evergreen, have large leaves, and thick barks. Adaptation to fire, however, is not only found in mature plants. Many seeds of savanna plants tolerate temperatures of over 100°C without their germination capacity being diminished (see Tamayo 1962; Vogl 1964, among others). Beadle (1940) showed in Australia that the seeds of *Acacia decurrens, Angophora lanceolata, Eucalyptus gummifera, Hackea acicularis* and *Casuarina rigida* can be stored for over four hours at 100° to 110°C and at low humidity without their germination capacity being seriously impaired. The characteristic species of the Brazilian campos cerrados, *Curatella americana,* germinate — as do many other campo species — particularly well after having been subjected to temperature shock treatment. The grass *Trachypogon montufari* present in the llanos of Venezuela, tolerates temperatures of up to 135°C, the frequentlv occurring tree *Byrsonima crassifolia*

Figure 66 *Adansonia digitata* in the dry savanna of Senegal (west of Dakar, July 1974).

Figure 66b Savanna in the vicinity of Korhogo in the northern area of the Ivory Coast (November 1979).

350°C and *Hyptis suaveolens* 770°C (Vareschi 1962). The gallery forests present along rivers in the moist savannas are spared from fires.

The savannas are the areas where the swarming of migratory locusts occurs (Krebs 1972). The dynamics of the migratory locust populations are

Figure 66c Moist savanna (foreground) with denser gallery forest (background) (50 km south of Ngaroundere in northern Cameroon, 3/28/1979).

dependent on the change in the conditions of the savanna biome. The spread of numerous pathogens in the case of humans is controlled by the belt-shaped arrangement of the savanna biomes. Louse relapsing fever at the beginning of this century spread within the African savanna strip. Because of the information already available about its ecological potential, the direction of migration and infestation could be predicted. The carriers of the South American chagas disease are the redubiids (*Triatoma infestans* and *Panstrongylus megistus*), which, ecologically speaking, are closely related to the campo cerrado biome.

The old-world savanna formations contain numerous big game species. Ungulates dominate the scenery in the African and Indian savannas. In Africa (Fig. 67), the savanna fauna includes, among the primates, the green monkey (*Cercopithecus aethiops*), the hussar monkey (*Erythrocebus patas*) and the Senegal galago (*Galago senegalensis*); among the canines, the side-striped jackal (*Canis adustus*), strand jackal (*C. mesomelas*), and African hunting dog (*Lycaon pictus*); among the verrids, the striped weasel (*Ictonyx striatus*), white-tailed mongoose (*Ichneumia albicauda*), dwarf mongoose (*Helogale parvula*), and banded mongoose (*Mungos mungo*); among the hyaenas, the spotted hyaena (*Crocuta crocuta*), striped hyaena (*Hyaena hyaena*), brown hyaena (*Hyaena brunea*), and aardwolf (*Proteles cristatus*); among the felids, the caracal (*Felis caracal*), lion (*Panthera leo*), and cheetah (*Acinonyx jubatus*). The order Tubulidentata, with its only species, the aardvark (*Orycteropus afer*), is as numerous on the savanna biomes as are the African Equidae (African wild ass, *Equus asinus;* grevy's zebra, *E. grevyi;* steppe zebra, *E. burchelli*), the Rhinocerotidae (black rhinoceros, *Diceros bicornis;* white rhinoceros, *Ceratotherium simum*), giraffe (*Giaraffa camelopardalis*) as well as numerous antelopes and gazelles (eland, *Taurotragus oryx;* lesser koodoo, *Tragelaphus imberbis;* horse antelope, *Hippotragus equinus;* hartebeest, *Alcelaph buselaphus;* topi, *Damaliscus korrigum;* white-bearded gnu, *Connochaetes taurinus,* white-tailed gnu, *C. gnou;* gerenuk, *Litocranius walleri,* dibatag, *Ammodorcas clarkei;* impala, *Aepyceros melampus;* Grant gazelle, *Gazella granti;* Thomson gazelle, *G. thomsoni;* crown duiker, *Sylvicapra grimmia;* dwarf saiga, *Rhynchotragus kirki*). The otter *Bitis arientans,* too, is a typical African savanna species. Numerous African sparrows are also found (for example, *Mirafra africanoides, M. cordofanica, Nilaus afer*), as well as the fire finches *Euplectes orix, E. franciscanus,* and *E. nigroventris* (Fig. 68).

The Indian savannas are much poorer in species than those of Africa, although the origin of the fauna of the African savanna is to a large extent oriental, as has been proven by the fossil findings in Siwalik, for instance.

The Australian savannas are dominated by marsupials, parrots, parakeets, and finch species. In Australia, the emu (*Domaius novaehollandiae*) replaces the ostrich.

A physiognomic analysis of the South American cerrado vegetation shows that all types of transitions from open forest to grass fields can occur.

Figure 67 The African savanna biomes are characterized by big game rich in species and individuals which shape the landscape. *(upper left)* Changing annual precipitations are responsible for the migrations of the gnus *(Connechaetes taurinus),* Ngorongoro Crater, 1972; *(center left)* and fluctuations of the topi antelopes *(Damaliscus korrigum),* Tsavo National Park, Kenya, 1972; *(lower left)* impalas *(Aepyceros melampus),* Krüger National Park, 1981; *(upper right)* elephants *(Loxodonta africana),* South Africa, 1983; *(center right)* kudus *(Tragelaphus strepsiceros),* Namibia, 1981; eland *(Taurotragus oryx),* Ngorongoro Crater, 1972.

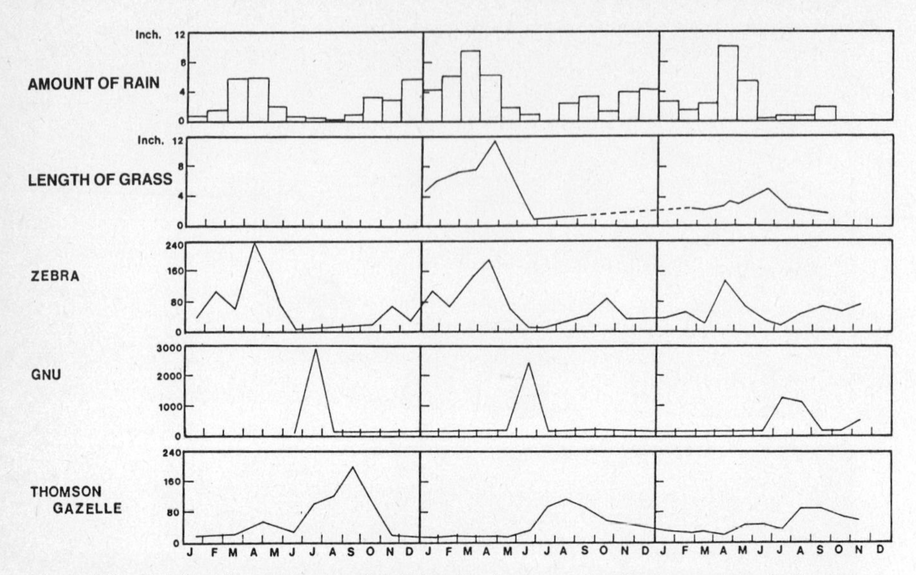

Figure 68 Relationships between rainfall, length of grass, and animal succession in the African savannas (according to Bell; from Remmert, 1973). The zebras first graze the tall grass, followed by the gnus and Thomson gazelles.

The cerrado presents a vegetation mosaic which is evidently strongly determined by the distribution of nutrients in the soil. In the campo cerrado (cerrado = dense, closed), at least three strata can be distinguished:

1. Low (3 to 8 m high), usually evergreen trees with thick barks which, like trees in the Scandinavian coniferous forests, sometimes exhibit spiral growth. Forest islands (Cerradão) occur locally (trees growing to heights of 8 to 18 m).
2. Low, mostly evergreen, large-leaved bushes (pillar cacti are absent).
3. Herb and grass fields which are easily broken up during the dry season (large proportion of ephemeral species).

The cerrados contrast with the campos limpos (limpo = pure, free) which can be defined as treeless grassy landscapes. All transitional phases between campo limpo and cerrado are present; some of them have their own names (such as campo sujo; containing sporadic tree and shrub growth). The largest part of the Central Brazilian campos cerrados (1.5 to 2.2 million km^2) is located within the tropical, summer humid, area with seven to eight humid months and an excess of 1,300 mm of precipitation (Fig. 69).

In close proximity to the ground water, trees may grow to a height of 12 to 15 m, and in narrow, water-carrying valleys, mostly evergreen gallery forests may develop, generally of the evergreen gallery type. The ecology of the campo cerrado formation has been studied very intensively during the past few years.

In spite of similar numbers of plant species in the individual campo sujo, campo cerrado and cerradão types, there are noticeable differences in the tree, shrub, and herbage strata (Goodland 1970) (Table 51).

Figure 69 The campo cerrado biome of Central Brazil (Pousada do Rio Quente; Goias, 1969).

So far, 774 shrub and tree species have been identified for the cerrados of Brazil (Heringer, Barosso, Rizzo, and Rizzini 1976). They belong to the genera which, in addition to their endemic, Central Brazilian stock, are closely related to the forests of the Serra do Mar and of Amazonia. The herbaceous ground flora, on the other hand, is composed primarily of campo limpo species (Heringer *et al.* 1976).

As long ago as 1943, Rawitscher, Ferri and Rachid realized that it was not lack of water that determined the physiognomy of the cerrado. In the

TABLE 51 NUMBER OF TREE AND SHRUB SPECIES IN THE CENTRAL BRAZILIAN CAMPO (according to Goodland 1969)

Number of species	Campo sujo min.	Campo sujo mean	Campo sujo max.	Campo cerrado min.	Campo cerrado mean	Campo cerrado max.
Trees	19	31	43	18	36	52
Herbs	42	60	79	42	53	72
Shrubs	1	5	9	1	4	6
		96			93	

Number of species	Cerrado min.	Cerrado mean	Cerrado max.	Cerradão min.	Cerradão mean	Cerradão max.
Trees	26	43	60	40	55	72
Herbs	18	47	59	21	42	60
Shrubs	1	4	6	0	3	7
		94			100	

Cerrado Emas (in the vicinity of Pirassununga, State of Saõ Paulo), they observed that at precipitations of 1,300 mm, the water storage at a soil depth of 20 cm equalled a reserve of three years, that the root systems of many of the cerrado trees (for example *Andira humilis, Anacardium pumilum*) reached a depth in excess of 10 m, that the fissure openings of the evergreen, large-leafed species examined were open all day long and that mechanisms which counteracted transpiration were mostly absent (in contrast to the plant species of the chaco or the caatinga).

Shortly afterwards, Ferri (1944) analyzed — also in the Cerrado Emas — the transportation of several campo plants and reconfirmed the fact that water is not the decisive factor in the distribution of the cerrado. Examinations by Rachid (1947) showed that the opening rhythms of some of the species could be correlated to their root depth. Andrade, Rachid-Edwards and Ferri again confirmed in 1957 that the water supply did not provide an explantion for the xerophytic phenotype of the three characteristic groups of the cerrado (perennials, ephemerals and grasses).

In 1958, Arens developed a hypothesis whereby the xeromorphic foliage was a "pseudo-xeromorphism" which could be explained by the oligotrophic balance of nutrients in the cerrado soils. Thus, in his opinion, it was the absence of important nutrients which, in the final analysis, supplied the basis for the xerophytic phenotype (Arens 1958, 1963).

In his dissertation on the subject of cerrado type in Minas Gerais, Goodland (1969) analyzed various campo types at 110 sites and determined that the nutrient supply in the soil increased continuously from the campo sujo to the cerrado, while the aluminum content decreased (58% in the campo sujo, 35% in the cerradão). Goodland (1969, 1970, 1971) determined in additional analyses that aluminum occupied a key position in the nutrient supply of the cerrado plants. The fact that many cerrado plants are resistant to high Al-doses (among the Vochysiaceae for instance, the genera *Qualea, Vochysia, Salvertia*) and that they store Al at high ppm (for instance, *Neea, Strychnos, Miconia, Psychotria, Antonia, Rapanea, Roupala, Rudgia, Palicourea;* Ferri 1976) is remarkable.

Since the physiognomic phenotype of Al-plants exhibits similarities with the cerrado plants, Goodland (1971) explained the phenotype of the cerrado formation by the oligotrophy of the soil which is directly controlled by the presence of aluminum (Goodland and Pollard 1973). In this connection, the cerradão analyses performed by Ratter, Askew, Montgomery, and Gifford (1976) in the northeast of the Mato Grosso, are of interest. They described two cerradão types according to their dominant plant species.

Type 1 is characterized by mesotrophic and type 2 by dystrophic soils. It is interesting to note, however, that their pH-values are between 5.2 and 6.5, i.e., above those of typical cerrado soils (below 5.2, Freitas and Siveira 1976). The calcium values, too, are higher; the Al-concentrations, on the other hand, are lower (in some cases zero). The authors analyzed similar cerradão types in Minas Gerais and Goias and were able to identify both cerradão types in these states (Pleistocene forest refuges in the cerrado?). Malavolta, Sarrugue and Bittencourt (1976) examined in detail the physiological effects of aluminum and manganese at low pH-values, as is the case in cerrado soils, and concluded that an agrarian use of many cerrado soils is not possible without buffering the aluminum.

It has only been in the past few years that the microbiology of the soil has been observed more closely (Drozdowicz 1976). The results so far obtained show that the decomposition of cellulose in the soil takes much longer than under the trees of the rainforest.

In contrast to the African and Indian savannas, the campos cerrados have very little big game. Frequent animal species in this South American habitat include *Melanopareia torquata* belonging to the rhinocryptides, the seriema (*Cariama cristata*), reminding one of the African bustard, nandu (*Rhea americana*), wild dogs *Lycalopex vetulus* and *Chrysocyon brachyurus,* snakes *Lygophis paucidens, L. lineatus dilepis, Crotalus durissus collilineatus, Chironius flavolineatus, Bothrops moojeni, Leimadrophis peocilogyrus intermedius* and *Epicrates cenchria crassus,* and anurans *Bufo paracnemis* (Bufonidae) and *Hypopachus muelleri* (Microhylidae). The closest relatives of the campo cerrado fauna occur in the North Brazilian caatinga area, in the Chaco, in the isolated high altitude campos, in the araucaria forests of the State of Parana, in the isolated campo islands in the Amazonian rainforest, in the Llanos of the Orinoco, and the coast and highland savannas of the Guyanas and Venezuela. The Pleistocene fossil history of the campo cerrado, which so far has been known to us only from the Lagoa Santa, shows that this formation also existed during the Quaternary period. The occurrence of monotypic genera limited to the campo cerrado would also indicate that the cerrado represents a habitat which existed before any human interference took place. Furthermore, many typical rainforest species completely avoid the campo cerrado. A large number occur in the disjunct areas containing mostly specifically or subspecifically differentiated populations of the forests of Amazonia and of the Serra do Mar. Even in the case of species capable of flying it can be proven that the isolated populations were no longer engaged in a gene exchange during the Recent period.

In the north eastern Brazilian caatinga, the long arid period is the decisive factor for the physiognomy of the plants. Strong evaporation and winds — the NE, E, and SE trade winds being predominant — negligible and irregular rain (less than 400 mm) with great insolation (in the interior of the caatinga, up to 3,200 hours of sunshine per year) are characteristic. The caatinga can be clearly distinguished from the campo cerrado by its xeromorphic character. Many plants are equipped with distinctive features to limit respiration losses.

The real caatinga can easily be distinguished from dry forest types, river forests, wax palm groves, and dry steppe lawns. A caatinga may be compared with the briar forests in other hot and dry tropical countries. Some trees remain strikingly green even during the arid period. These include the

Joazeiro (*Zizyphus joazeiro*) and the shrub *Capparis yco* with its leathery leaves. Pillar cacti, which are absent in the campo cerrado, such as *Cereus jamacaru, C. suquamosus* or the "Xique-Xique" *Pilocereus gounellei,* the "Barrigudos" *Cavanillesia arborea* and palm tree (*Copernicia cerifera, Cocos schizophylla, C. comosa*) dominate.

The caatinga contains many physiognomic and ecological parallels to, and also genetic relationships with, the Argentinian-Bolivian Chaco (Figs. 69a and 69b)

5.1.3 The Steppe Biomes

In contrast with the savanna, the steppes undergo marked changes between the warm and cold seasons. Steppe biomes therefore correlate with the grasslands of the extra-tropics, although they are closely related to the grasslands of the tropical high mountains (for example, the Puna grasslands in the Andes; cf. High Mountain Biomes).

In the Palaearctic, the steppes reach from the mouth of the Danube River to the Altai and the Saur Mountains in a closed belt. In mountainous central and eastern Siberia, they are limited to the lower-lying basin areas. Steppe islands occur in isolated instances in the forests of the Yakut Republic and at various elevations in the Eurasian Mountains. In the north, as far as the Ural Mountains, the Eurasian steppe borders on deciduous forest biomes (broad-leaved species), followed by birch and aspen forests as far as the River Yenisei in western Siberia, and by *Larix* and *Pinus* forests in the eastern Transbaikal region. The forest steppes occur in the form of macro-mosaics consisting of deciduous forest stands and meadow steppes representing the transition between the steppes and forest biomes. Lavrenko (1970) divided the Eurasian steppes into a western group influenced by the ocean, and an eastern group. The western steppes are dominated by spring geophytes (*Ornithogalum, Crocus, Tulipa, Poa bulbosa*), and annual ephemerals, such as *Arenaria serpyllifolia, Bromus, Holosteum,* and *Valerianella.* These are absent in the eastern group.

The steppe islands located in the evergreen forests in the Lena Valley near Yakutsk occupy a special position, since, at rainfall levels of 150 to 200 mm, they can be characterized by numerous endemics.

Chernozems have A-C profiles with an A_h-horizon over 50 cm thick, consisting of marl, which is dark grey when dry and nearly black when wet. The high proportion of organic material (in the Russian chernozem sometimes in excess of 10%; in Central Europe and North America 2 to 4%) is ultimately attributable to two factors:

(a) Invertebrate species (Oligochaetes, among others) and vertebrates (particularly rodents) in sometimes high population densities populate the A_h-zone and leave behind them correspondingly large amounts of excrements and burrows;

(b) The seasonal change from warm summers to, in part, extremely cold

Figure 69a Dry forest in the Serra de St Ana (Falcon Peninsula, Venezuela, 2/17/1979).

Figure 69b Castinga at the foot of the Serra de St ana (Falcon Peninsula, Venezuela, 2/17/1979).

winters stops the biochemical and bacterial decomposition processes. ıt of

Both of the above groups of factors make possible the development of nitrogen-rich humic acids present in the form of insoluble Ca- and Mg-salts separated into flakes. Thus, a biologically optimal soil type is created (mostly on loess or marl) with a high degree of saturation of Ca^{2+} and some Mg^{2+}, a neutral pH, loose crumbing soil structure and good aeration inherent in soils with the greatest yields. After having been rinsed by the melting snow waters in the spring, $CaCO_3$ is concentrated, below a depth of 50 cm, as a finely-distributed, "often mycelium-like $CaCO_3$" (Ganssen 1965). Therefore, the more arid the climate, the higher the location of the lime zone. The potential fertility of the soil and the agrarian output potential which would normally be expected as a result, however, are reduced by the steppe climate prevailing in many regions. Extremely dry summers with devastating dust storms and early winters are limiting factors on the grainfield expansion undertaken in many areas. The relict black soil islands in Central Europe (for example, around Magdeburg, Halle, Erfurt, Upper Rhine Valley) originated during the climatic optimum (ca. 5,000 B.C.). It can be concluded from the existing fossil burrows in today's forest regions built by obligate steppe rodents (for instance, the suslik), that in some cases, significant shifts between forest and steppe biomes took place during the postglacial period.

Since the soil type is greatly influenced by the macro-climatic conditions, it is not surprising that a close connection also exists between soil and steppe types. The water supplies in the soil and evaporation favor the predominance of grasses over forests. Natural fires and those started by humans maintain the competitiveness of grasses on better nutrient supplied soils.

Productivity and biomass depend on the individual steppe cycle and its seasonal sequence. Many authors continue to emphasize this seasonal change of the steppes which is usually accompanied by a noticeable change in color of the vegetation. The following change of aspects applies to the Transbaikal steppes which are dominated by Mongolian elements:

1. Second half of May = blue carpet consisting of *Pulsatilla turczaninovii*
2. Mid-June = White blossoms of *Stellera chraemaejasme*
3. July = Most colorful phase, caused by the yellow *Tanacetum* blossoms
4. August = Blue blooms of *Scabiosa fischeri*
5. September = All elevations colored a bright red due to the dying off of *Tanacetum* leaves

In humid years, the above-ground phytomasses of the herb deficient feathergrass steppes amounts to 4,530 to 6,250 kg/ha compared with 710 to 2,700 kg/ha in dry years (Walter 1970). The underground phytomass, on the other hand, remains practically unchanged. It usually exceeds the above-ground phytomass many times over. From a phytomass of 23.7 t/ha in the meadow steppes, the underground portion can amount to 84%, and for the phytomass of a feathergrass steppe, it can be as much as 91%.

Seasonal change (summer/winter) has far-reaching effects on the life of animals living in steppe regions. Their reaction to the winter in the Eurasian steppes lasting five to six months is winter hibernation or migration. As with animals living in the tropical savannas, they generally possess excellent faculties of vision and smell. Migratory animals usually form herds.

The avifauna is characterized by seasonal migrations. This applies equally to the tawny eagle (*Aquila rapax*), steppe harrier (*Circus macrourus*), saker (*Falco cherrug*), steppe partridge (*Perdix dauricae*), demoiselle crane (*Anthropoides virgo*), great bustard (*Otis tarda*), steppe peewit (*Vanellus gregarius*), Pallas's sandgrouse (*Syrrhaptes paradoxus*), desert jay (*Podoces panderi*), Mongolian lark (*Melanocorypha mongolica*), and rose-colored starling (*Sturnus roseus*).

The rodents (*Marmota, Citellus*) build immense hill and burrow systems. The bobac (*Marmota bobak*) constructs lodges with a height of up to 1.5 m and a diameter of up to 16 m. These earth cones are populated by a special vegetation. Other rodents (*Citellus*) build their U-shaped burrows in close sequence (up to 35,000 individuals/km^2) and thus loosen the soil.

The hamster (*Cricetus cricetus*), a representative of the western Palaearctic steppe fauna — which is also widely distributed in the cultivated steppes of forest regions — also belongs to the burrowing animal species. The same applies to steppe lemmings (*Lagurus lagurus*), which populate the steppes from the Ukraine to Mongolia, and the mole vole (*Ellobius talpinus*). Chigetai wild asses and saiga antelopes are particularly well adapted to the extremely dry steppes with their dust storms in the summer and their snow storms in the winter. The large antechambers — elongated in the form of a trunk — in the nose of the saiga antelope filter the inhaled air and retain the dust. The young animals possess mimetic adaptations to the burrows of the suslik and bobac. They existed in the Late Pleistocene steppes, together with the megaloceros (*Megalocera giganthus*).

The North American steppes (prairies) extend in the rainshadow of the Rocky Mountains from 54° latitude north (northern Saskatchewan and Alberta) to the Mexican highlands. South of 30° latitude they change into the prosopis savannas. Prairies also occur in the isolated basins of the Rocky Mountains. As with the East European steppes, the greatest rainfall occurs during the summer months. In accordance with their grass stand associations, they can be divided into three zones from east to west:

(a) Long grass prairie with high precipitation (660 to 815 mm), well moistened soil (potential forest land) containing *Andropogon scoparius, Stipa spartea, Sporobolus herterolepis, Agropyrum smithii,* and *Koelleria cristata.*

(b) Mixed prairie (precipitation 535 to 610 mm) with grass stands of *Andropogon scoparius.*

(c) Short grass prairie (Great Plains) with low precipitation (405 to 485 mm) and short grasses (*Boutelous gracilis, Buchloe dactybrides,* among others).

The depth of the roots and length of the blades decrease from the long grass to the short grass prairie; they also follow the chessboard-like arrangement of the soil types and the climatic north–south gradients (Bryson and Hare 1974). Prairie chickens (*Tympanuchus cupido* and *T. pallidicinctus*) the burrowing owl (*Speotyto cunicularia*) which is also widespread in the South American savannas, granivore sparrow species (*Ammodromus bairdii, Calamospiza; melanocorys, Spizella pallida, Rhynchophanes maccownii, Calcarius ornatus*), burrowing rodents (for example, *Perognathus flavescens, Spermophilus franklinii, S. tridecemlineatus, S. spilosoma*), the pronghorn (*Antilocapra americana*), bison (*Bison bison*), and rattlesnake species (*Sistrurus catenatus, Crotalus atrox, C. viridis*) are vertebrates characteristic of the North American prairies. In contrast with the northern hemisphere, the steppes of the southern hemisphere occupy smaller areas. The eastern Patagonian steppe belt of South America, in the rainshadow of the Andes, is a dry steppe containing tufts of hard grasses and dwarf shrubs.

The South American pampa (between 32 and 38°S, the agricultural core of Argentina, was a particularly disputed steppe; many scientists contended it had originally been forest land, destroyed by Indian fires. Grass species (*Stipa neesiana, Panicum sp., Paspalum sp., Bromus unioloides, Poa lanigera, Eragrostis lugens,* etc.), dominate. At higher groundwater levels, *Paspalum quadrifarium* forms huge tufts reminiscent of the *Carex* clumps present in the northern hemisphere. In the southern hemisphere, this form of growth is called "tussock". As the southern latitude increases, it becomes more and more dominant (tussock grasslands; cf. Tundra Biomes, Section 5.1.8).

The relatively high precipitation in the pampa (1,000 mm in the northeast, 500 mm in the southwest) is compensated for by the potential evaporation. Shallow lakes containing soda and having no drainage and which usually dry up during the summer months, are further indications of the semi-arid climate of this steppe. "The negative water balance in the humid pampa amounts to about 100 mm, and in the most arid parts of the pampa up to 700 mm" (Walter 1970). Woody plants (*Celtis spinosa*) grow on these well-drained soils. The large humus zones (to 1.5 m) are reminiscent of black soils and prairie soils and show no indication of early forest vegetation.

Charles Darwin (1832), during his ride on horseback from Bahia Blanca to Buenos Aires, was the first to question the causes of the lack of forest in the pampa and distinguished very clearly between the original landscape of the pampa, which he deemed treeless for evolutionary-genetic reasons, and the present potential for plant growth in the pampa; these findings have often been neglected by later scientists. More recent biogeographic analyses of the Neotropic distribution centers show that the pampa represents its own distribution center of a steppe fauna (Müller 1973) and thus supports the assumption that the pampa constitutes original grassland.

Just as certain pathogenic organisms and disease carriers affecting man are associated with tropical rainforests and savannas, so there are disease symptoms associated with steppes in the belts of both northern and southern hemispheres (Jusatz 1966). Rabbit fever, whose agent is *Pasteurella tularensis,* is a steppe disease. It is transmitted by ticks and mites to rodents, rabbits, and sheep which form a natural reservoir. The occurrence of tularemia is restricted to continental climates with natural steppes or with steppes caused by excessive cultivation of the soil.

5.1.4 Desert Biomes

Cold and arid deserts are areas which demand exceptional forms of adaptation (Maloiy 1972, Edney 1977).

5.1.4.1 The Arid Deserts In the New World, deserts and semi-deserts exist in the southwestern part of North America (Sonora Desert, Mojave Desert, Death Valley, Lower California), along the South American Pacific coast of Peru (4°S) and Chile, as well as in northwestern Argentina. In the Old World, the most extensive desert areas are in the Sahara, on the Arabian peninsula, in Asia (Gobi, for example), India, southwest Africa, and western Australia. In the case of arid deserts, it is not always possible to distinguish clearly between semi- and full deserts: semi-deserts surround the arid cores or are interspersed with them in a mosaic pattern in climatically or edaphically favourablc locations (Table 52).

The physical environment conditions of the desert biomes — most widespread in the continental arid belt of the northern hemisphere in the Old World — are characterized by 11 to 12 arid months, a precipitation usually less than 150 mm per year, sharp contrasts in temperature between day and night (+56°C to −40°C), high evaporation caused by the desert winds (for instance, simoom, harmattan, ghibbi, sirocco) and high insolation as well as a predominance of mechanical weathering. The various types of desert can be identified on the basis of the predominant substratum (rock or stone desert = Hamada; gravel desert = Serir; clay desert = Sebcha, Takyr; sand desert with dunes = Erg and Barchans; salt deserts), on which varying types of life forms are dependent. The temporal and spatial distribution of precipitation, however, is without doubt the factor having the most drastic effect on plant growth and animal life. Characteristic zonal soils in the tropical and subtropical dry belts of the earth are: the real desert soils (including quicksand soils, dunes,

TABLE 52 ARID DESERTS OF THE WORLD (according to Cloudsley-Thompson 1977)

Desert	Area (km²)
Sahara	9,100,000
Australian Desert	3,400,000
Turkestan Desert	1,900,000
Arabian Desert	2,600,000
North American Deserts (incl. Great Basin, Mojave, Sonora, Chihuahua)	1,300,000
Thar Desert (India)	600,000
Namib and Kalahari Deserts	570,000

Takyr soils) and semi-desert soils (including brown semi-desert soils, sierozem soils, reddish-brown soils). Solonchak and sononetz soils are often present as intrazonal soils. Increasing aridity is coupled with a decreasing tendency toward soil formation.

In the vicinity of groundwater, the desert scene often changes abruptly (oases). Some species, which, with their deep roots reach the groundwater (for example, tamarisk), form wind-pruned scrub which represents a habitat absent from the remaining desert region. Areas having a uniform appearance are often interspersed, in a mosaic pattern, by habitat islands which are sometimes only explainable by the effects of exposure or relief. This applies particularly in the case of the isolated high mountains in the Sahara, whose vertical vegetation succession illustrates the tropical and subtropical climatic influences. Flora, fauna, and humans exhibit specific forms of assimilation to the environmental conditions of the deserts.

In the case of plants, it is particularly remarkable how with most species, it is less a question of the emergence of a protoplasmic drought resistance than a regulation of vegetation density and root systems to make optimal use of the available rainfall. This means that there are far fewer species than in the tropical rainforests, and those that are present are distributed across the surface in clearly distinct zones.

Stocker (1962) pointed out the ecological asymmetry within the arid tropical belt. On the equatorial side of deserts, the scrub savanna biomes often expand — in the form of scattered trees — right up to the edge of the desert, while towards the regions with a moderate climate, only semi-desert and steppe biomes containing no trees or, at best, subshrubs, are encountered (cf. Lauer and Frankenberg 1977).

Plant species which must survive in totally arid zones (for instance, the Saharan species *Aristida pungens, Anabasis aretioides, Genista saharae, Artemisia herba-alba*) possess specialized ecophysiological adaptations to the water conditions (dehydration) of their environment.

Rain plants (for example, *Mesembryanthemum, Mollugo*) are capable of unfolding their blossoms after a short moisture contact. Poikilodydrous plants, i.e. plants capable of drying out, exist in the desert regions, particularly among the algae, lichens and mosses. However, impressive examples of this capacity can also be found in the higher flowering plants which probably acquired it secondarily. *Chamaegigas intrepidus* is a water plant growing in the shallow basins of the arid southwest African granite mountains which, after the rainy period, has to survive, often for months, in totally dry puddles. As soon as the hollow is filled after a rainfall, the shrivelled-up leaves quickly swell and, within a short period of time, increase to ten times their original length. A few days later, the floating leaves and blossoms rising above the water surface develop. The bush *Myrothamnus flabellifolia,* endemic to southwest Africa, survives the dry period of several months at its site with folded-up leaves, which although so dry that they can be crumbled into a dust-like powder, still exhibit a very slight CO_2 production (Ziegler 1974). After half an hour's rain, the leaves unfold and look fresh and green.

Xerophytic, succulent, sclerophyllous, halophile, and perennial plants (often with root systems reaching down to the groundwater — tamarisk roots can be up to 30 m deep) are further types of life-form characteristics of the desert biomes.

The animal life is adapted not only to the physical desert conditions, but, in particular, to the temporal and spatial characteristics of the desert vegetation. It is, therefore, not surprising that different phylogenetic, related groups in deserts which in many cases are widely separate from each other, are represented by similar convergent types of life-form. Sand burrowers among the lizard species, side writhers among the snakes, bipeds among the desert rodents, desert coloration of numerous birds, wing reduction in beetles, strong development of salt glands and a predominance of nocturnal activity above the ground surface in the case of mammals are only a few of the particularly striking characteristics of desert animals. In addition, there are special physiological advantages which are specifically aimed at minimum water consumption (among other things, large capacity for urine concentration, insignificant water loss through evaporation, negligible water content in excrement). Generally, the circadian rhythm is adapted to the daily cycle of temperature and air humidity. Many species adapt their entire seasonal development to the phenology of the desert conditions. In Egypt, the adult *Adesmia bicarinata* appears at the end of October and by March reaches its population peak. During the summer months, however, it is present exclusively in the larval and pupal stages. The migration of *Schistocerca gregaria,* occurring in the peripheral deserts, is determined by the phenology of the vegetation. Remarkable morphological structures serve as a protection against transpiration. Many poikilothermic species solve their problem of temperature regulation by means of special behavioral strategies. Desert snakes, in contrast to desert lizards, usually cannot endure temperatures beyond 42°C. The North American desert rattlesnake *Crotalus cerastes* dies at a temperature of 41.5°C, while lizards living in the same habitat survive at temperatures of 45 to 47.5°C (Cloudsley-Thompson 1977).

Among the desert rodents, the granivores are dominant; their biped mode of locomotion suggests their adaptation to travelling great distances in search of food. Leaf-eating rodents are less frequent (Kenagy 1974, Daly and Daly 1973), such as the North American *Dipodomys microps* which prefers to feed on the highly saline leaves of *Atriplex confertifolia,* and the Saharan *Psammomys obesus* which specializes in chenopodiaceae. The dentition of *Dipodomys microps* is adapted to the scraping of the highly saline outer layers of leaf tissue. The lower rodent teeth are broad, flat at the front, and chisel-shaped. The relatives of *Psammomys obesus* and *Dipodomys microps* are predominantly granivores.

Typical desert fauna of the Sahara include the jerboas (*Jaculus jaculus, Gerbillus campestris, Gerbillus nanus, Gerbillus gerbillus, Gerbillus pyradmidus*), the fennec, the rare sandcat (*Felis margarita*), the mastigure *Uromastix* (with salt glands), the side-winder vipers (*Cerastes cerastes* and *C. vipera*), the common skink *Scincus scincus* ("poisson de sable"), the desert larks (*Ammomanes deserti* with desert coloration), the wingless desert locusts, the long-legged, usually wingless tenebrionids, and the desert isopod (*Hemilepistus reaumuri*).

In most cases, the desert and semi-desert species of North Africa penetrate as far as the Indian arid regions. Species and genera which are attributable to such a distribution type are designated Saharo-Sindhian species. These include, among others, the saw-scaled viper *Echis carinatus* occurring from Algeria to Sri Lanka, and psammophile sandcat (*Felis margarita*).

The Saharo-Sindhian fauna and flora must be viewed as a biogeographical entity. Characteristically, the semi-deserts of Rajasthan contain numerous genera present in the Sahara and Arabian deserts (among the mammals, for instance, *Gerbillus*).

The Palaearctic-Ethiopian desert belt encompasses the largest part of the arid areas of the biosphere. According to Walter (1968), a subdivision into six large regions can be made.

1. The Sudano-Sindhian deserts, including the Sahel zone south of the actual Sahara, the coastal strip along the Arabian Sea, and the Thar Desert (Sindh) in Pakistan and India;
2. The Saharo-Arabian deserts, whose central areas receive, at best, negligible amounts of precipitation during the winter months;
3. The Irano-Turanian deserts which include the plateaux of northern Iran and the southern part of the Aralo-Caspian lowlands;
4. The Kazakhstano-Dzungarian semi-desert and desert regions whose transition areas to the east European–south Siberian steppes, reach eastward from the Caspian lowlands of the lower Volga via Lakes Aral and Balkhash;
5. The Central Asian desert regions of Mongolia and northern China;
6. The high mountain deserts of Tibet, which, in the form of cold deserts, are located at altitudes of about 4,000 m.

In contrast to the Kazakhstano-Dzungarian deserts, the Central Asian desert regions receive their precipitation from the east during the summer (Chinese monsoon region). Winter and spring are correspondingly dry. This applies to the Ordos north of the Great Wall of China (Hwang Ho), Alashan, and Baishan, as well as to the Tarim Basin, Takla Makan, Tsaidam Basin and Gobi. The winters are very cold, dry, cloudless, and usually snow-free in the Gobi, the average elevation of which is 1,000 m. The development of vegetation only starts with the onset of the summer rains.

In the southwest African Namib Desert which is divided — by the Swakop — into a southern region dominated by drifting sand dunes, and a northern region (desert of stones), we encounter — in addition to the monotypical endemic genera — close relatives of the Saharo-Sindhian fauna. The miracle plant of the Namib Desert, *Welwitschia mirabilis,* is known to have existed during the Miocene; this is a Gymnosperm which received its name in honor of its discoverer, the Austrian physician Welwitsch (1859). By means of deep, 2 m long, tap roots, a short trunk, two leathery perennial leaves which continue to grow only at their base, and dioecious "flowers" in cone-like inflorescences, it has adapted to life in the desert (cf. Schulze, Ziegler and Stichler 1976) as have the silicolous and window plants (Lithops), the sand dwelling reptiles (*Palmato-gecko rangei, Ptemopus garrulus, Aporosaura anchietal, Mesoles cuneirostris, M. reticulatus, Acontias lineatus, Bitis peringueyi;* for additional examples, see Low 1972, Mertens 1972), the reptiles living in the stone deserts (*Eremias undata, Rhoptropus afer, R. bradfieldi, Cordylus namquensis*) and the highly specialized Tenebrionidae (Holm and Edney 1977) (Fig. 70).

The sand dunes of the Namib Desert, drifting in the wind, are free of vegetation. Only drifting *Salvadora* leaves and grasses (*Stipagrostis gouatostachys, Aristida sabulicola*) can occasionally be found. Organic substances are covered by dunes as "organic depots". Locusts (*Anacridium investum*), flies and ants (*Camponotus detritus*) again and again penetrate the dune areas from the more densely vegetation covered peripheral locations. This is also true for various vertebrates, among which the burrowing gecko (*Palmatogecko rangei*) is particularly worth mentioning. The Tenebrionidae *Cardiosis fairmairei* and *Onymacris laeviceps* belong to the detritus eaters processing the drifting organic material, and they can be encountered "during the daytime under the hottest sun" (Kühnelt 1976). At dusk, *Lepidochora argenteogrisea* (Tenebrionidae) becomes actively engaged in eating detritus; it populates the sand in its hundreds. Other Tenebrionidae, on the other hand, appear on the surface only at night time (*Lepidochora porti, L. kabani*), together with Lepismatidae and dune termites.

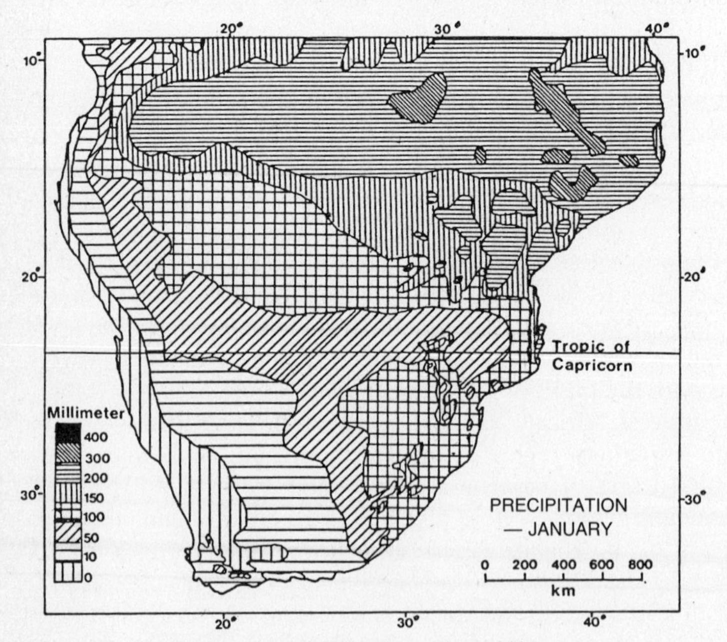

Figure 70 Distribution of precipitation in southern Africa.

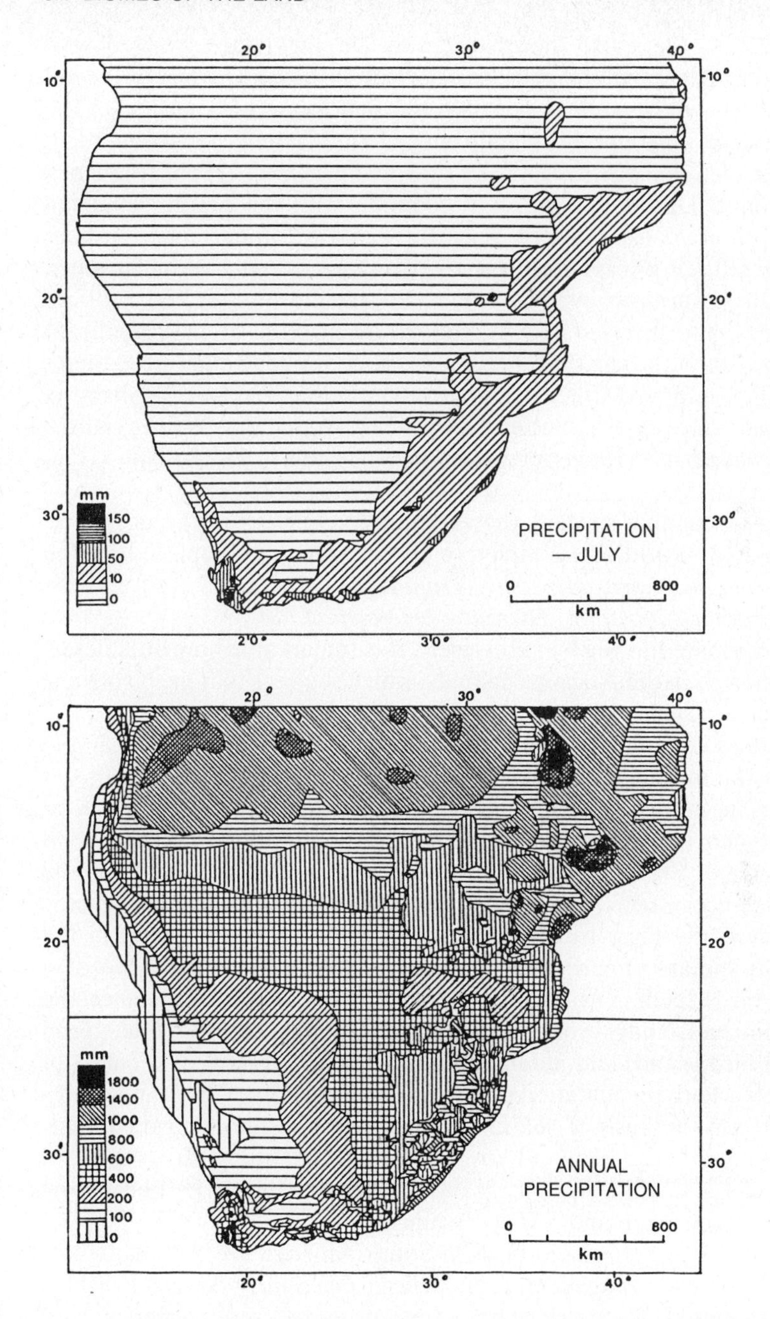

These detritus eaters are hunted by carnivorous species, among which the Solifugae (*Eberlanzia flava*), spiders (*Caesetuis deserticola*) and larvae of the Tabanidae are noteworthy. The territorial lizard *Aporosaura anchietae* occurs on many dunes; it is capable of burying itself very quickly into the sand. Also the Namib golden mole (*Eremitalpa granti namibensis*) lives as a sand swimmer in the loose dune substrate. On the dunes covered with *Aristida* and *Naras,* phytophagous insects (for instance *Onymacris plana, Hetrodes, Rhabdomys pumilo*) appear as the first group of consumers. The locust *Anacridium investum,* the Namib chameleon (*Chamaleo namaquensis*), the lizard *Meroles cuneirostris,* and the snake *Bitis peringueyi* are frequent.

The Namib Desert has much in common with the South American Atacama. Both are coastal deserts, caused by cold ocean currents. Onshore winds discharge their precipitation as they blow over the cold ocean currents, whereupon they warm up over land and thus remain unsaturated. In both coastal deserts, the dreaded fog formations occur without any significant precipitation (in South America — Garuas). A continuous plant belt, therefore, is absent and wide areas are free of plants. In places, Tillandsia (Bromeliaceae) and cactus species (*Eulychnia, Copiapoa, Philocopiapoa, Mila, Borzicactus, Weberbauer-ocerceus*) occur in the form of islands (Fig. 70).

Among the animal species, the Agamidae occurring in the desert regions of the Old World are completely absent. They are replaced by the Iguanidae (*Ctenoblepharis adspersus, Tropidurus peruvianus,* etc.) and the Teiidae (*Dicrodon guttulatum, Dicrodon heterolepis*).

When compared to the Namib Desert, it is remarkable how little desert adaptation there is in the Atacama animals, which suggests that the Neotropic desert biomes are relatively young.

While the Humboldt Stream is the cause of the desert conditions in the Atacama, in the marine area, it is a zone of very great phyto- and zoomass production. The immense phytomass (algae) development results in the diverse fauna rich in individuals, which is based, through various food chains,on the plant biomass. Via the aquatic birds, numerous food chains end up in the arid terrestrial ecosystems. For the aquatic bird populations living in Pacific South America, the huge fish populations (sardines, among others) in the Humboldt Stream are of particular significance. The excrement produced by the aquatic birds feeding on small fish, on some islands (for instance the Peruvian Chincha Islands) produces hills up to 60 m high (guano). Guano can be found on bird islands and mountains along the subtropical west coasts of the continents where precipitation is low (particularly Peru, Chile, and southwest Africa). On the basis of coloration, age, and phosphoric acid content, one can distinguish between red (fossil) guano containing 20 to 30% of phosphoric acid and white (recent) guano (containing 10 to 12% phosphoric acid, 10 to 12% nitrogen and 3% potassium).

The most important guano birds in South America are the guano cormorant *Phalacrocorax bougainvillii,* the pelican *Pelecanus occidentalis* which is widely distributed along the coasts of South and Central America, and several gannet species (*Sula nebouxii, S. variegata, S. dactylatra*). On the southern island of the Chincha group alone, the colony of *P. bougainvillii* consists of about 360,000 individuals which have settled on an area of 60,000 m^2 (3 nests per m^2). The penguin *Spheniscus humboldti* builds its nesting cavities right into the guano.

The Australian deserts (Eremaea) are positioned at the same geographical latitude as the Kalahari. The southern border is represented by the Nullarbor Plain extending for 700 km in an east–west direction, consisting of karstic, Tertiary limestone. Acacias dominate the peripheral zones while in the sand and clay plains, the saltbush semi-deserts containing the two key species *Atriplex vesicaria* (saltbush) and *Kochia sedifolia* (blue bush) are predominant. Hard-leaved hedgehog grass tufts including *Triodia pungens* and *Plectrachne schinzii,* exhibit a mosaic distribution. The mulga scrub favoring the growth of woody plants, *Acacia aneura, Cassia* and low *Eucalyptus* species are dominant. The characteristic Eremial, Australian fauna includes *Antechinomys laniger, A. spenceri, Lagorchestes hirsutus,* the bird species *Polytelis alexandrae* and *Stipiturus ruficeps* and over 70 reptile species. The agama lizard *Amphibolurus isolepsis* populates the desert of Western and Central Australia and feeds mostly on ants, small arthropods, and plants. From an ecological point of view, the species is strictly limited to sandy soils and is absent from the interspersed rocky plateaux. This ecological specialization is characteristic of most of the Australian desert reptiles (Pianka 1972).

Apart from the holarctic relationships, the North American desert biomes show numerous connections to South America. Various plant communities are represented in the Sonoran Desert. The Carnegiea-Encelia-community is dominated by *Cerciduim microphyllum,* the red-flowering *Ocotillo Fouquieria splendens, Acacia constricta, Celtis pallida, Opuntia* species, *Ferocactus wislizenii, Jatropha cardiophylla, Prosopis juliflora, Ephedra trifurca, Heteropogon contortus, Hibiscus coulteri, Lippia coulteri,* and *Selaginella arizonica;* in the Larrea-Franseria community, the creosote bush *Larrea tridentata* which can be leafless during the dry period, and soft-leaved Composites (*Franseria*) and *Opuntia* species are characteristic; in the *Prosopis velutina*-zone, *Prosopis* species (with groundwater roots 20 m deep), *Acacia gregii,* and *Celtis pallida* are present; and, finally, in the vicinity of depressions where water is available, there exists a *Populus-Salix* community.

Morafka (1977) compiled a monograph of the Chihuaha Desert, in which he defines the habitat of reptiles and amphibians and their adaptations to plant biomes which define the character of the landscape.

5.1.4.2 The Cold Deserts The Arctic habitats extend from the tundra biomes towards the poles; these habitats owe their existence and productivity to marine food chains. The role of the terrestrial primary producers is insignificant. The animal world is a characteristic coastal and ice floe fauna.

This tendency becomes noticeable in the Arctic tundra biomes. Water fowl and beach birds are dominant and have been investigated in Arctic Alaska by Pitelka (1974), among others.

Arctic Cold Deserts The carnivorous species dominate among the semi-terrestrial mammals. This is also true for the circumpolar-distributed polar bear (*Ursus maritimus*). Apart from the human being, the polar bear is the end consumer in Arctic food chains.

The Atlantic walrus (*Odobenus rosmarus rosmarus*) which feeds almost exclusively on molluscs was partly responsible for opening up the far north, and due to the exploitation of the seemingly inexhaustible supply of walruses led to geographic research expeditions of world-wide significance. Since — in contrast with the ringed seal (*Phoca hispida*) — the walrus is unable to breathe through holes in the ice sheet, the animals retreat during winter to the ice-free edges of the Arctic. The bearded seal (*Erignathus barbatus*), which often appears sympatrically with *P. hispada* and which is intensively hunted because of its skin, is also able to blow breathing holes into the dense ice sheet and prefers to remain in the shallow areas of the ocean. The harp seal (*Pagophilus groenlandicus*) is restricted to the North Sea all the year round, but it undertakes long migrations, during the course of which some animals may stray as far as England or the Elbe delta. Frequently, these migrations proceed in the direction of the principal food animals. The snow-white young animals are hunted for their pelts. Some aquatic bird species also deeply penetrate the Arctic (Salomonson 1967).

A daily rhythm of environmental factors exists on land and in the moors and lakes of the regions close to the Pole. It can be suppressed by certain weather conditions; however, all the animals analyzed so far are able to perceive this daily rhythm and to synchronize their internal clock to the rotation of the earth (Remmert 1965). At nearly 80°N (Spitzbergen), a strict daily rhythm can be observed for birds as well as for insects.

Antarctic Cold Deserts Ocean depths of over 8,000 m (east of the Scotia Arc) surround the Antarctic continent which measures 12.4 million km^2, and is covered by ice of an average thickness of 2,000 m. Under the constance influence of westerly winds, the highly stratified water masses circulate around the Antarctic in an easterly direction. Submarine cold water currents flow from this rotating disc towards the equator. The top stratum of the water which is 70 to 200 m thick, i.e. the Antarctic surface water, originates in the Antarctic divergence, a narrow zone in the southern Antarctic Ocean where the water located right below the top layer rises between two divergent strips of the surface water, the rise being caused by the prevailing winds blowing from opposite directions. Westerly winds are dominant north of the Antarctic divergence; in the south, easterly winds with an onshore tendency push the surface waters in a westerly direction. Low temperatures and low salinity (caused by the ice thaw) are characteristic of the surface waters. The influence of relatively warm (1 to 3°C), huge, deep currents rich in nutrients and up to 2,000 m in depth, however, is remarkable. These deep currents are pushed under the Subantarctic and Antarctic surface waters, wherupon they surface along the Antarctic divergence and form the basis for a rich flora and fauna. Immense pack-ice zones (in winter, the distance between the northern boundary and the mainland base is about 800 km), covering nearly 19 million km^2 of ocean surface, constantly change their boundaries during the course of the seasons. During the summer when ice thaws in the south, huge pack-ice islands break off and form floating platforms for penguins and seals.

The information available with respect to the productivity of the Antarctic surface waters is contradictory. The zooplankton is estimated at 8 to 10 g dry mass/m^3. The Euphausiidae, however, have not been included in this estimate; among these, *Euphausia superba* growing to a length of 7 to 8 cm is able to supply nearly 30 g/dry mass/m^3 in the northern Antartic. Of particular ecophysiological interest are Antarctic fish devoid of any haemoglobin. They include the ice fish *Chaenocephalus aceratus* (Chaenichtyidae) whose gills and blood are colorless. A sufficient supply of oxygen is transported in the blood plasma.

Because of the great decline in radiation during the winter months, the climate is extremely severe. The lowest temperature, −88.3°C, was recorded at the eastern Antarctic Station Wostock (cold pole of the Earth). The Antarctic is surrounded by a low pressure trough caused by the temperature contrast between the continent and the ocean. As a result of these low pressure areas, 340 storm days per year are recorded in Adélieland.

Nearly every year, some Antarctic bird species travel to the South American coast, particularly in winter. On Isla Grande (Tierra del Fuego), the penguins *Aptenodytes patagonica, A. forsteri, Eudyptes cristatus, E. chrysolophus,* and *Spheniscus magellanicus* during breeding time, are often observed (the latter is regularly washed ashore along the Brazilian coast from Rio Grande do Sul to Bahia). Representatives of these Antarctic bird species also live along the Australian and New Zealand coasts. Representatives of the sheath bill (Chionididae) reach the southern tip of South America and Australia. Reminiscent of pigeons in their behavior, *Chionic alba* regularly go to Tierra del Fuego and Argentina, whereas the second species, *C. minor,* travels to the south of Australia and Tasmania. Both species feed on carrion, young birds, and the eggs of penguins.

5.1.5 The Sclerophyll Biomes (Biomes of the evergreen hard-leaf forests)

Evergreen hard-leaf forest formations follow the arid subtropical deserts in a northerly or southerly direction along the western side of the continents (Schmithüsen 1968); during the summer months, these hard-leaf forests are completely under the influence of the subtropical highs bringing hot, dry weather, whereas in winter, they are subjected to the cyclonic rains of the moderate zones.

Mild, wet winters and dry, warm summers are thus typical of this habitat which, in the western Palaearctic, southern province of the Cape of Good

Hope, Central Chile (south of 30°S), and Southern Australia can be characterized by a remarkable ecological regularity in spite of its isolated location and genetically varying structure. Terra rossa soil is frequently present (as in the Mediterranean region) which can be interpreted as the relic of former more humid climatic conditions.

This terra rossa is a soil rich in clay (up to 90%) and iron (up to 10%) which can form on carbonate rock and which possesses an A-B-C profile whose B-zone is characterized by the brilliantly red coloration giving the soil its name. It develops from lime and dolomite rock in subtropical, humid climatic conditions. Because the high $CaCO_3$-content, coupled with the hardness of the lime, it leaves a larger residue of solution only after a fairly long period of time, even under very humid climatic conditions, which is why terra rossa is usually regarded as being very old. Hard-leaved shrub formations, such as the Mediterranean *maquis,* the Californian chaparral, and Australian sclerophyllum scrub which may consist of multi-layered *Acacia carpophylla* thickets (Brigalow-Scrub) or umbrella-shaped *Eucalyptus* species (mallee), are relics of originally widely distributed open forest formations with large numbers of clearings, which, due to anthropogenic influences, have turned into one of the most widely destroyed formation on earth.

After closely analyzing 34 relic areas, Braun-Blanquet (1936) described the Quercetum ilicis galloprovincialis (so named by him) as the natural forests of southern France:[1] (Figs. 71 and 72).

1. Tree stratum: 15 to 18 m high; closed; *Quercus ilex.*
2. Shrub stratum: 3 to 5 m high, containing *Buxus sempervirens* (2), *Viburnum tenuis* (1), *Phillyrea media* (1), *Phillyrea angustifolia* (+), *Pistacia lentiscus* (+), *Pistacia terebinthus* (+) *Arbutus unedo* (+), *Rhamnus alaternus* (+), and *Rosa sempervirens* (+).
3. Lianas: *Smilax aspera* (2), *Linicera implexa* (1), *Clematis flammula* (1), *Hedera helix* (1), and *Lonicera etrusca* (+).
4. Herbaceous stratum: max. 30% coverage; *Ruscus aculeatus* (2), *Rubia peregrina* (1), *Asparagus aculeatus* (1), *Carex distachya* (1), *Viola scotophylla* (+), *Asplenium adiantumnigrum* (+), *Stachys officinalis* (+), *Teucrium chamaedrys* (+), and *Euphorbia characias* (+).
5. Moss stratum: sparsely developed; *Drepanium cupressiforme* (1), *Eurynchium circinnatum* (1), *Scleropodium purum* (1), *Brachythecium rutabulum* (+), and others.

In contrast to the Mediterranean maquis, the vegetation in the Californian chaparral (Spanish chaparra = bushy oaktree) reaches heights of only 4 m which, however, is not to be interpreted as a degradation characteristic, but rather is attributable to the low rainfall (around 500 mm). The most frequent plant species is the rosaceous *Adenostoma fasciculatum* (Knapp 1965), in addition to the bush-shaped oaks (*Quercus dumosa,* for example), the shrub-shaped Papaveraceae (*Dendromecon rigidum*), *Rhus* species,

[1](1) = dispersed, (2) = in small groups; (+) = covering less than 5% of the area.

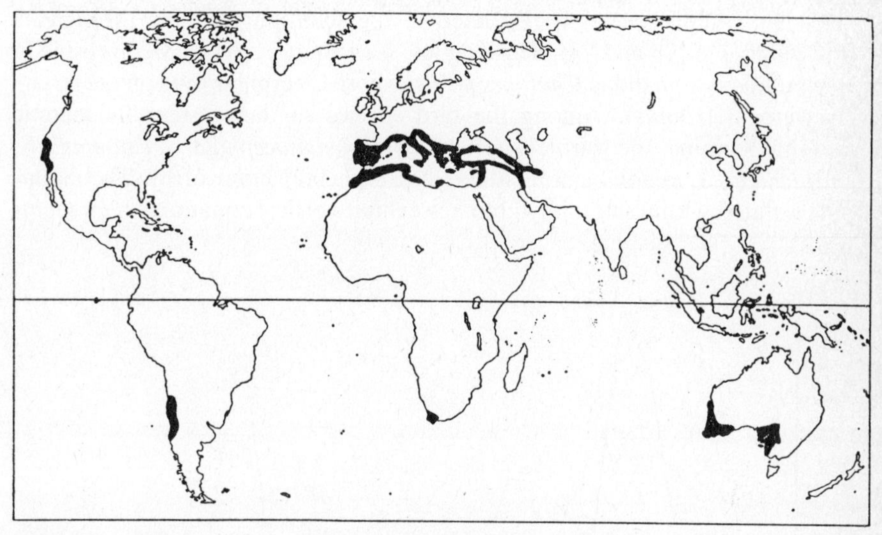

Figure 71 Regions of the world with a Mediterranean climate (according to Thower and Bradsbury, 1977).

Figure 72 The cultural history of the sclerophyll biomes is synonymous with the cultural history of the olive tree (*Olea europaea;* here, an old plantation on the Island of Crete; April 1978). Its present distribution, for which man is responsible, marks only a potential sclerophyll region.

Pasania densiflora, Castonopsis chrysophylla, Myrica californica, and *Arbutus menziesii.*

It is only recently that the fauna of the sclerophyll biomes has been comparatively analyzed (Cody and Walter 1977, among others). A natural ecological connection with the vegetation can easily be established, where a particular animal species is restricted to the sclerophyll biome through its food plant (Figs 73 and 74). This applies, for instance, to the Mediterranean distributed Nymphalidae *Charaxes jasius,* whose caterpillar lives on the strawberry tree (*Arbutus*). Among the bird species of the western Palaearctic sclerophyll biome, the warbler species (*Sylvia melanocephala, S. cantillans, S. conspicillata, S. undata,* among others) are especially noteworthy. The mountains isolated within sclerophyll biomes exhibit vertical zonation which seems

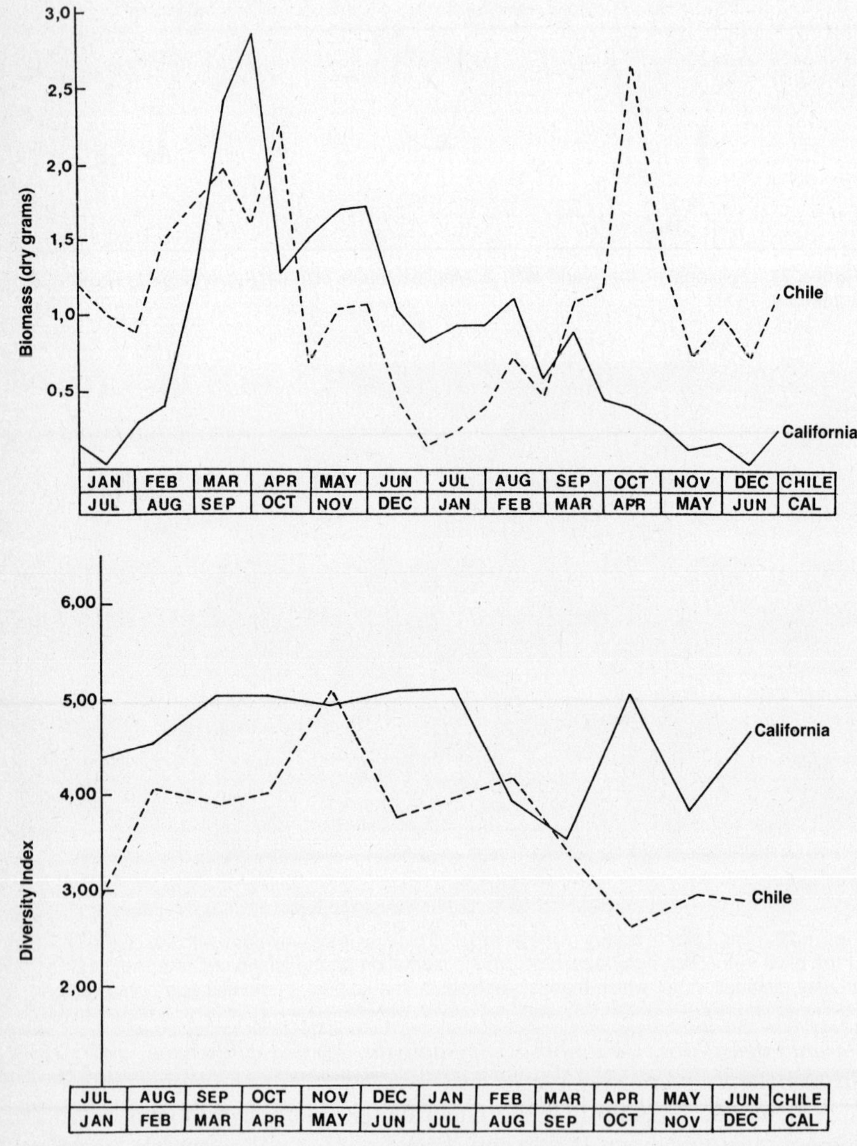

Figure 73 Annual biomass (top) and diversity index (bottom) in the Californian and Chilean sclerophyll biomes (according to Thower and Bradsbury, 1977).

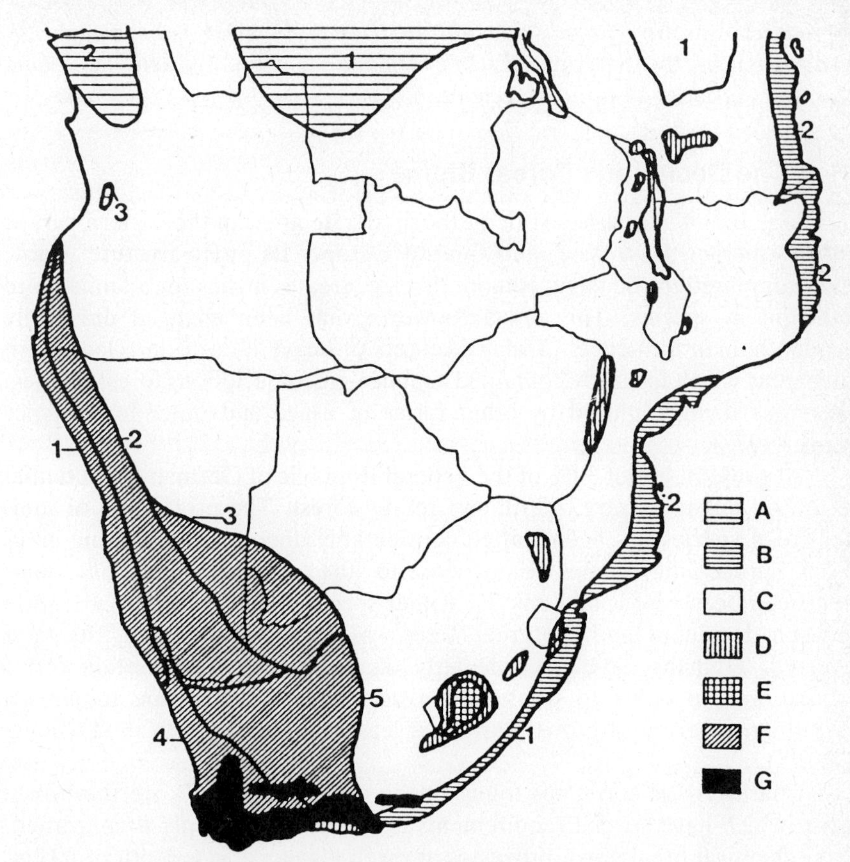

Figure 74 Phytogeographical division of Southern Africa.
A: Guinea–Congo region; A1: Congo zone; A2: Nigeria–Cameroon zone.
B: Coastal region (Indian Ocean); B1: Tongaland zone; B2: Zanzibar zone.
C: Sudan–Zanzibar zone; C1: Eastern zone.
D: Afromontane region.
E: Afroalpine region.
F: Caroon–Namib region: F1: Namib zone; F2: Namaguland zone;
F3: Southern Kalahari zone; F4: West Cape zone; F5: Caroon zone.
G: Cape region.

to make a distinction appropriate between humid and arid mountains. Within the region of the Sierra Nevada (southern Spain), the lowest zone is of *Quercus ilex* which is followed by a narrow belt of *Q. pyrenaica.* On the north side, the latter is divided by a small strip consisting of *Sorbus* and *Prunus* species (mountain forest of the Sierra Nevada) from a dwarf-shrub heath containing *Cytisus, Genista,* and *Erica. Festuca* lawns, at the foot of the peaks (southside), accompanied by a belt of low thorny dwarf shrubs, at an elevation of 3,000 m represent the transition to the Alpine grassy and herbal fields which, depending on their isolated location, are characterized by a number of endemic species (*Pseudochazara hippolyte, Plebicula golgus*). In addition, populations occur which may also be present in other Central European mountains and at lower elevations in the north. These include butterflies of the species apollo (*Parnassius apollo;* Sierra Nevada subspecies *P.a. nevadensis*), of the Satyridae *Erebia hispania,* and the Lycaenidae *Jolana jolas* and *Agriades glandon* (species *züllichi*).

5.1.6 The Deciduous Forest Biome

This large biome is characteristic of the moderate areas in the eastern part of North America, East Asia, and Central Europe. Its phytostructure which, when compared to the taiga, is more diverse, creates niches for a fauna more abundant in species. This fauna, however, has been changed drastically through human influences. Today's centers of heavy industry are located in wide areas which formerly contained summer-green deciduous forest biomes. These are being replaced by other forms of usage and other forest types (Mayer 1977).

At present, about 30% of the Federal Republic of Germany, a potential deciduous forest country, is still covered by forest. The proportion of summer-green species has been shifted considerably under anthropogenic influences. Spruce and fir trees occupy close to 40%, pine and larch 25%, summer-green beech 24%, and oak 8% (other species = 5%). In 1961, the ratio between deciduous and evergreen trees was 1:2; 150 years ago, the ratio reversed. Humans have a particularly damaging effect on certain forest populations. In order to flourish, the deciduous forest biome requires a warm-humid period of growth lasting at least 4–6 months, and mild winters (Fig. 75).

Characteristic soil types found in the deciduous forests are the brown earths which have special requirements as far as water supply is concerned. Their chemical breakdown processes are well balanced to a depth of 60 cm. Clay and iron compounds exhibit surprisingly constant values when compared with the values for a podsol profile in the taiga. The planting of evergreen forests on brown earths may lead to a raw humus deposit, stronger surface drainage, acidification and podsolization processes. These processes also have an effect on certain plant species of the herbaceous flora. Thus, for example, the growth of the following species is promoted on a Luzulo-Fagetum site after it has been stocked with a spruce forest: *Dryopteris carthusiana, Vaccinium myrtillus, Carex pilulifera, Galium hercynicum, Epilobium angustifolium, Unium hornum, Rubus idaeus, Deschampsia flexuosa, Senecio fuschsii, Agrostis tenuis,* and *Digitalis purpurea.*

It may generally be stated that the growth of acid tolerant coppice plants, ferns, and certain mosses can be promoted by means of afforestation with evergreen trees. Species such as *Polygonatum verticillatum, Milium effusum, Poa chaixii,* or *Festuca altissima,* however, recede.

Brown earths show an A_h-B_v-C-zoning. The humic A_h-zone is colored brown-grey and is only rarely deeper than 20 cm. It slowly changes into an ochre-brown B_v-zone whose coloring is caused by the Fe-oxides created during the decomposition of silicates containing Fe and which, in contrast to the A_h-zone, is not covered by the black-grey color of an organic substance.

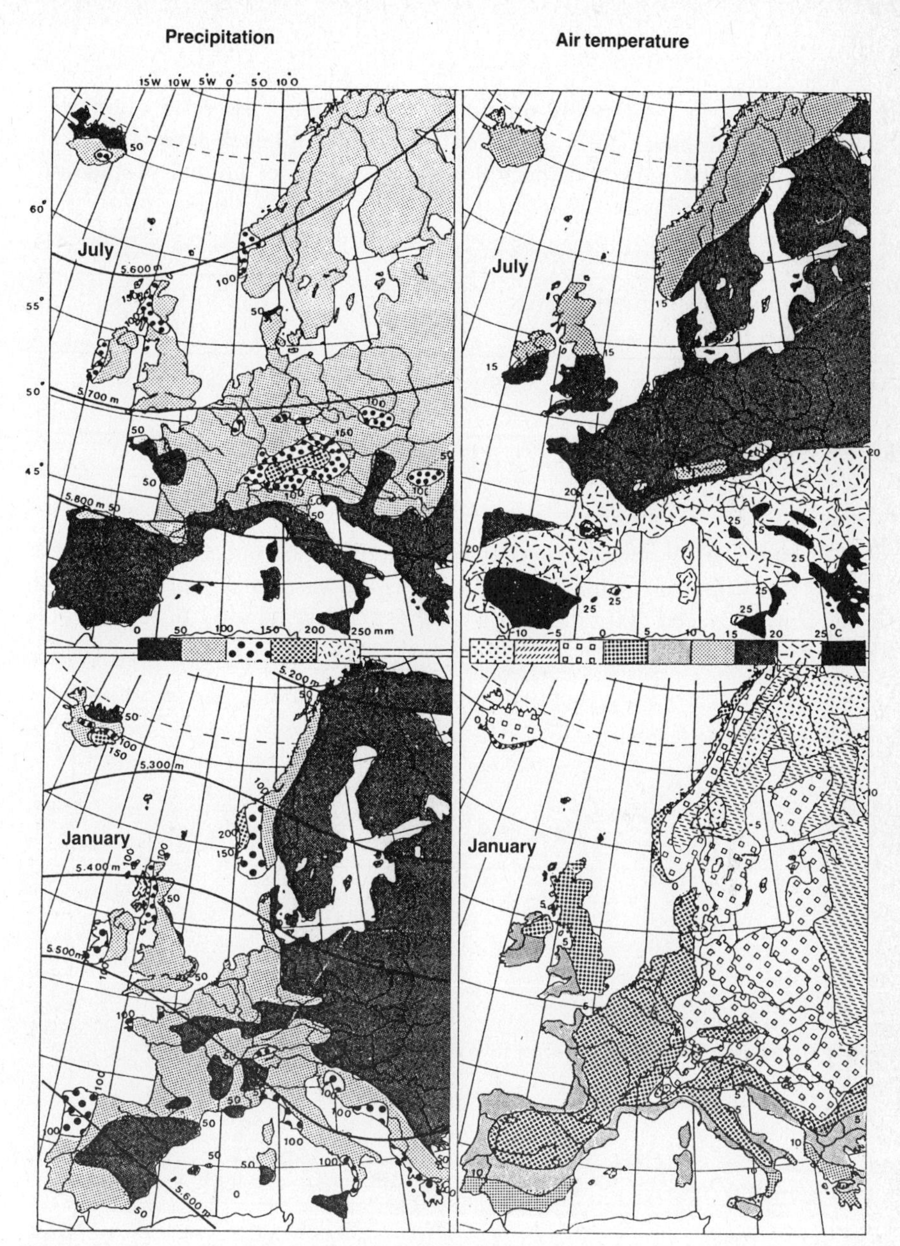

Figure 75 Temperature and precipitation distributions in Central Europe during January and July.

The thickness of the B-zone varies between 20 and 150 cm. The transition to the C-zone (which is not brown) often occurs without sharp delineation. Two basic forms can be distinguished:

(a) Eutrophic brown earth on a rich substratum

(b) Oligotrophic brown earth with a poor (acid) substratum

The latter usually develops on Ca- and Mg-deficient rocks. Brown earths originate in moderately humid climates. Phytomass production was examined by numerous authors in Central European beech forests. Müller and Nielsen (1965) indicated the following quantities of annual production (tons of dry matter/ha and year) in the case of a 40-year old beech forest:

Total production by assimilating leaves = $23.5\ t \cdot ha^{-1} \cdot a^{-1}$

Losses through respiration: leaves = 4.6 t; stems = 4.5 t; roots = 0.9 t; total = $10.0\ t \cdot ha^{-1} \cdot a^{-1}$

Annual leaf production (2.7 t); stems (1.0 t); litter and roots (2.9 t); total = $3.9\ t \cdot ha^{-1} \cdot a^{-1}$

Above-ground wood production 8.0 t; underground 1.6 t; total = $9.6\ t \cdot ha^{-1} \cdot a^{-1}$

Walter (1971) points out that the standing phytomass of the forest increases continuously with age and that in the case of 50-year old forest stands it may exceed 200 t/ha, and with those 200 years old, more than 400 t/ha (Fig. 76, Table 53, Fig. 77).

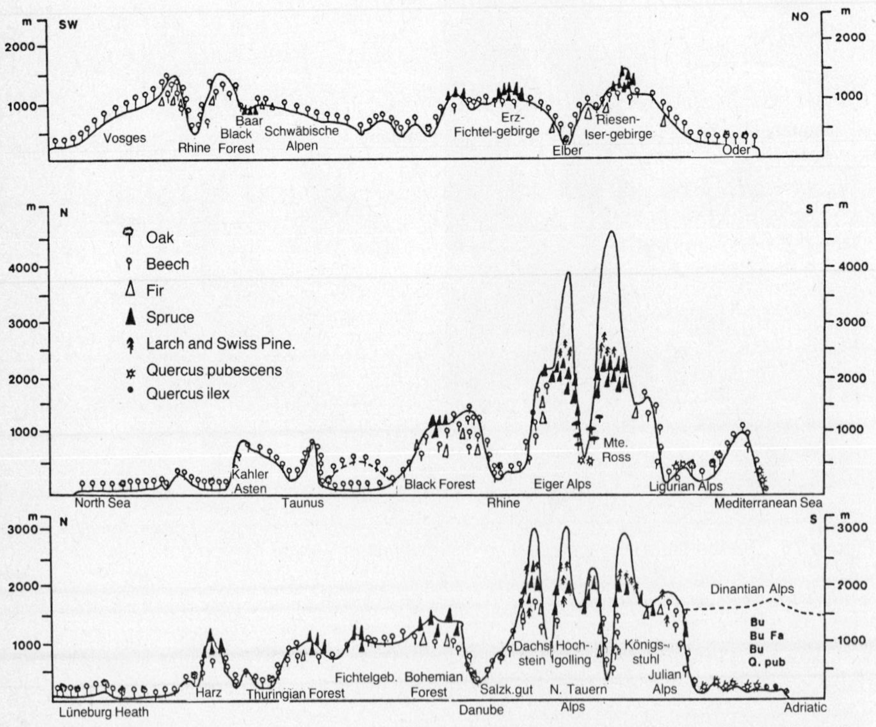

Figure 76 Diagram showing vegetation profiles across Central Europe (according to Haeupler, 1976).

TABLE 53 ABOVE-GROUND BIOMASS OF A 120-YEAR OLD WESTERN EUROPEAN DECIDUOUS FOREST BIOME (according to Duvigneaud 1962)

Trees	Leaves	4 t/ha
	Branches	30 t/ha
	Trunk	200 t/ha
Undergrowth		1 t/ha
		235 t/ha
Birds		1.3 kg/ha
Large mammals		2.2 kg/ha
Small mammals		5.0 kg/ha
		8.5 kg/ha

Density and biomass of the fauna in the litter are considerably greater than in the boreal coniferous forests (Cornaby *et al.* 1975, Funke 1977, Schauermann 1979).

The annual rhythms of the deciduous forest biome and the related ground-level light conditions result in varying species compositions (Fig. 78).

Some of the best-known spring flowering plants among the still leafless or freshly green Central European deciduous forests are: the celandine (*Ranunculus ficaria*), wood anemone (*Anemone nemorosa*), primrose (*Primula elatior*), arum (*Arum maculatum*), and lily of the valley (*Convallaria majalis*).

Depending on the ecological similarity of the disjunct deciduous forest

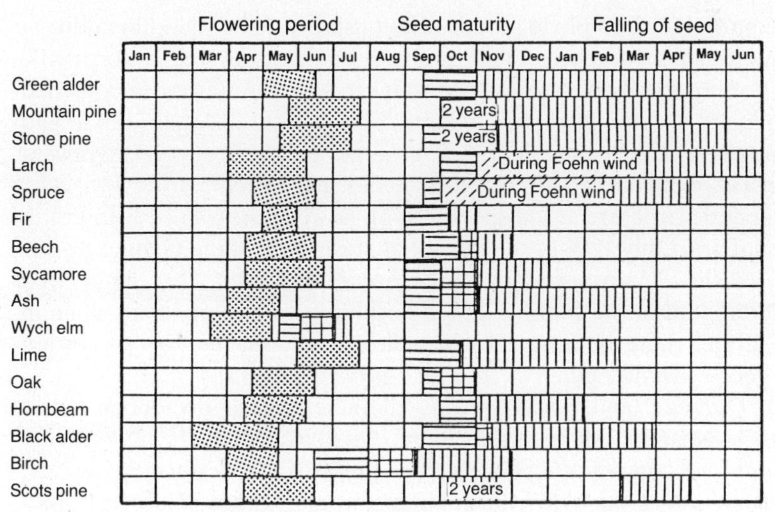

Figure 77 Flowering period, seed maturity, and seed fall of the most important Central European forest trees (accoring to Leibundgut, Rohmeder, and Schreiber, from: Mayer, 1977). The seasonal weather and local site factors (elevation, local climate) can lead to clear shifts in seed and fruit fall.

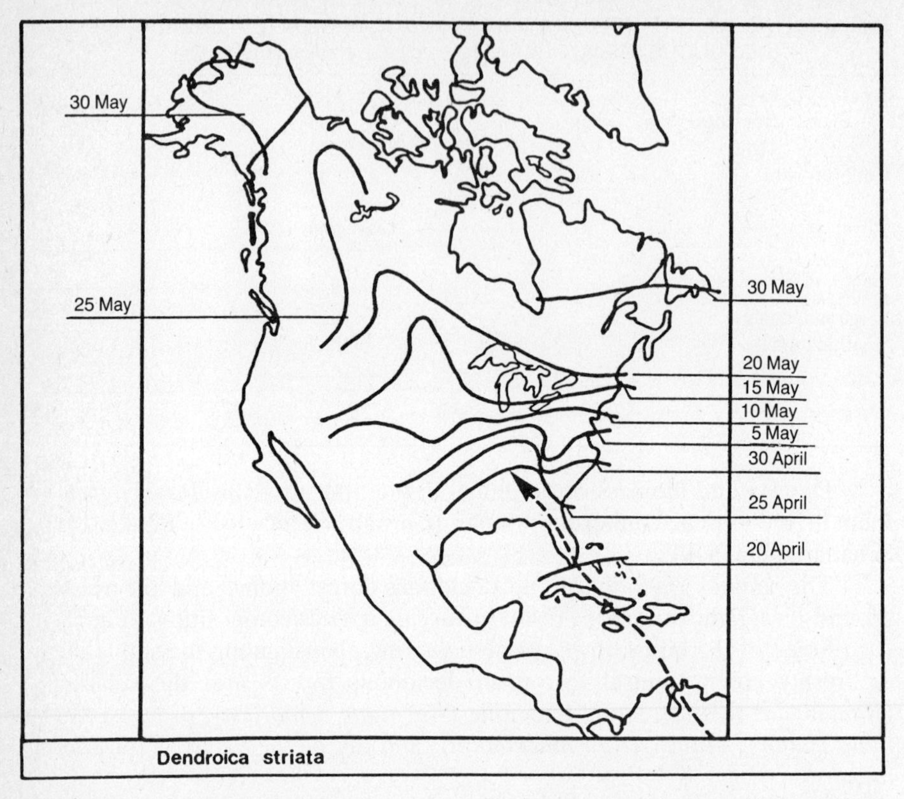

Figure 78 Arrival dates (phenological start of spring) of *Dendroica striata* in North America.

biomes, remarkably close phylogenetic relationships can be found in addition to the numerous convergences. This applies to animals as well as to plants. The moss flora of the Appalachian Mountains in North America is composed of more than 353 species of which 196 (55.3%) also occur in Japan. Japan and western Europe exhibit parallels in their vegetation classifications (Miyawaki and Tüxen 1977, Pignatti 1977) and their ecological structures.

The genetic structure at the species level, however, is totally different in the two countries. Only 12% of the Pteridophytes, 1.6% of the Gymnosperms and 9.7% of the Angiosperms in Japan occur in Europe. No single tree species is dominant in the summer-green deciduous forests of Japan while in western Europe, there is nearly always a dominance on the part of certain species (*Fagus sylvativa, Quercus petraea, Alnus glutinosa*).

Acer, Quercus, and *Fagus* species characterize the deciduous forest biomes of the Nearctic (*Acer saccharinum, Quercus rubra, Q. macrocarpa, Fagus grandiflora,* among others) as well as those of the western Palaearctic (*Q. petraea, Q. robur, Fagus sylvatica,* among others), China (*F. longipetiolata, F. engleriana*), Taiwan (*F. hayatae*), and Japan (*F. sieboldii, F. japonica*).

While new habitats for small animals were also created by human activities and numerous species owe their present existence in Central Europe

Type of site		cool humid			dark				dry warm			bright	
Sites examined	8.2	2.2	8.1	2.1	12.2	9.2	9.1	17.2	17.1	7.1	17.3	12.1	7.2
Stenecological forest species													
Patrobus atrorufus	21	5	–	–	–	–	–	–	–	–	–	–	–
Agonum ruficorne	15	6	–	–	–	–	–	–	–	–	–	–	–
Trechus quadristriatus	1	13	–	–	–	–	–	–	–	–	1	–	–
Pterostichus nigrita	6	2	–	–	–	–	–	–	–	–	–	–	–
Bembidion mannerheimi	3	2	–	–	–	–	–	–	–	–	–	–	–
Agonum assimile	5	13	21	1	–	–	–	–	–	–	–	–	–
Pterostichus niger	83	9	1	22	1	–	–	–	–	–	–	–	–
Nebria brevicollis	15	–	12	–	–	–	–	–	–	–	–	–	–
Eury-ecological forest species													
Abax ater	181	55	98	24	228	36	48	27	54	184	123	324	92
Carabus problematicus	82	11	26	25	14	52	6	24	52		81	16	3
Carabus purpurascens	6	3	1	15	49	–	–	35	6		38	29	1
Carabus nemoralis	3	3	1	–	2	–	–	–	1	2	9	4	3
Cychrus attenuatus	10	2	4	–	4	9	11	6	18	–	27	3	–
Pterostichus niger	83	9	1	22	1	–	–	5	–	–	14	–	–
Pterostichus oblongopunctatus	6	4	4	–	–	–	5	–	3	–	3	1	–
Pterostichus madidus	1	–	10	–	–	–	–	–	–	23	–	4	4
Differential species of the Fagetalia (Löser, 1971)													
Abax ovalis	–	2	1	–	14	–	2	–	18	–	–	171	–
Molops piceus	–	1	2	1	4	–	–	–	–	15	–	13	–
Abax parallelus	8	3	17	–	6	2	8	1	21	29	22	13	26
Pterostichus cristatus	47	126	102	3	1	–	–	–	–	–	–	–	–
(Tr. laevicollis)	2	2	1	1	1	1	1	1	1	–	1	4	–
Species of clearings													
Carabus coriaceus	–	–	–	–	1	1	–	–	–	1	–	3	21
Harpalus latus	–	–	–	–	–	–	–	1	1	–	1	1	2
Bembidion lampros	–	–	–	–	–	–	–	–	–	–	1	–	6
Carabus arcensis	–	–	–	–	–	–	–	–	–	–	12	–	–
Notiophilus biguttatus	–	–	–	–	–	–	–	–	–	–	–	–	2
Trechus secalis	–	–	–	–	–	–	–	–	–	–	1	–	–
Amara lunicollis	–	–	–	–	–	–	–	–	–	–	1	–	–
Pt. anthracinus	–	–	–	–	–	–	–	–	–	–	–	–	1

Figure 79 Characteristic ground beetles (Carabidae) in 13 forest sites (8.2–7.2) in the vicinity of Saarbrücken. The forest sites (natural forest units) were defined on the basis of their macro- and microclimates as well as their ground vegetation. The cool humid sites (8.2, 2.2, 8.1, 2.1. 12.2) are populated by stenecological forest species. Eury-ecological forest species are present on all sites.

or Northern America exclusively to humans, numerous other vertebrate species (lynx, wildcat, wolf, bear, Old World otter, beaver) are in the process of disappearing or have already been extirpated (e.g. the aurochs).

The ecological relationships of some deciduous forest animals have been examined particularly thoroughly in the case of beech forest species (cf. among others, Funke 1977).

On the basis of numerous snail species, Ant (1969) was able to subdivide individual beech forests. Thus, for instance, the occurrence of *Abida secale, Clausilia parvula,* and *Cepaea hortenis* is linked to the conditions prevailing in beech forests. Thiele (1977) was able to show similar links in the case of ground beetles. The biotype of *Abax ater* and *A. ovalis* exhibits forest conditions close the ground in the "natural forest units" of the Saarland. The link between individual forest species among the carabids and summer-green forests has been confirmed (Müller *et al.* 1975) (Fig. 79). 13 sample sites were classified into cool humid and dark to dry warm and bright. Comparing — within these surfaces — the occurrence of stenecological and eury-ecological forest species as well as the different species of the beech forests (Löser 1972) and types of clearings, it becomes apparent that the individual sites can be determined by means of characteristic species combinations.

Deer (*Capreolus capreolus*) "occur, by nature, considerably more frequently as inhabitants of the edges of forests or of regenerating forest than they did in primaeval forests thousands of years ago... deer must choose easily digestible, concentrated nourishment, for example the buds of deciduous trees, budding greens, acorns, beechnuts. It is only in this manner that they are able to assimilate enough energy per time period in their omasum, leaving a sufficiently large surplus for growth and milk production" (Ellenberg 1977).

The summer-green deciduous forests of the southern continents (*Nothofagus* forests), differ — ecologically and in their species composition — from those in the northern hemisphere. Although the *Nothofagus* forests of Chile, for example, with their monotypic amphibian genera *Telmatobufo, Batrachyla, Hylorina, Calyptocephalella,* and *Rhinoderma,* as well as their endemic bird genera *Sylviothorhynchus, Aphrastura, Pygarrhichas,* and Enicognathus, possess a basic stock which is related to the remaining South American fauna, they have their own history since the Miocene as has been proven by the fossil record. *Calyptocephalella* has been known to exist in the southern part of South America since the Oligocene.

Two monotypic marsupial genera of the Chilean *Nothofagus* forests, *Rhyncholestes* and *Dromiciops,* can be traced back as far as the Eocene (*Rhyncholestes*), and the Oligocene and Miocene (*Dromiciops* and its relatives) in Patagonia.

5.1.7 The Boreal Coniferous Forest Biome

Cold climates in the winter, with temperatures falling to −78°C (eastern Siberia), short, but warm summers with continuous day conditions, and a growing season lasting from three to five months, characterize these areas with perenially frozen ground in the northern part, and podsol soils as well as bogs and swamps. The boreal evergreen biome runs around the pole south of the Arctic tundras in the form of a belt 1,500 km wide, interrupted by the oceans. The number of days with mean temperatures of above 10°C is below 120. The growing season (= number of days with an average temperature of at least 5°C, per year), consisting of 105 to 110 days along the Alpine forest limit is of about the same duration as it is along the forest limits in northern Fennoscandia (Walter 1970, Holtmeier 1974). The latter, which forms the northern border towards the tundra biome (boundary of warmth deficit) is characterized by a cold season lasting more than 8 months and by only 30 days of daily mean temperatures in excess of 10°C. Climatic fluctuations have a marked effect on the frequency of seed years and thus control productivity and natural regeneration. Contrary to a widespread assumption, biomass production does not increase due to the conditions caused by long days. The CO_2-production decreases markedly under permanent light (when compared

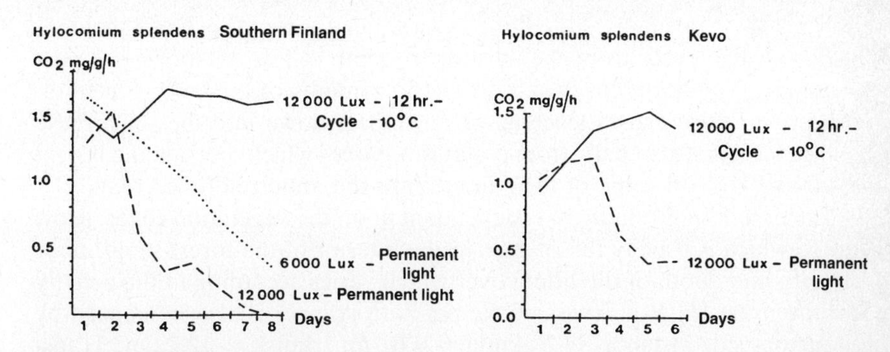

Figure 80 Comparison of the CO_2-production of *Hylocomium splendens* in southern and northern Finland (Kevo) under permanent light conditions and in the 12-hour cycle. The mosses growing under permanent light reduce their CO_2-production after the third day (according to Kallio and Valanne, 1975).

to the 12 hour rhythm) (Fig. 80). The phytomass (per ha) increases from north to south, although oceanic or continental climatic influences make possible exceptions to this rule through modifications to genetic structures (Table 54).

The average annual yield amounts to 5.5 t of organic matter, about 3 t of which result from the growth of wood (Table 54). The low annual temperatures and low evaporation and the resulting vegetation structure (plants with nutrient deficiencies) encourage the formation of a podsol soil. In the boreal coniferous forest biome, small-scale, zonal podsolic soils (podsol = Russian name for ash soil) often alternate with peaty and gley soils. The soil profile of typical podsol soils is $O–(A_h)–A_e–B_h–B_s–C$ and they are formed by a layering of Fe and Al, together with organic substances.

Subclassifications are made according to the type of B-zone (iron podzol, humus podzol, among others) and transition forms (for instance, brown earth podzol, gley podzol).

Under the abiotic conditions of the boreal coniferous forest biome, litter is decomposed only partially due to the limited activity of the ground biota. Organic complex formers and reduction processes are more prominent in the soil solution. They liberate Fe and Al, where especially Al — at low pH-values (below 4) — becomes available to the plants (cf. Campo Cerrado,

TABLE 54 PHYTOMASS (LIVING AND DEAD MATERIAL; C kg/ha) FROM PINE AND BIRCH FOREST SITES IN FINLAND (KEVO) AND NORWAY (HARDANGERVIDDA); (according to Kjelvik and Karenlampi 1975)

	Kevo (69°22′N, 19°3′E) Pine forest	Kevo Birch forest	Hardangervidda (60°22′N, 7°25′E) Birch forest
Living phytomass	35047	11035	34310
Living + dead phytomass	38650	12075	35140
Roots	7324	3541	11340

section 5.1.2). On humic sites, humus hardpan forms. The Fe_2O_3-content increases continuously from the surface to depth. Close relationships exist between the vegetation cover and the biotic contents of the soils. Precipitation, however, also causes leaching of substances down into the soils; these substances originate, in part, from pollution sources which, outside the boreal pine woods, give off some of their fumes into the atmospheric currents.

The extent of groundfrost is dependent upon the vegetation cover, snow coverage which is usually full of gaps in the interior of the forest stand, poor insolation, and depth of the litter cover which varies according to the activity of decomposers. The thickness of the permafrost layer is also influenced by these factors (for instance, in Yakutia = 9.01 m; Irkutsk = 12.22 m; Transbaikalia = 22.78 m; Amur = 22.8 m).

Coniferous trees (*Picea, Abies, Pinus, Larix*), birch (*Betula*), poplar (*Populus*), alder (*Alnus*), and willow (*Salix*) are the tree species forming the forest stand. Oceanic and continental climatic influences lead to distinct biogeographic differentiations. In the area of oceanic climates, the boreal pine forests usually border onto the summer-green deciduous forests with a wide transition zone. As the continentality increases, arid steppes and semideserts form the border biomes. In Scandinavia, *Betula pubescens* ssp. *tortuosa* (the discussion with respect to the species status of *tortuosa* has not yet been resolved) represents the northern forest limit. The northernmost occurrence of *Pinus sylvestris* is located in the Stabbursdal (70° 18′N), west of the Porsangerfjord (Holtmeier 1974).

The growth forms of coniferous species are notable, particularly on exposed sites. In Fennoscandia, *Pinus sylvestris* assumes spruce-like growth forms on some sites. *Picea excelsa* occurs in slim columnar forms. Many trees exhibit twisted growth, often bizarrely covered by lichen. As it approaches the forest limit, *Betula* shows an increasing tendency to form multiple trunks.

In the herbaceous flora (higher plants), the number of species diminishes as the northern latitude and altitude increases. Chamaephytes and hemicryptophytes become dominant.

The polar forest limit is influenced not only by abiotic, but also by biotic and anthropogenic factors. Fungus diseases (*Phacidium infestans,* for instance) affect pines, and the infestation manifests itself by noticeably red needles. The susceptibility of *Pinus* is particularly great on sites located away from natural stands (cultivations). By chewing on the trees, the reindeer (*Rangifer tarandus*) counteracts the expansion of forest growth. Thus, large numbers of reindeer cause far-reaching destruction of birch, willow and pine forests. Mass reproduction of insect species can locally destroy entire forest stands. The development cycles of the Holarctically distributed green inchworm, *Oporinia autumnata,* seriously affecting the beech forest limits in northern Norway as well as in Finland, has been particularly well examined. In the boreal coniferous forest biome, the mosaic pattern of dryland spruce-pine-birch-mixed forests, large lakes and bogs is striking. A knowledge of this mosaic is important for the correct interpretation of areas of the boreal forest species. Of these, there is a great diversity. Although their areas can be correlated with the coniferous forest biome, they exhibit a typical habitat island distribution. Numerous species limited to the remnants of humid regions and heaths in Central Europe and which are in continuous decline there, have conjunct propagation areas in the boreal coniferous forest biome. This applies, for instance, to the crane (*Grus grus*) as well as to *Colias paleano* and the Holarctic elk (*Alces alces*). Via the moors and lakes, some amphibians and reptiles also penetrate far into the pine woods. In Scandinavia, these include the ovoviviparous forest lizard (*Lacerta vivipara*), the northern viper (*Vipera berus*) which, however, occurs only as far as 67°N, the moor frog (*Rana arvalis;* to about 69°N) and the common frog (*R. temporaria;* to the North Cape); in North America, the garter snake (*Thamnophis sirtalis*) and wood frog (*R. sylvatica*). In addition, there are species whose life cycle restricts them to the tree and herb species of the forests. These include the fir parrots (*Loxia curvirostra*), occurring in North America, which as invasion birds, can appear far beyond their breeding area. Their relative, the binder parrot, *L. Leucoptera* is even more strictly confined to the Holarctic pine forests. In addition, these also include the Siberian jay (*Perisoreus infaustus*) widely occurring in the Scandinavian forests and its Nearctic representative, the gray jay (*P. canadensis*), the Holarctic pine grosbeak (*Pinicola enucleator*), which is the size of a starling, the bramble finch (*Fringilla montifringilla*), rustic bunting (*Emberiza rustica*), Siberian tit (*Parus cinctus*), redwing (*Turdus iliacus*), bluebonnet (*Tarsiger cyanurus*), wood warbler (*Phylloscopus borealis*), waxwing (*Bombycilla garrulus*), Holarctic three-toed woodpecker (*Picoides tridactylus*), and its relative *Picoides arcticus,* limited to North America. Several owl species — some of them active during the daytime — are typical representatives of the taiga fauna. Among these are also the Palaearctic Ural owl (*Strix uralensis*), penetrating also into the cities, as well as the Holarctically distributed owls *Strix nebulosa* (Lapland owl), *Surnia ulala* (hawk owl) having a falcon-like appearance, and *Aegolius funereus* (Tengmalm's owl) which, however, penetrates the summer-green deciduous forests along the southern periphery of its western Palaearctic habitat. The Palaearctic sable (*Martes zibellina*) is a forest animal living predominantly in spruce and *Pinus cembra* forests, whose skins were exported during the 3rd century B.C. to the 17th century from the Scythian Empire across the Black Sea to adorn the czar's crown. In the boreal pine forest of North America, the sable is replaced by the American marten (*Martes americana*) and fisher (*M. pennanti*). These close connections between northern Eurasia and North America are also observable in insects and in plant species and thereby confirm the general rule that the relationship between the continents grows closer with increasing northern latitude.

The ground fauna consists of only a few groups of species (mites, insect larvae, threadworms, primary insects, etc.). Snails do exist; they only play a minor role, however, being absent in many coniferous forests. Horntail, sawfly, gall wasp, and ants (*Formica, Camponotus*), on the other hand, are

TABLE 55 HIGHER PLANTS IN SOUTHERN NORWAY, DEPENDING ON ALTITUDE AND DISTRIBUTION TYPE (according to Dahl 1975)

Altitude limit (m)	Number of species	Amphi-Atlantic (%)	Circumpolar (%)	Eurasiatic (%)	Endemic (%)
>2000	29	51.7	37.9	10.3	0
1800–1999	39	35.9	53.8	10.3	0
1600–1799	75	32.0	58.7	8.0	1.3
1400–1599	63	23.8	47.6	27.0	1.6
1200–1399	138	14.5	37.0	48.6	0.7
1000–1199	109	11.0	36.7	52.3	0

characteristic representatives among the insects. Numerous forest pests (cf. Kloft, Kunkel and Ehrhardt 1964) are associated with the coniferous forest belt.

The seasonal change in living conditions is the deciding factor for a large number of different modes of adaptation. For invertebrates long periods of hibernation (diapause), for insect-eating birds and mammals (reindeer, amongst others) migration; and birds and mammals who feed directly on primary producers (including seed), survive the taiga winter by staying put (rough foot hens, amongst others). Animal population levels annually change; the numbers of waxwings and crossbills, for example, depends on the nutrient output of the forests.

Some kinds of nordic boreal coniferous forests often have related, isolated populations at corresponding altitudes in the Alps, Caucasus and Himalayas. Hidden, however, behind this boreoalpine distribution are ecological adaptation strategies which are fundamentally different. The ring thrush (*Turdus torquatus*) has a spatial disjunction. However, if its propagation area is investigated more closely it is soon recognized that the species in Scandinavia avoids the closed coniferous forest, although the bird is found in much greater numbers of woodlands around the Atlantic fjords. In the Alps it is found on the outskirts of forests. The pine jay (*Nuci fraga caryocatactes*) lives in the Alps at the coniferous forest level (like its North American counterpart *N. columbiana*) and is an important distributor of *Pinus cembra;* in the northern part of its distribution it inhabits summergreen forests and in the coniferous forest itself its counterpart is the Siberian jay (*Perisoreus infaustus*). The mealy redpoll (*Acanthis flammea*), in its northern habitat, lives in birch and alder forests; in Iceland and Greenland it also populates the tundra; and in the Alps it occurs in the pine as well as deciduous forests.

5.1.8 The Tundra Biomes

The name tundra is derived from the Finnish "tunturi", meaning "forestless hill". Looking at these "tunturi" in Finland, it is not difficult to recognize that some significant differences exist in the Arctic tundra. In Fennoscandia, the forest limit in the area of the North Cape — apart from a few regions — represents an altitude limit. It is only on the Kola Peninsula that it changes

Figure 81 In Fjell, Norway (June 1978).

into a real northern limit at sea level (Fig. 81).

In addition to abiotic features (frost, aridity, icy winds, permanent day or permanent night conditions, short vegetation phase, permafrost, etc.), a multitude of biotic factors (competition with lichens, mosses, and higher plants; man, reindeer, tree parasites, among others) determine the Arctic forest limit and, thereby, the transition from tundra to boreal pine forest. The transition, therefore, is generally not abrupt; it takes place in a forest-tundra belt of varying width, dominated by birch trees (oceanic influence) or larch and spruce trees (continental influence). Wide palsa moors are frequently present in this zone. These are frost cores of varying thickness which, because of the dense humus and moss layer, do not thaw, even in summer, and which are surrounded by moor areas which in turn rest on a nanopodsol soil. In these regions, ice wedges — cone-shaped structures, formed by the frost effect in the presence of abundant organic substances — are frequently present (Semmel 1977, Washburn 1973). Large and abrupt temperature fluctuations as they occur in the high mountains are not characteristic of the tundra biome during the summer months. For this reason, many tundra insects — in contrast to the fauna of the high mountains — are unable to survive large differences in temperature. In the northern Siberian tundra, the number of days with average temperatures of above 0°C fluctuates between 188 and 55. Since the low temperatures reduce evaporation, the conditions in the tundra — despite rainfall levels of around 200 mm — are humid. Summers are short and cool (0°C annual isotherm) and the ground at depth remains constantly frozen. The depth of the frost is not only dependent on abiotic, but also on biotic factors. In spite of the lack of precipitation in the tundra (northwest Spitzbergen about 400 mm; northwest Canada about 200 mm per year), open water surfaces and sodden flat moors form particularly during the snow-melting season. Their existence is due to the low evaporation and obstruction of the water discharge, caused by the permafrost. They often do not dry up until the end of the growing season. Animal and plant communities are dependent on this periodicity. With increasing latitudes, arctic communities consisting of generally circumpolar-distributed species groups dominate (cf. Hulten 1970, Thannheiser 1976).

The Holarctic tundra biome is characterized by a predominance of mosses, lichens, and heathlands abundant in dwarf shrubs. It extends from the subpolar region of northern Eurasia, along the coasts of Greenland, over large areas of Iceland, and along the edges of the North American Arctic (Wielgolaski and Rosswall 1972) and in northern Siberia it occupies an area of 3 million km^2.

Tundra biomes are absent in the southern hemisphere, although the most southern tussock grasslands exhibit many similarities. This is apparently also true of many plant communities on moors in the southern hemisphere. The forest limit on the southern continents runs along the southern tip of South America (56°S) north of the Falkland Islands, and, in the South Atlantic Ocean, retreats to about 40°S. Tristan da Cunha, at 37°S, has one tree species. Grounds with frost patterns and low pH-values occur. Crytophytes often assume the storage of important elements.

Nitrogen fixation is directly controlled by the temperature, humidity, and light. One lichen species, *Solorina crocea,* fixes nitrogen below 0°C.

The proportion of fungi and bacteria in the soil is high. A multitude of different *Penicillium* species, whose species composition can change from year to year (Hays and Rheinberg 1975), are found in the litter of individual plant species, for instance, of *Rubus chamaemorus, Empetrum hermaphroditum,* or *Dryas octopetala.*

With mass reproduction of lemmings the surface vegetation is sometimes destroyed because roots are actually eaten as well; this can have serious repercussions for the thaw depth of the ground. Of special significance are the great variations in radiation distribution (polar day, polar night). As a result of the consistently low temperatures, the period of development for poikilothermic animals is considerably extended, and plant growth accordingly delayed. The widespread lichen *Cladonia rangiferina* grows about 1–5 mm per year, and it must be assumed that in the tundra this pasture land needs at least 10 years to regenerate. In accordance with the small growth rate, primary production, when compared with other biomes, is also small. Of importance, however, is the degree of exploitation of this production by consumers. Although most kinds of animals and plants are forced to use the short summer months as a period for reproduction there are species whose development takes place over several years. In this category is *Braya humilis* whose buds can appear two years before the actual period of bloom. Since photosynthesis in certain cryptogams occurs at 0°C, they dominate this border area. Vertebrates which are typically found in the tundra are the reindeer (*Rangifer tarandus*), blue hare (*Lepus timidus*), polar fox (*Alopex lagopus*), musk ox (*Ovibos moschatos*, which at present is only to be found in Greenland, Spitzbergen (introduced) and Norway (introduced)), lemmings (upon which the presence of certain nordic birds of prey depends (*Buteo lagopus, Nyctea scandiaca,* amongst others)), snow grouse and snow buntings.

The snowy owl is the non-migratory bird of the tundra. Reindeer in many locations only appears in summer and in winter migrates southwards in immense herds. It has a lasting effect on the cryptogam vegetation and it is assumed that the extensive lack of lichens in reindeer pastures in central Spitzbergen is due to the high density of reindeer.

The circumpolar-distributed Arctic fox (*Alopex lagopus*) penetrates beyond the tundra into the icelands of the Arctic. Two color breeds, the "blue" and "white fox" are so named because of their polymorphic winter coat, the white fox dominating the Arctic area. Its closest relative is the corsac (*Alopex corsac*), which lives in the steppes between the Volga and Mongolia and in isolated populations in Manchuria.

A representative of the forest tundra is the Holarctic white hare (*Lepus timidus*) which, in the form of an Arcto-Alpine species (cf. High Mountain Biomes for the Arcto-Alpine distribution, section 5.1.9) populates the dwarf pine belt in the Alps, also lives in the Arctic tundra of Greenland and Taimyr Peninsula. It shares the white winter coat of a smaller Nearctic spruce forest species which, because of the heavy hairgrowth on its hindlegs, is named the skating hare (*Lepus americanus*). Among the tundra birds, the non-passeriformes are dominant. In the "tunturi areas" of Finland and in the Norwegian tundra, faunal elements of varying biome affiliations — corresponding to the small alternating areas in the landscape units — have been determined.

Mass reproduction of lemmings can result in a reduction of the tundra production in excess of 20%. In general, the invasion cycles of the lemmings recur every 3 to 5 years (MacLean, Fitzgerald, and Pitelka 1974). Apart from the abiotic factors (air temperature, snow cover) and vegetation, predatory animals also have a significant impact on these cycles. Examinations of the winter nests of the North American *Lemmus trimucronatus* and *Dicrostonyx groenlandicus* confirm that the white weasel (*Mustela nivalis*), which also occurs in Central Europe, has a controlling influence on lemming cycles (Pitelka 1973).

Also among the Lepidoptera, there are circumpolar-distributed tundra species (among others, *Colias nastes, Colias hecla, Oeneis verna*). The abundance of mosquitoes in the tundra during the summer months is characteristic. This is explained by the numerous water ponds well-suited for their reproduction and by the fact that the females can also survive on plant juices (i.e., without sucking blood).

According to the hypothesis established by Remmert (1972), homoiothermic vertebrates, because of their consumption and fertilization impact, represent a significant reason for the survival of the tundra.

"Their destruction would be equivalent to the destruction of the tundra.

Thus, the fate of the tundra is also decided in winter along the European coasts, the locations to which the wild geese of the Arctic migrate."

5.1.9 The High Mountain Biomes

Above the closed forest limit, we find the high mountain biomes whose extent varies with geographic location, vertical change of atmospheric factors, and the vegetation whose altitudinal distribution is influenced by these elements. The term "high mountain" as described by Troll (1955) applies to these biomes. With increasing altitude, the air generally becomes colder, purer, thinner, and, viewed in absolute terms, drier, while insolation and wind velocity increase. Level (= altitude) and relief effects (= sunny and shady side, windward and leeward), however, mean that absolutely no uniform climatic conditions prevail in high mountain areas. Their reciprocal penetration often results in mosaic-like distribution patterns of the taxa existing there. Based on the coexistence of completely different vegetation types above the forest limit, it can be concluded that climatic contrasts occur closely side by side which can often be attributed to the differences in relief.

Sunrise in the high mountain areas sees a radiation level that is approximately double the intensity of that in the lowlands. Global radiation maxima in June, July and August are exceeded, around mid-day (over 2.2 $cal.cm^{-2}.min^{-1}$). There exists, therefore, a much greater difference in the high mountains between places in the sun and places in the shade than similar differences in the lowlands. In bright conditions where cloud has broken up, solar radiation coupled with high cloud radiation of such intensity results that it exceeds even extra-terrestrial levels. In mountainous regions the intensity of vertical solar radiation is considerably greater than in the lowlands. At the same time daily and annual levels are also much more consistent. In the snowless vegetation period the average circumpolar radiation figures (= onto a spherical receiver from the sun, sky and reflecting background; incident radiation included) for the lowlands and high mountain areas are closely related; in glacier regions (high incidence of global radiation) the intensity is greater all the year round. The same is true for the intensity of ultra-violet radiation which is of direct importance in many life processes. The decrease in temperature, noticeable in summer at increasing altitudes, is modified by different relief effects and differences in the weather (e.g. mountain sides act as heat collectors). In the foothills of the Alps mean annual temperatures of 8–10°C are recorded, at 2,000 metres 0°C and at 3,000 metres −6°C. Above 2,000 metres no summer days have a daily maximum of over 25°C. The number of frost days increases from 100 days in the lowlands to 300 days at 3,000 m altitude. Ground temperatures, on the other hand, decrease more slowly with increasing altitude, and the annual mean temperature of the ground in high mountain areas is much warmer than that of the air; above 2,800 m, the ground is constantly wet or covered with ice and snow.

Precipitation increases with altitude, although here too, relief, topographical and geographical location play an important role. The Alpine villages of Oetz and Tessin have recorded — although situated at nearly the same elevation (≈700 m) — extremely different precipitation, i.e. 700 and 1,800 mm, respectively.

Looking at the tropical high mountains (Fig. 82), the elevation-dependent, hydric classifications differ from the above. Frequently, the lower cloud limit (first condensation level) is located between 1,500 and 1,800 m, followed by the second condensation level at 2,700 to 3,500 m. This "supplies humidity to the tropical mountain biomes, and causes rain to fall in them, producing the life forms which have been described several times by Troll" (Lauer 1976) (Fig. 82). In the Central Alps, the snow limit, i.e., the zone in which the winter snowcover no longer melts on the flat surfaces, is met at 3,200 m. Along the northern edge of the Alps, this limit is located between 2,400 and 2,700 m.

Local winds (valley, slope, and mountain winds) are of particular importance; in the Alps, they can reach horizontal speeds in excess of 200 km/hr and vertical movements of up to 2 km. Wind-sensitive plant species, for example, the alpenrose *Rhododendron hirsutum* (calcicole), and *Rhododendron ferrugineum* (calcifuge), belonging to the Ericaceae, avoid areas which are highly wind-exposed. In the lee of rock bluffs, in cold air pockets, and in valley locations, snow accumulations (snow patches) often last through

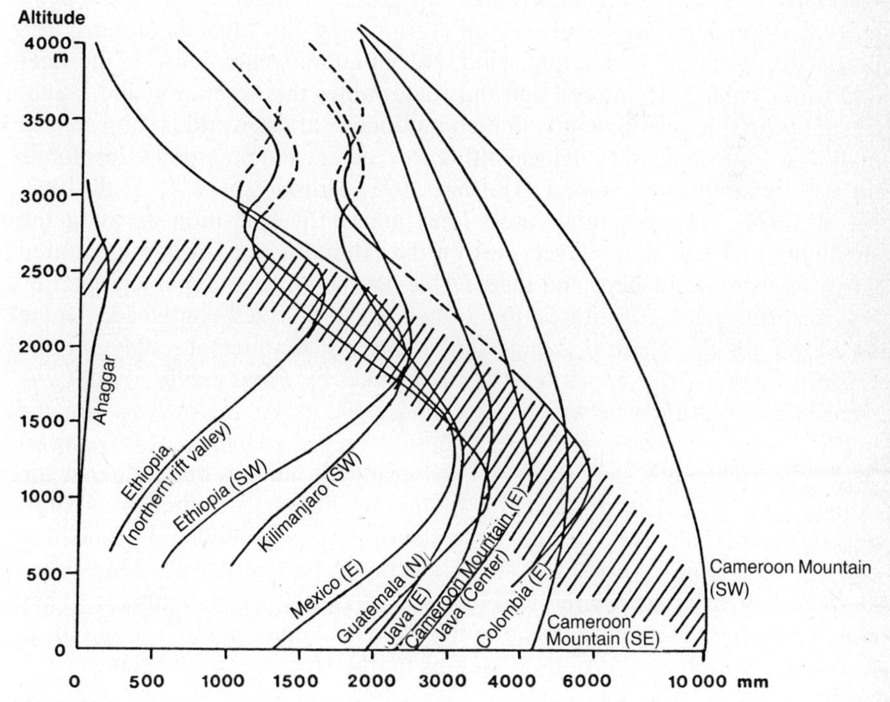

Figure 82 Annual precipitation in various tropical mountain escarpments in relation to altitude (hatched area = elevation of the maximum amount of annual precipitation; according to Lauer, 1976).

the growing season and are characterized by their own particular flora and fauna.

Some prominent characteristics are inherent in the soils of the high mountains of moderate latitudes (Franz 1976) (Fig. 82a). Because of the low temperatures and long vegetational dormancy period, the soil formation generally slows down correspondingly. The capacity for soil formation generally decreases as altitude increases, while increased erosion phenomena counteract the formation of mature soil profiles.

Physical processes (frost shattering, solifluction, aeolian movement, slope wash, among others) are dominant in soil formation when compared with chemico-biological factors. These often result in scree with only slightly weathered primary minerals and in soils rich in coarse waste. Thawing processes close to the surface over a frozen substrate lead to periodic water retention which in turn results in pseudo-gley like soil type. The humus in the high mountain soils is particularly rich (up to more than 80% of organic substance). Humus, consisting predominantly of the excrements of ground animals swells very easily and, when frozen, may lead to a volume increase in the hydrated soil, forcing the stones out of the fine soil, and thus leading to the formation of soils with frost patterns. Aeolian accumulations can be noted in all high mountain regions. Podsol soils are present in a lesser thickness under the Alpine grass heaths than the nanopodsols (cf. The Tundra Biomes, section 5.1.8). The individual density of ground animals seems to depend more on soil humidity than on soil warmth. In the high mountains, the interaction between insolation, wind and mountain relief controls the heat and water balance in the soil and thus determines the vegetation and fauna. Because the tree line in nearly all high mountains of the world is subjected to manifold anthropogenic and local influences, actual and potential forest limits do not generally correspond (Haffner 1971, Holtmeier 1974, Troll 1962, Meyer 1974). The potential forest limit marks the transition between the subalpine and low-alpine levels (within the Alpine region), wherein stunted trees may still occur here and there. This belt is, however, characterized by dwarf shrubs and constitutes the transition to the self-contained Alpine meadows. In the Central Alpine and relatively continental valleys of the Eastern Grisons, the upper tree line is formed by *Pinus cembra* and *Larix decidua* at an altitude betweeen 2,100 and 2,300 m. As the oceanic climatic influences increase towards the west, these are replaced by *Picea abies*. Apart from man and abiotic site factors, it is generally the animals that influence the Alpine tree line. The productivity of the vegetation is limited to a short vegetation period. Larcher and his collaborators examined the production ecology of Alpine dwarf shrub stands on the Patscherkofel near Innsbruck (Larcher 1976), particularly *Vaccinium* and *Loiseleuria* heathlands. They concluded that the climate affecting the plants is influenced by varying orographic and edaphic conditions as well as by the structure of the stands.

Figure 82a Mountain grassland (above) and mountain forest (below) above the tree line on Cameroon Mountain (Cameroon, 3/15/1979).

The latter controls the micro-climate (= bioclimate) pertaining to the stand, which can differ greatly from the relief-dependent terrain climate. Thus, for instance, the *Loiseleuria* shrubs growing along windswept edges exhibit a much warmer, more humid and calmer micro-climate than the rhododendron bushes occurring in sheltered hollows. *Calluna* stands evaporate relatively large amounts of water and their behaviour with respect to the heat balance is comparable to that of a meadow well supplied with water; the behaviour of the *Loiseleuria* stands, which exhibit less transpiration, is more like that of forests.

The transition from forest to the open terrain of the meadows and of the dicotyledon cushions represents an important zoogeographical transition. Although there are numerous lowland species which, due to a great ecological potential, can penetrate far into the high mountains oreal, survival there requires numerous physiological adaptations. In the Alps, the annelid worm *Dendrobaena octaedra* is still present above the tree line (to 3,500 m), *Lumbricus rubellus* (to 3,000 m), and *Octolasium lacteum* (to 3,000 m); and the snails *Ariantha arbustorum* (to 3,000 m) and *Pryamidula rupestris* (on lime rocks to 3,000 m). The wolf spiders (*Lycosidae*) have been found to exist at 4,300 m. The pseudo-scorpion *Obisium jugorum* occurs up to 3,000 m, the centipede *Lithobius forficatus* and *L. lucifugus* (about 2,500 m), the bristletail *Machilis tirolensis* (to 3,800 m), the grasshoppers *Gomphocerus sibiricus* and *Decticus verrucivorus* (to 2,600m), the tiger beetle *Cicindela gallica* (to 2,700 m), the carabids *Nebria gyllenhali* and *Techus glacialis* (to 3,500 m), and the icewater scavenger beetle *Helophorus glacialis* (in snow puddles, up 3,200 m) in the Alps reach altitudes of over 2,500 m. The same applies to the leaf-eating beetle *Chrysochloa gloriosa,* the bumblebees *Bombus alpinus* and *B. lapponicus,* the snow flea *Boreus hiemalis* (to 3,800 m), the lacewinged fly *Chrysopa vulgaris,* the geometrid *Psodos alticolaria,* the hawkmoth *Macroglossum stellatarum,* the Alpine salamander *Salamandra atra,* the Alpine newt *Triturus alpestris,* the common frong *Rana temporaria,* the common lizard *Lacerta vivipara* (to 3,000 m), the adder *Vipera berus* (on southern slopes to 3,000 m), the bat *Pipistrellus savii,* and the water shrew *Neomys fodiens.* The common propert of all of these species is that they also occur at considerably lower altitudes, some as low as sea level.

Apart from the above, there are species in the high Alps which belong strictly to this zone. The snow glass snail *Vitrina nivalis,* the glacier crab spider *Xysticus glacialis,* the millipede of the northern limestone Alps *Leptoiulus saltuvagus,* the Alpine *Hypsoiulus alpivagus,* and the Alpine crane fly *Oreomyza glacialis* belong to this faunal community, as does the geometrid *Dasydia tenebraria,* the owls *Anarata nigrita* and *Agrotis fatidica,* the Alp apollo *Parnassius phoebus,* the tent caterpillar *Malacosoma alpicola,* spotted butterfly *Melitaea asteria,* the blue Lycaenid *Albulina orbitulus,* the black butterfly *Erebia gorge,* and the ice black butterfly *E. pluto* flying as high as 3,000 m on the Grossglockner. Among the vertebrates, the following species are particularly remarkable: the rock ptarmigan *Lagopus mutus,* the dotterel *Eudromias morinellus* occurring on the short-grass meadows, particularly in the Eastern Alps, the snow finch *Montifringilla nivalis,* the Alpine white hare *Lepus timidus,* the snow mouse *Microtus nivalis,* the marmot *Marmota marmota,* the chamois *Rupicapra rupicapra* which has adapted to the rocky regions, and the Alpine ibex (*Capra ibex*). Because of their island-like distribution and isolated location, the high mountain biomes — in contrast to the tundra biomes — are not a self-contained genetic entity. Manifold ecological differences and a history which is by no means uniform for the entire high mountain biome, created the mosaically distributed differentiation centers. It is, therefore, not surprising that even the Alpine high mountain biome cannot be treated as a self-contained entity. This is illustrated, among other things, by the allopatric dispersal pattern of the ground beetle *Nebria bremii, N. atrata,* and *N. fasciatopunctata. N. bremii* is an Alpine species generally living close to summer snow patches; it is found in the Alps west of Innsbruck. The ecologically related *N. atrata* is completely absent in this region. It occurs in the Hohen Tauern eastward to the Rottenmanner Tauern and stays south of the Salzach and the Enns. *N. fasciatopunctata* which also occurs in lower regions, lives in the foothills of the southeastern Alps between Friesach in the northwest and Cilli (Yugoslavia) in the southeast. The butterfly *Arctia cervini* is an endemic of the Central Alps, while the beetles *Amara spectabilis* and *A. alpestris* are endemics of the southern limestone Alps, and the tower-shaped snail *Cylindrus obtusus* has its dispersal center in the limestone mountains of the Eastern Alps (locally also on rocks in the Central Alps which are rich in lime). The dispersal areas of the weevils *Otiorhynchus foraminosus* (western part of the Eastern Alps), and *O. auricapillus* (eastern part of the Eastern Alps), which occur nearly allopatrically, also illustrate the strong differentiation of the Alpine high mountain biome.

Several beetle species are endemic to isolated mountain rocks. In the extreme east of the Central Alps, the ground beetles *Nebria schusteri, Trechus grandis,* and *T. regularis* are found only on the Koralp. Numerous parallel examples exist for the plants. It is worth noting that the abundance of endemics, particularly the wingless mountain endemics, the cave dwellers, and the terricolous blind beetles increases, the closer we get to the glaciated peripheries of the Southern and the Eastern Alps.

Numerous true cave animals existing today in the caves of the Dobratsch (southwest of Villach) presumably survived the last glaciation on this site.

The original cave beetles attain their greatest species diversity in the Karawanken in Austria. Only a few cave Carabidae have been found so far north of Drau and Gail; this also applies to the terricolous blind beetles.

Due to their winglessness and their sensitivity to aridity and high temperatures, the mountain endemics (including cave dwellers) and subterranean blind beetles in their distribution areas generally show little change since Pleistocene times.

The relationships between the high mountain and tundra biomes do not by any means hold everywhere, so in the present state of the art no attempt is made to treat high mountain and tundral fauna as belonging to a single biome. This is true in particular of the Mediterranean Alps (Varga 1974), of the Andes (Vuilleumier 1970, Müller 1973, 1976), of the high mountains of Africa (Moreau 1966, Eisentraut 1973), of the Himalayas (Dierl 1970, Swan 1970), and the high mountains of New Guinea (Mayr and Diamond 1976).

The climatic differences between the high mountain of the tropical and the temperate latitudes are also significant. The genetic relationships between the Eurasian tundras and the high mountain biomes are in many cases based on historical causes. At present, there are numerous species possessing a disjunct distribution whereby one of the populations belongs to the tundra and the other to the high mountain biome. In some cases, it can be proven that species of the European high mountain and tundra biomes lived in Central Europe during the last glaciation, thereby permitting the development of a mixed fauna during the ice age (Thienemann 1914).

Relicts of this glacial mixed fauna in rare cases survived on the cold sites of the Central Alps or, after the climate improved, completely disappeared. Naturally, the same is also true for the vegetation.

The mountain avens *Dryas octopetala,* particularly widely dispersed on limestone in the higher parts of the Alps and the Abruzzi during recent times, had occurred in the lowlands of Central Europe during the last glaciation (the Dryas Period). The area fragmentations of such taxa occurred during the postglacial period are called Arctic-Alpine area disjunctions. A good example of this disjunction type is represented by the group of relatives of the snow grouse (*Lagopus mutus*) (cf. Tundra Biomes, section 5.1.8). The invertebrate species in particular supply a wealth of examples (Holdhaus 1954). Arctic-Alpine areas are known even for spiders (Thaler 1976).

Darwin suspected that the Arctic-Alpine dispersal type must have come about by means of area shifts during the last glaciation. Whether the disjunct populations are sub- or semispecifically differentiated, it is probably true that this mode of development applies in general.

From the temporal sequence of the glacial and postglacial biome shifts it can be concluded that the Boreo-Alpine disjunction must be younger than the Arctic-Alpine one (see Fig. 83).

A downward shift of the high mountain biomes during the Pleistocene can be proven world-wide. This process is substantiated by numerous pollen-analytical and geomorphological examples, as well as by the proofs of the Pleistocene snow limit depressions (Heine 1976, among others). On Kilimanjaro, the depression of the snow limit is put at 1,300 m, in the Andes at 1,400 m, in the Bismarck-Chain of New Guinea at 1,000 m and on Mt. Kenya at 1,100 m. While in the Transylvanian Alps, the values were 1,900 m, in the High Tatra 1,600 m, and in the Riesen Gebirge about 1,200 m, in the Black Forest they are 900 m and in the British Isles 600 m.

The fact that these vertical shifts had significant phylogenetic consequences was shown by Mayr and Diamond (1976), among others, in the case of the montane avifauna of New Guinea and the Solomons.

Ecological conditions in the tropical high mountains differ fundamentally from those of the extratropical mountains. Of decisive importance is the

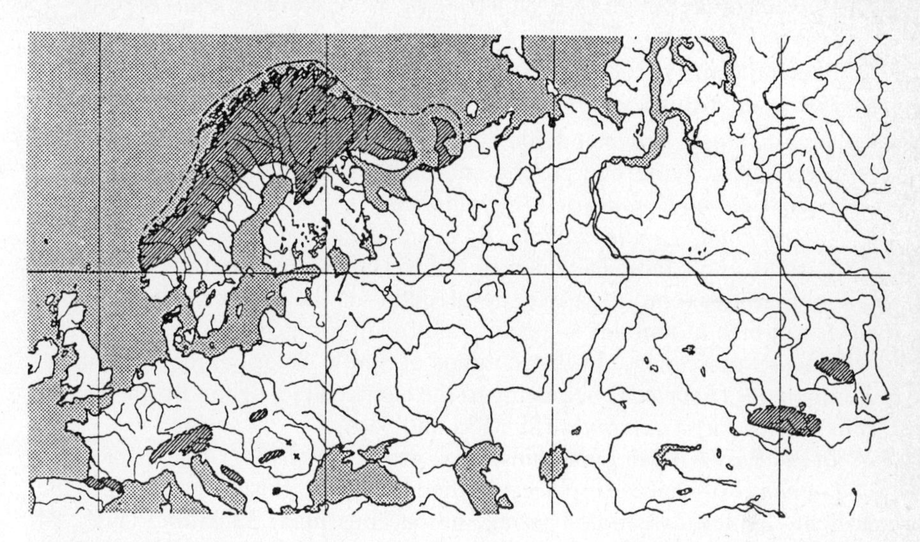

Figure 83 Dispersal of the butterfly *Erebia pandrose,* as an example of the Arctic-Alpine distribution type.

fact that the tropical high mountains have a daily climatic rhythm and that the modification of precipitation can lead to a mosaic of totally different montane biomes. The taxa in this biome must be adapted to the frequency of the short-term frost changes.

The neo-tropical zones known as **Paramos** have been particularly intensively studied over the last 100 years. Cleef (1977) restricts Paramos to neo-tropical areas, and excludes 'aro-alpine' and high mountain zones despite their similar forms of vegetation (cf. Hedeberg 1973): "Paramos are open vegetations, generally occurring above the upper forest-line in the mountains of the humid tropics of Latin America."

Paramo formations thrive in areas with rainfall levels between 750 mm and 2,500 mm (occasionally as much as 3000 mm) in the high mountain ranges of Costa Rica and Panama, the Eastern Andes in Peru and Bolivia, and the high Andes in Venezuela, Colombia, Ecuador and Northern Peru. The flora of the highlands in the isolated mountains between the Orinoco and the Amazon, and in the Guyanese Tepuis, have features in common with the flora of Andean Paramos.

Cuatrecasas (1958, 1968) distinguishes three types of Paramos:

1. *Subparamo.* From 3,000 m to 3,500 m, defined as an intermediate form of vegetation between the Andean mountain forests and open Paramos, distinguished in their lower levels by the presence of composite and ericaceous bushes, and in the upper levels by *Arcythophyllum nitidum* (Rubiaceae) and *Gaylussacia saxifolia* (Ericaceae).
2. *Paramo grassland* (= 'Paramo propiamente dicho' — genuine paramo) between 3,500 m and 4,100 m (reaching 4,300 m locally), with tussock

grasses (*Calamagrostis*), between which the rosettes of the composite Espeletia are particularly striking. There are occasionally woods of *Polylepis* at this level.

3. *Superparamo,* between 4,100 and 4,750 m (up to the snow line), characterized by patterned ground and plant species of the genera *Draba, Lycopodium* (*L. saururus, L. erythraeum, L. brevifolium, L. rufescens, L. cruentum, L. attenuatum*), *Alchemilla, Poa,* and *Agrostis.*

Various plant families which in the lowlands are generally known by small, herbaceous representatives, developed — in the northern Andes and in the African high mountains — pachycaul growth forms, giving the landscape its characteristic scenery. In the Paramos of South America, these are man-high espeletias (*Espeletia*) belonging to the Compositae, or the Bromeliaceae of the species *Puya raimondii* (Fig. 83a and 83b).

In the East African mountains, there are tree-*Senecio* (Compositae) and tree-*Lobelia* (Lobeliaceae) species; in the Himalayas, this group includes the succulent, laniferous candle-bearing, herbaceous plant *Saussurea* (Figs. 84 and 85). In spite of the isolated location of the Neotropical Paramos, their fauna usually consists of two large groups of relatives. The first group includes species which also populate the open biotopes at sea level either in the northern hemisphere or in the south of the southern continents and which, in the Paramos, traverse the tropical lowland conditions unfavorable to them. There is a wealth of genera and species, among the vertebrates as well as the invertebrates, belonging to this group, from the butterflies of the genus *Colias* to the duck genus *Anas.* The second group is represented by an endemic Paramo fauna poor in individuals as well as in species. Of the 758 bird species living in Costa Rica, for instance, only 21 particularly differentiated subspecies occupy the isolated Paramos of the high mountains of Talamanca.

The number of in endemics increases noticeably in the far-flung Paramos of the northern Andes. Cervine species (*Pudu mephistophiles*), tapirs (*Tapirus pinchaque*), rodents (*Thomasomys paramorum*), marsupials (*Marmosa dryas*), predatory animals (*Tremarctos ornatus, Nasuella*), and numerous birds (for instance, *Hemispingus verticalis, Cisthothorus meridae, Schizoeaca fuliginosa, S. griseomurina, Asthenes virgata, Cinclodes excelsior, Chalcostigma olivaceum, Nothoprocta curvirostris*), populate this biotope which the small hummingbird *Oxypagon guernii* never leaves all year long. The proportion of tropically related groups is much lower than in the lowlands. The iguana of the genus *Phenacosaurus,* whose closest relatives occur in the tropical lowlands, populates the Colombian Paramos in two species (*P. heterodermus, P. nicefori*). *P. heterodermus,* the most abundant species in Bogota (Fig. 85a), has — like the other species — a prehensile tail with which it anchors itself in the *Espeletia* or the *Rubus* bushes. The small *Euspondylus brevifrontalis* is oviparous in spite of the high altitude at which it occurs (above 3,500 m).

Because of the high air humidity, a large number of amphibians were also able to conquer the Paramos. Among these, the endemic Paramo frog *Niceforonia nana,* which was only described in 1963 and whose ecology is still to a large extent unknown, is particularly worth mentioning.

The floristic and faunistic relationship between the Andean Paramos and Patagonia as well as the sub-Antarctic islands is extremely close (cf, among others, Cleef 1977, Müller 1976, 1977). This is in complete contrast with, for instance, the tropical African and Malayan high mountains. The

Figure 83a *Espeletia* meadows in the Paramo of Sumapaz, Colombia (1979)

reason for this is that the Ethiopian as well as the Oriental high mountains are much more isolated and the Andes were able to function as an immigration path for the sub-Antarctic taxa since the beginning of the Pleistocene. In

Figure 83b *Puya* (above) and *Polylepis* forest (below) in the Paramo of Sumapaz, Colombia (1979).

South America, there exist — apart from the Andean system — biogeographically highly interesting and generally speaking greatly isolated high mountain biomes. These include, in addition to the Guianas with their numerous Tepuis, the Alpine campos of the Serra do Mar and of the Minas Gerais (Brazil, cf. Magalhaes 1956).

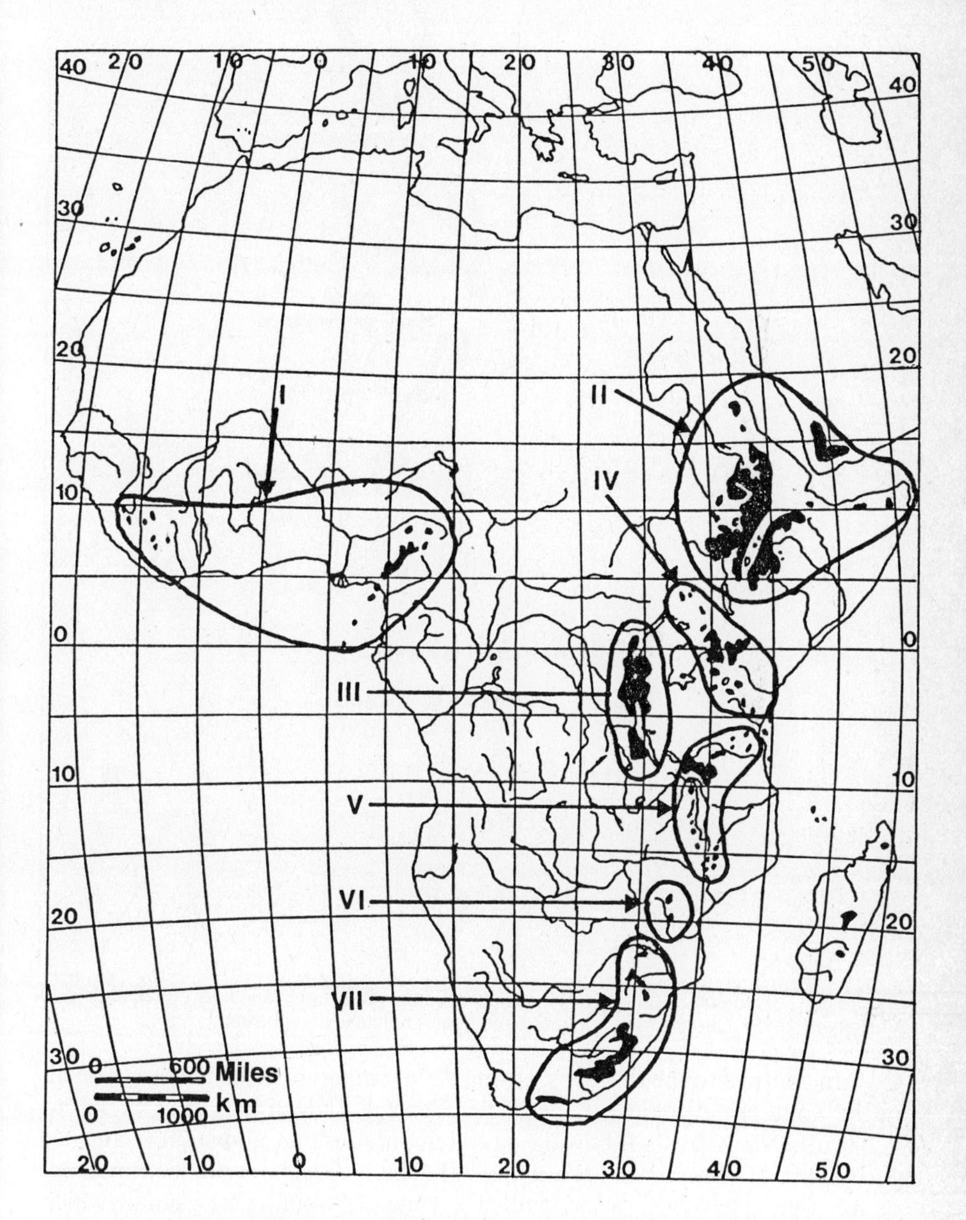

Figure 84 Dispersal of the 'Afromontane' biomes in Africa and Madagascar. I = West African Region; II = Ethiopian Region; III = Kivu-Ruwenzori Region; IV = Usambara Region; V = Uluguru-Mlanje Region; VI = Chamanimani Region; VII = Drakensberg Mountain Region.

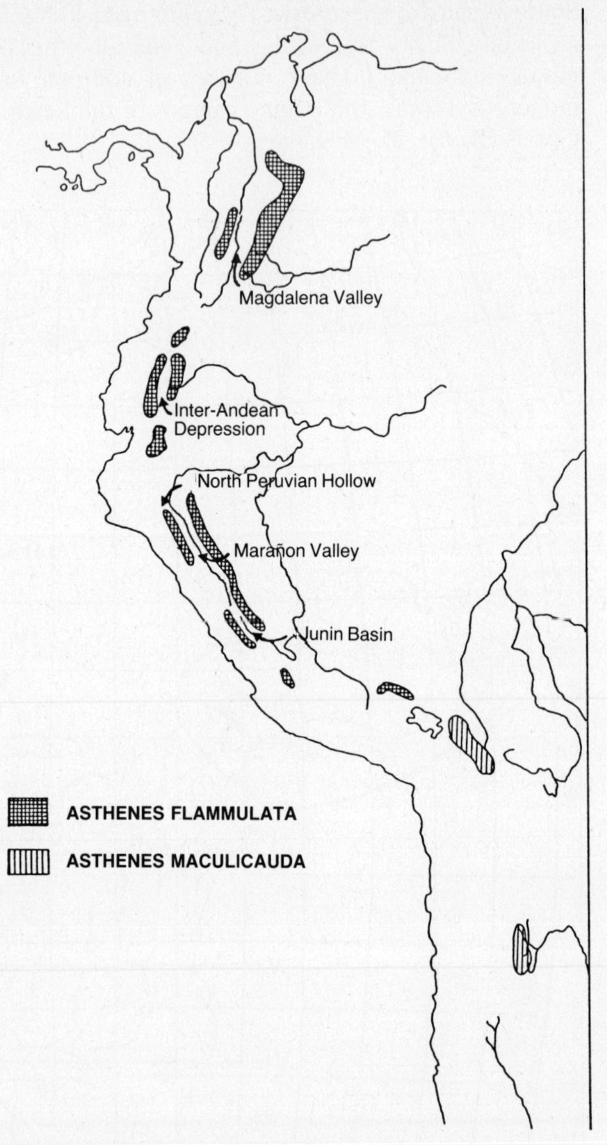

Figure 85 Dispersal of the High Andean bird species of the genus *Asthenes* (according to Vuillemier, 1975). Geographical barriers have been particularly emphasized.

In contrast to the Paramos, strong daily temperature fluctuations prevail in the Southern American Puna of the Central Andean highlands (3,300 to 4,700 m). Night frosts following daily temperatures of above 30°C are not exceptional (Cabrera 1968, Werner 1976). The seasonal development process of the Puna, however, is dependent on the distribution of rainfall which clearly diminishes from north to south. Frost-resistant Ichu grasses of the genera *Calamagrostis, Festuca,* and *Stipa* are dominant over vast areas, inter-

Figure 85a *Phenacosaurus heterodermus,* in the vicinity of Bogota (Colombia) where the arboricole Iguanide is frequently found in blackberry bushes.

rupted by hard-cushion and rosette plants or dwarf shrubs with foliage rich in resin. This uniformity may fundamentally change in small areas due to a deviating flow of water in the ground. Humid Puna types alternate with drier ones.

The stony and rocky mountain sites are populated by various briars (*Tetraglochin, Junellia, Mulinum*). Whereas dry valley floors, river banks and edges of saline area, in the presence of ground water close to the surface, are populated by *Parastrephia* subshrubs, the drift-sand areas and salt-free sand surfaces are covered by a fabiana-pennisetum community, which, with increasing salt concentrations, is replaced — depending on the substrate — by *Sporobolus* or *Distichlis* species.

In the region of the Andean Puna in northwest Argentina, Ruthsatz (1977) distinguishes four altitudinal zones:

1. The subnival zone (above 4,900 m) with yearly mean temperatures at or below 0°C, frequent night frosts and cryoturbation processes. The vegetation is limited — in the form of islands — to special microclimate sites. Cushion plants, evergreen dwarf shrubs and hemicryptophytes with a vigorous root system are predominant. The dwarf shrub and herb meadows of this Alpine cold desert have been described as the Wernerion-pseudodigitatae community.
2. The High Andean zone (between 4,100 and 4,900 m) with yearly mean temperature between 0 and 6°C, in its upper region (above 4,500 m) is characterized by hard-leaf thicket grasses forming a Festucion-

orthophylla community. In the lower region, the remains of *Polylepis tomentella* coppices can be found. Soligenous mires consisting of hard cushion plants occur here (Wernerion pygmaeae).

3. The Puna zone, at yearly mean temperatures between 7 and 10°C, includes vast Alpine plateaux, low hill ranges and the upper slopes of the Central Andean valley between 3,200 and 4,100 m altitude. Daily variations of 40°C at 2 m above ground level are frequently encountered. The predominant vegetation formations and the open shrub semi-deserts generally consist of deciduous shrub species, whereas evergreen species are rarer. They are classified as Fabianion densae. Meadows consisting of dune grasses (*Sporobolion rigentis*) occur on the saline drifting sand soils, whereas in the areas of the brackish lakes and saline areas which are fed by ground water, saline meadows (Salicornio-Distichlidion humilis) are present.
4. The pre-Puna zone is limited to the deeply incised Central Andean valleys with yearly mean temperatures between 12 and 15°C. The slopes and valley floors are covered by open semi-deserts consisting of succulent shrubs which have been defined as the Cassion-Trichocereion community.

The plants of the Puna have had to adapt to a lack of water prevailing practically all year long, to the intensive radiation and evaporation, great daily temperature fluctuations, salt enrichment and erosion-susceptibility of the soils, the strong winds prevailing at this altitude and last, but not least, the presence of a herbivorous fauna. The latter has adapted to the varying productivity of the vegetation (dry Puna: about 2,000 kg of surface, phytogenic dry weight/ha and year, humid Puna: about 7,000 kg of surface phytogenic dry substance/ha and year). In contrast to the root eaters, including the widely-dispersed comb rat of the genus *Ctenomys,* large herbivores, such as the Darwinian ostrich *Pterocnemia pennata;* or the Camelidae *Lama vicugna* undertake local migrations to the Alpine plateaux. Large, rock-dwelling rodents such as *Lapidium* and *Chinchilla* prefer to graze the coppice grasses present in their habitat, whereas small rodents such as species of the genera *Abrocoma* and *Octodontomys* have adjusted — as have several bird species — to a diet of grains. A strict food specialization, however, has not been confirmed. It can be shown that many species which in the lowlands were food specialists became omnivores in the Puna.

Of the 153 high Andean bird species described, 35 species are limited to the Puna highlands. The species exhibiting reduced or shortened wings, and well-suited either for quick running (*Pterocnemia pennata*), excellent swimming (such as the endemic Titicaca diver *Centropelma micropterum*) or fast flying, deserve special mention. The proportion of endemic ovenbirds is particularly large. In addition, there are finches, pipits, tataupas, plovers, and pigeons. The two South American flamingos *Phoenicoparrus andinus* and *P. jamesoni,* live in the saline lakes of this Alpine habitat. Some of the small bird species developed remarkable adaptations to the Puna. Their population density — with a few exceptions — is small. The tendency to avoid the fast changes in weather by means of migration appears to be insignificant. The lakes of the Puna plateau having no outlet contain amphibians which, in the case of the *Telmatobius* species which lives in the water throughout the year, for instance, are represented by 13 species (of which three are endemics to Lake Titicaca).

The reptiles have also adapted to this habitat. Five species of the sand lizard genus *Liolaemus,* which is widely dispersed in the southern part of South America, occur here. The species in the vicinity of the beaches, however — for instance, the Brazilian *Liolaemus occipitalis* — are oviparous (cf. Resting Biome, section 5.1.10), while the species of the Puna are viviparous. This is also true for *Ctenoblepharis jamesi* occurring in the same habitat and having a parallel in the Alpine reptiles of East Africa and the Himalayas. The great number of frost days in the tropical Alpine mountains and the fact that freezing generally takes place in a very thin surface layer leads most reptiles to utilize crevices located at a lower elevation or the burrows of rodents.

Close relations exist between the Puna animals and those of the Patagonian cold grasslands. Thus, the Darwinian ostrich in the Puna region occurs in the subspecies *garleppi,* whereas in Patagonia it exists in its nominal form. Similarly, the closely-related species of the Puna tataupa, *Tinamotis pentlandii,* lives in the Patagonian lowlands (*T. ingoufi*). Additional examples of this type of relationship can be cited in the case of starlings and of some mammals (for instance *Lama guanacoe*).

The history of development of both the Paramo and the Puna is young. It is very closely tied to the genesis of the massifs in which they occur today and to the Pleistocene climatic development (Haffer 1974, Müller 1976, 1977, Van der Hammen 1974).

During the last glaciation, the Andes from Chile to Costa Rica wore ice caps which covered large areas of today's Paramo and Puna. An analysis of numerous pollen profiles showed that during the Pleistocene, in eastern Colombia a subtropical climate prevailed at an altitude of 500 m. Today, the transition from the tropical to the subtropical climate takes place between 1,500 and 1,700 m elevation. Similar data are available for other tropical high mountains; this is the reason why we may assume that as recently as 11,000 to 12,000 years ago (end of the last glaciation), the high mountain communities were located at considerably lower elevations than they are today. It was only at the time when the climate improved that the Paramos and the Punas migrated to their present sites which had become free of ice. Since they have been subject to the stronger insolation of the high mountain habitats only since this shift in elevation, we may assume that many of the dispersal patterns and differentiated species of the Puna and Paramo faunas are very young. This also applies with respect to earlier history. Whereas the rise of the northern Andes to their present level occurred at the turn of the Plio-Pleistocene, i.e., about 2 to 4 million years ago, that of the Central Andes of Peru and Bolivia occurred only during the Miocene. This means that during the early Tertiary no habitats existed for the South American Puna and Paramo organisms.

5.1.10 The Restinga or Coastal Forest Belt Biome

Restinga (Portuguese = reef or sandbank) in Brazilian linguistic usage means "shore or coastal thicket". Jakobi (1977) interprets this as "the tropical, wind-sheltered coastal forest belt of the supralittoral in a hilly landscape leading from the wide sandbeaches with a dune ridge to the coastal timber forest." Although with this definition he separates the 'Restingal' from the dune zones, he nevertheless recognizes that "in view of the characteristic connection to the Paranaean Coast, the two must be treated together" (Jakobi 1977, p. 49).

In his analysis of the coastal forest belt of Rio de Janeiro, Dansereau (1947) pointed out that this habitat could be truly understood if it were possible to ascertain the often quickly-changing influences — in small areas — of saltwater, fresh water, dunes and their terrestrial systems which are very closely interlinked with the coastal forest belt.

Without a knowledge of the numerous fresh water lakes (Lagoas) and swamps, of the often immense lagoons (exhibiting a fluctuating salinity), the sandbars running parallel to the coast, and vegetation-free, drifting sand dunes, it is not possible to understand ecologically the coastal forest belt of today.

The proximity of the sea has a particular effect on annual temperature variations. The direction, frequency, and force of the winds form and reform the habitat. Sand storms are no rarity in the South Brazilian coastal forest belts (Fig. 86).

Beach ridges and dunes follow the Brazilian coast, interrupted by granite and gneiss rocks of the Serra do Mar jutting out into the Atlantic and by the estuaries of the river systems draining into the Atlantic. In the northeast (Maranhão) and in the southeast (Rio Grande do Sul), they reach huge proportions. East of Sao Luis (Maranhão), an almost 25 km wide coastal strip begins "in which hundreds of low beach ridges, fortified by vegetation, extend inland, covered only by low brush..." (Hueck 1966). Drifting sand dunes, completely devoid of vegetation, which, as they advance, bury shrubs, jut out of the flat landscape; farther to the east, however (for instance in the delta area of the Paraiba), they are broken up by river mouths and vast mangrove swamps.

In the region of Cabo Frio (Guanabara), Ule (1901) distinguished between two coastal forest belt types (heath and myrtle coastal forests), dominated by Ericaceae, Melastomaceae, Eriocaulaceae, and, in the open areas, by cacti. In the myrtle coastal belts, they are replaced by tree-shaped Myrtaceae.

Treeless sites are occupied by the dwarf palm *Diplothemium maritimum* and the ground Bromeliaceae whose cisterns are populated by — among others — *Corythomantis brunoi.* Ule (1901) described an additional, more mature succession phase, i.e., the *Clusia* coastal forest belt, which is dominated by the *Clusia* equipped with support roots and growing to a height of

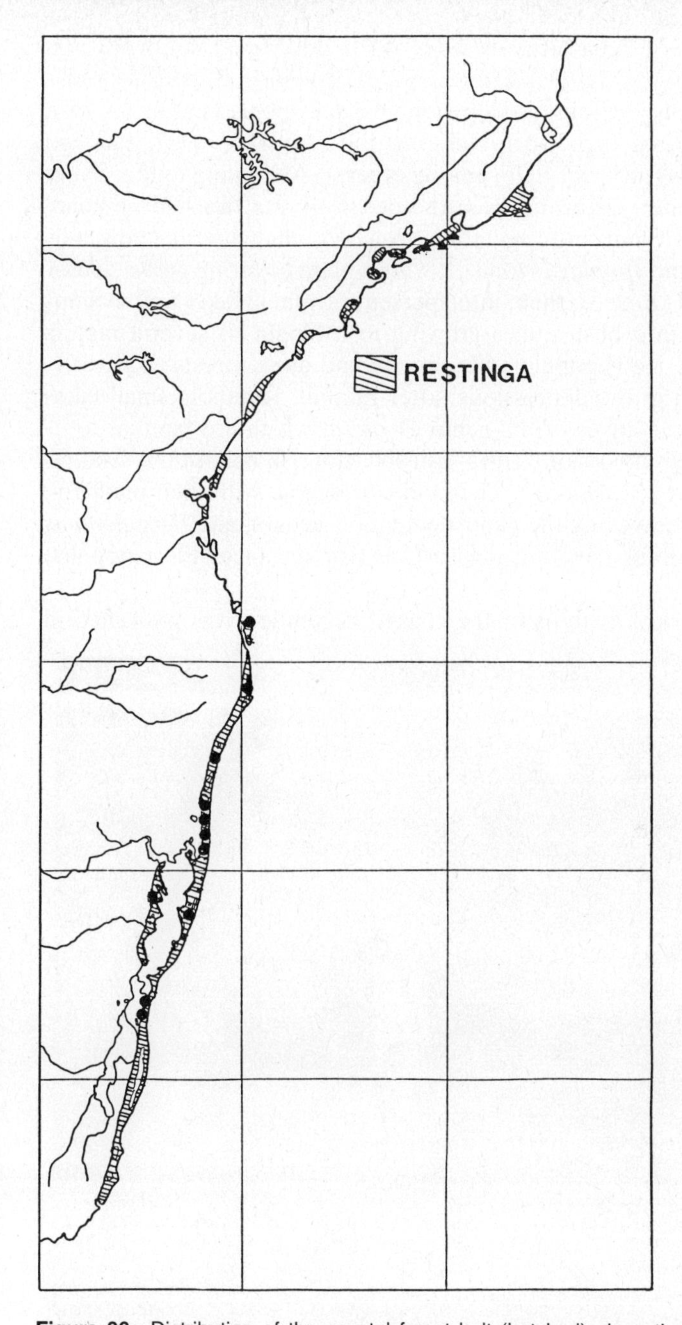

Figure 86 Distribution of the coastal forest belt (hatched) along the southeastern Brazilian coast and the dispersal of the sand lizard genus *Liolaemus* (according to Müller and Steiniger, 1977). Triangles = *Liolaemus lutzae;* dots = *Liolaemus occipitalis;* squares = *Liolaemus wiegmannii.*

between 3 and 10 m. The coastal forest belts of Rio de Janeiro, south of Cabo Frio, have been analyzed by Dansereau (1947). The basic classification developed by him can also be found — albeit increasingly poorer in species — in Santos, Santa Catarina, and Rio Grande do Sul (Andrade 1967, Lamego 1940, Rawitscher 1944, Rizzini 1976, among others). According to this classification, the first zone — starting from the ocean — is a 'washed up zone' consisting of sand without any vegetation (zone 1), changing into low hills covered by grasses and *Ipomoea* (zone 2), whose plant covering grows denser as it advances inland (zone 3); then, interspersed with small lakes and swamps (zone 4), changing into bush dunes growing to a height of several meters (zone 5), and whose lee is populated by woods and dune forests (zone 6).

The vegetation in the depressions, after rainfall, results in small lakes forming. The representatives of the genus *Drosera,* which are familiar to us from the low and high moors of Central Europe, flourish here in the sand, as does the bladderwort (*Utricularia*). The sand *Utricularia,* with their bladders, trap small representatives of the sand depression ecosystem. *Paepalanthus* species sometimes occur covering the surface (for instance, Florianopolis) (see Fig. 87).

The first historical analysis of the coastal vegetation was published in

Figure 87 Restinga (coastal forest belt) Biome on Florianopolis Island (Brazil; September 1976).

1954 by Rambo. He established that the "Quaternary coastal region of Rio Grande do Sul, southern Brazil, measuring 30,000 km^2, contains no endemics on the site." He concluded "that the entire phanerogamous flora of the coast of the Rio Grande — and the same also applies to its extension southward to St. Catarina and to the northeast of Uruguay — immigrated from the neighboring areas." Independent of the evolutionary genetic conceptions he derived, he recognized that the flora of Rio Grande do Sul could be meaningfully interpreted only by the assumption of Quaternary fluctuations in the vegetation (cf. also Rambo 1954; Smith 1964). The first summary of the coastal fauna in the Rio Grande region was compiled by Gliesch (1925). He also described the resting places of the eared seal *Otaria byronia* and *Arctocephalus australis* on the Ilha dos Lobos near Torres which today is visited only sporadically by isolated individuals.

As in the case of the flora, the fauna, also, exhibits a marked north–south gradient. While bird species such as the crown sparrow *Zonotrichia capensis,* the parasitic cowbird *Molothrus bonariensis,* the pipit *Anthus correndera,* the tyrants *Pitangus sulphuratus* and *Tyrannus melancholicus,* the falcons *Milvago chimachima* and *Caracara plancus,* and the burrowing owl *Speotyto cunicularia* occur from the north to the south of the coastal forest belt, the mocking thrush *Mimus saturninus* is characteristic of the coastal forest belt of Rio Grande do Sul. All of the above bird species, however, share the fact that in addition to the coastal forest belt, they also populate other biomes. This also applies to the snail hawk *Rosthramus sociabilis* occurring as far as Florida, which hunts snails (Ampularias) living in the isolated lagoons and swamps and whose brilliantly yellow-red egg pouches floating above the water surface are visible from afar. The swamp- and water-bird fauna, rich in species, is particularly impressively represented in the wildlife preserve of Taim in the south of Rio Grande do Sul. Also present are the Parastacidae (for instance, *Parastacus pilimanus*) which, in the southern hemisphere, take the place of the Holarctic river crayfish (Astacidae). In addition, among the invertebrate and vertebrate species, there are some specialists which ecologically speaking are closely restricted to the coastal forest belts and whose phylogeny evidently proceeded parallel to the history of this biome.

This is true, for example, for the Brazilian tiger beetle genus *Cicindela,* also occurring in Central Europe, as well as for the moth genus *Ecpanteria* (Arctiidae), whose caterpillars feed on *Senecio* species in the coastal forest belt and which betray their presence by prominent tracks in the sand. Just like the burrowing sand rats of the genus *Ctenomys,* or the toad *Bufo arenarum,* numerous arthropods do not appear on the sand surface until nightfall. These incude the nearly white Forficulidae *Labidura batesi,* as well as certain termite species. The three allopatrically dispersed Brazilian sand lizards of the genus *Liolaemus* (Müller and Steiniger 1977), are also typical coastal forest belt and coastal dune species.

5.1.11 The Mangrove Biome

The mangrove biome, at the boundary between the marine and terrestrial ecosystems, is inhabited by a fauna and a flora which have adapted to the tidal zone of the tropical oceans and which are poor in species, although, in many cases, extremely abundant in individuals. They are able to endure significant fluctuations in salinity, and have incorporated in their nutrient cycle the organic mud (as a typical substrate). In shallow bays protected from the breakers and containing rich marine and fluvio-marine accumulation material, the mangrove often forms vast forests. Examples of Maldivian and other coral islands as well as successful germination experiments with various mangrove plants in pure fresh water show that its development is not necessarily dependent on brackish water. Although the various plant species resemble each other physiognomically and — as long as they occupy a comparable ecological position within a stand — also physiologically (Chapman 1976), they often belong to different genetic groups. Widely-branched support roots anchor the mangle tree *Rhizophora* in the oxygen-deficient, but H_2S-rich mud. The above-ground, aerial roots (pneumatophores) of the *Sonneratia* species or the "kink roots" of the *Bruguiera* jutting out of the mud supply the plants with sufficient oxygen. Numerous species are viviparous. The embryo develops immediately out of the bloom, and then falling off the mother plant and boring itself into the soft mud, or being carried by the water to another site where it takes root very quickly. All species exhibit an amazing ability for survival at different salt concentrations. This results from the fact that individuals of the same species occur far upstream in the estuaries of the rivers, whereas others occur in pure salt water. The varying requirements to which the plants in the mangrove belt are subjected on the ocean as well as on the land side, lead to a specific dispersal of the individual plant species and thus to a characteristic mangrove structure. *Rhizophora* species penetrate the farthest into the Atlantic where they intercept the rolling breakers. In Southeast Asia, they are replaced by *Sonneratia* which erects a protective forest belt towards the sea and can grow to a height of as much as 25 m. At high tide, some of the trees are immersed up to their crowns in the water, while at low tide, even the prop root system is exposed.

Like coral reefs, the distribution of mangroves is determined by the course of cold and warm ocean currents (Fig. 88). Cold southern ocean currents (Benguela and Humboldt Streams) on the west coasts of South America and Africa shift their southern limits towards the equator while the warm East African Agulhas Stream makes it possible for the mangrove to flourish as far as Southern Madagascar.

On the Pacific side of South America, the mangrove extends as far as 40° latitude south; under the influence of the warm Brazil Stream, however, on the Atlantic east coast of the continent it extends as far as Santos and, in a fragmentary form, as far as Florianopolis (State of Santa Catarina). The mangrove penetrates farthest south on the Chatham Islands (at a latitude of 44°S). The distribution of plant genera in the mangrove belt shows that an

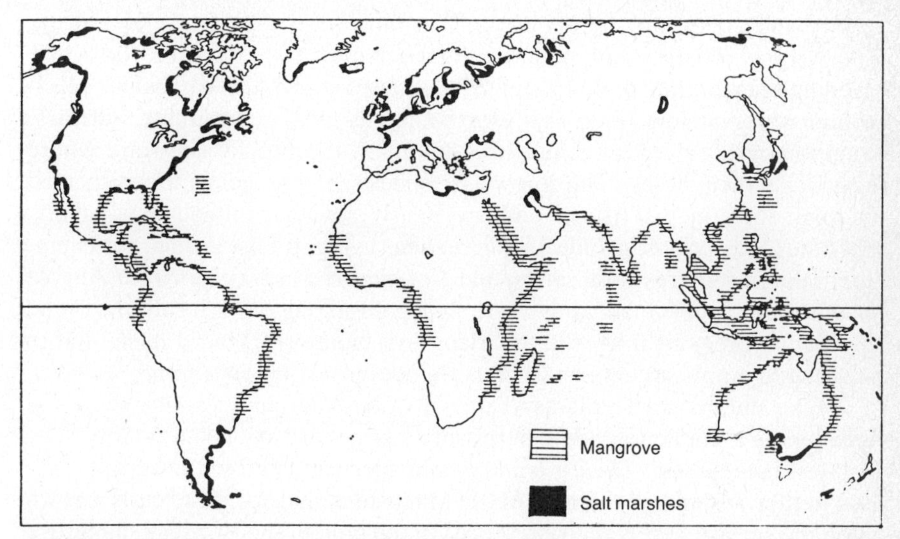

Figure 88 Global distribution of mangrove and salt marshes (according to Chapman, 1977).

eastern mangrove abundant in species can be distinguished from a western mangrove which is poorer in species. The border, on the one hand, runs through the African continent and approximately correlates with the phytogeographical separation between Palaeotropics and Neotropics.

Because of this Southeastern Asiatic species abundance, many authors assumed the Sunda Archipelago to have been the center of origin of the mangrove. It can, however, be shown through numerous palaeontological findings concerning other taxa that it is always dangerous to attempt to identify the original center of a biome with its present day center of greatest diversity.

The fauna of the mangrove zones consists of species of varying origins. It is possible to encounter tropical rainforest species as well as marine animal groups. Nearctic and Palaearctic swamp species wintering in the tropics search for their food among the crustaceans which are strictly confined to the mangrove biome.

The species' composition and daily periodic rhythm of the mangrove fauna are greatly affected by the vegetation structure and periodicity of low and high tide (Gerlach 1958, MacNae 1966, Müller 1977). Numerous species, particularly the sessile or semi-sessile mangrove oysters and balanids (for example, *Chthalmus rhizophorae*) living on the trunks of the *Rhizophora* have their most active phase at high tide, while the mangrove crabs *Uca, Sesarma* and *Cardisoma* search for their food at low tide. The Neotropic crab *Aratus pisoni* lives on the mangrove trunks above the high tide level and feeds on mangrove leaves. Like the mangrove crab *Goniopsis cruenta,* the mudskippers (for example, *Boleophthalmus, Scartelaos,* and *Periophthalmus*) are capable of an amphibian mode of life.

At low tide in the mangrove forests of southern Senegal, the *Periophthalmus* populates the open mud areas in thousands. At the first sign of danger, they flee into the open water or into the dwelling burrows of the fiddler crabs, if their own holes are too far removed from the site. During the breeding season, they build funnel-shaped burrows into the mud which reach ground water at low tide. The young animals grow up in these burrows. Numerous fiddler crabs (genus *Uca*) populate the muddy substrate where they make their holes. The known 62 species of this genus are dispersed predominantly in the tropics and are nearly always confined to a muddy substrate. Exceptions include *Uca tetragona,* living in East Africa, for example, in that it requires coral sands; and *Uca tangeri* which occurs from Angola to southern Spain (Guadalquivir). In Senegal (among other locations, in the Sine-Saloum region) it lives in the mangrove biome whereas in its habitat in southern Spain, it occurs in salt marshes overgrown by *Salicornia.*

Like the mangrove plants (Table 56), the *Uca* species which share the same biome are able to endure surprisingly large fluctuations in salinity. The shell-less callianassid, *Thalassin anomala,* which is related to the hermit crab, lives in tube systems in the mud of the Malayan mangrove. Mud casts thrown up on the surface betray their presence. In the mud of the Brazilian mangrove (for instance, near Santos, south of São Sebastião), one frequently encounters the branched-out dwelling structures of the giant whale's tongue *Balanoglossus gigas* (Enteropneusta) which can be up to 2.5 m long. The ducts consist of mud particles which are glued together by the animal's skin slime.

However, the mangrove zone is not only populated by an amphibian, frequently euryhaline, fauna. There are species which advance, at least from time to time, from the sea as well as from the rivers and from the rainforests of the terra firma.

Today's distribution of mangrove zones is only a momentary pattern of a long, historical process. The strict ecological link between the mangrove and certain environmental factors turns its fossil occurrences into indicators for the shifts in other large biomes on the earth during our most recent past (Müller 1973). Thus, during the last glaciation the mangrove oysters (for instance, *Ostrea arborea*) living off the roots of *Rhizophora mangle,* occurred as far as the Argentinian Rio Negro (today's limit: Florianopolis) and would lead us to suspect considerably warmer ocean temperatures in this area than at present. Because of a drop in the sea level at that point in time, South America became connected to the Falkland Islands by means of a peninsula protruding eastwards, with the result that the Antarctic cold water covered far less of the South American coast than it does today. Similar observations were made along the Colombian coast (Müller 1973). Pollen of the red mangrove (*Rhizophora*) have been used for some time for the reconstruction of past fluctuations in sea level.

5.2 THE MARINE BIOMES

When in 1924, Hardy described the food chains of the herring along the English east coast, he recognized — as did Möbius, in 1877 during a study of

TABLE 56 DISTRIBUTION OF MANGROVE PLANTS (according to Chapman 1976)

Families and genera	Number of species	Indian Ocean and Western Pacific	Pacific USA	Atlantic USA	West Africa
Rhizophoraceae					
Rhizophora	7	5	2	3	3
Bruguiera	6	6	–	–	–
Ceriops	2	2	–	–	–
Kandella	1	1	–	–	–
Avicenniaceae					
Avicennia	11	6	3	3	1
Meliaceae					
Xylocarpus	? 10	? 8	?	2	1
Combretaceae					
Laguncularia	1	–	1	1	1
Conocarpus	1	–	1	1	1
Lumnitzera	2	2			
Bombacacaea					
Camptostemon	2	2			
Plumbaginaceae					
Aegialitis	2	2			
Palmae					
Nypa	1	1			
Myrtaceae					
Osbornia	1	1			
Sonneratiaceae					
Sonneratia	5	5			
Rubliaceae					
Scyphiphora	1	1			
Myrsinaceae					
Aegicera	2	2			
Total	? 55	? 44	? 6	10	7
Other genera	35	19	9	7	4
Total	? 90	? 63	? 15	17	11

the oyster bank near Kiel — that a knowledge of the structure of marine biomes is an important prerequisite for a secure fishing industry in the future. Structure and function of marine ecosystems (Steele 1974), however, are not only of importance to marine organisms. Any great changes in the oceans also result in modifications of the terrestrial ecosystems. The oceans are weather regulators, transportation paths, a source of numerous nutrients and mineral substances and the end link in the ecosystemic pollution chain. The Atlantic Ocean alone contains about 20,000 times as much dissolved organic substances as the annual world-wide wheat harvest (Coker 1966); the substances' distribution is influenced by ocean currents, the circulation of nutrient salts, insolation, temperature and salinity.

About 3 billion tons of terrestrial material annually find their way into the oceans. Without compensatory movements of the earth's crust, therefore, the terra firma would erode in a short time. The major part of terrestrial fresh water emanates from evaporation on the surface of the oceans.

5.2.1 Physico-chemical Factors

The oceans are our largest heat reservoir. Here, the temperature fluctuations are considerably smaller than they are on land. Nowhere on the open sea are the amplitudes greater than 10°C during the year or greater than 1°C during the day. Below the euphotic zone penetrated by light, considerable annual fluctuations no longer occur. The salinity of the oceans generally fluctuates between 34 and 37‰. The Gulfs of Bothnia and Finland, however, only contain 2‰, the Red Sea, on the other hand, 40 to 46.5‰, the central Sargasso Sea 38‰. The salinity in the estuaries can fluctuate considerably (Cronini 1975). On average, the surface waters in the southern hemisphere are more saline than those in the northern hemisphere. The vertical fluctuations in salinity are remarkable. Their cause is evaporation which leads to an increase in the salinity on the ocean surface. This results in an increase in specific gravity. Surface water is transported off into the depths and water from the depth rises to the surface. Such vertical movements are particularly noticeable on the sill of Gibraltar between the Mediterranean and the Atlantic Ocean. The Mediterranean waters with a salt content of 38‰ flow in a sub-surface current into the Atlantic whose salinity at a depth of 600 to 1,200 m increases from 36‰ to 36.5‰. Atlantic water penetrates the Mediterranean in the surface water zone.

Sea water is a solution of heavier, non-volatile materials in easily evaporated water. Many physical factors of the sea water can be understood as a function of salinity and temperature (Fig. 89). The first sea water analyses of non-evaporated residues (Challenger Report) show the following chemical composition:

Sodium chloride	77.76%
Magnesium chloride	10.88%
Magnesium sulfate	4.74%
Calcium sulfate	3.60%
Potassium sulfate	2.46%
Calcium carbonate	0.34%
Magnesium bromide	0.22%

Other trace elements were first discovered — some of them in high concentrations — in marine organisms before they were established for sea water itself. Vanadium was found in the blood of ascidians and holothurians, cobalt in mussels and lobsters, nickel in snails, and lead in the ashes of various marine organisms. The high magnesium content in the ionic composition of the seawater is reflected in the magnesium content of marine animals (Table 57).

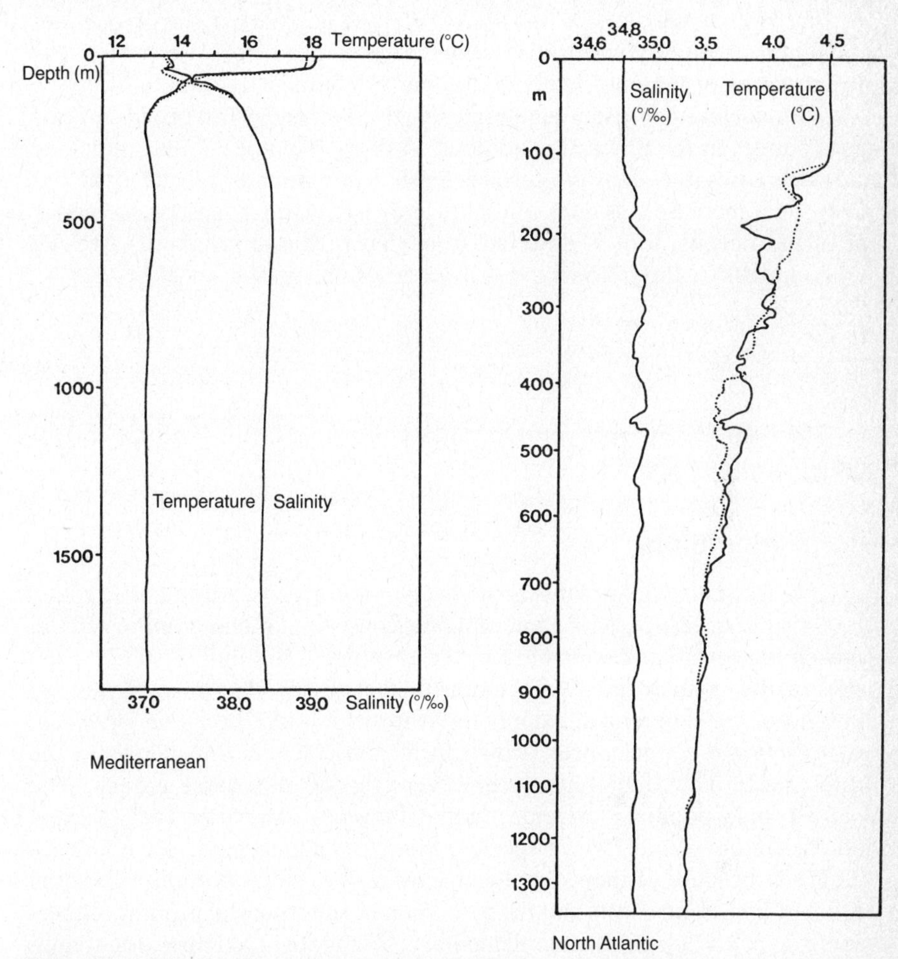

Figure 89 Vertical profile of temperature and salinity in the Mediterranean (left) and North Atlantic (right).

TABLE 57 MAGNESIUM CONTENT OF THE SEAWATER AND OF THE BLOOD SERUM OF MARINE AND TERRESTRIAL ANIMALS, IN RELATION TO THE CALCIUM CONTENT (Calcium = 100)

Seawater (Ocean)	311		
Marine animals		Terrestrial or fresh water animals	
Crustaceans	130–196	Human being	25
Echinoderms	240–253	Mammal	18–29
Molluscs	214–275	Chicken	19
		Crocodile	33
Homarus americanus	34.7	Tench	34
Gadus collarus (Codfish)	35.9	Roman snail	11
Pollachius virens (Green pollack)	47.0		
Sperm whale	713		

The composition of the silicified skeletons of many marine organisms and numerous physiological adaptation features are closely related to the concentrations of various elements (Malins and Sargent 1975).

With declining or rising temperatures, the cold and warm ocean currents effect changes in the physical conditions. A drop in temperature, on the one hand, decreases the density, surface tension and sound velocity, whereas specific heat increases, as does solubility and viscosity. The gaseous composition of the oceans, too, is different from that of the atmosphere. At 35‰ salinity and 10°C, the following dissolved gases are present in seawater:

Nitrogen = 64%
Oxygen = 34%
CO_2 = 1.6%

The high CO_2 proportion (50 times more than in the atmosphere) is, therefore, noteworthy.

5.2.2 Marine Biota

Thallophytes (unicellular plants forming colonies or beds, without true roots, vascular stem or leaves, with gametophores consisting of one or more cells all of which are fertile and which are never enveloped in a hull of sterile cells) make up the main form of vegetation in the oceans. Spermatophytes are represented by submerged halophytes (the eelgrasses *Posidonia oceanica, Zostera marina,* for instance), which form dense seagrass meadows in the littoral zone. The thallophytes, however, represent several groups. The Cyanophyceae populate extreme marine biotopes. They are the principal food for *Littorina* and *Patella* species. Many of their genera, occurring endolithically because of their dissolving activity, play an important role in the erosion of limestone coasts and the formation of small coastal profiles. Others assume a color which is complementary to the light composition (complementary chromatic adaptation); this is the reason why some species from depth have a reddish or purple color. The chrysophytes produce oil as an assimilation product (starch is absent). Planktonic and benthic forms are present which are usually autotrophic (there are also saprophytic species); their calcium or siliceous shells contribute greatly to the formation of sediments, and they serve as food for numerous plankton-eating fishes.

Among the Pyrophyceae, there are species which are responsible for marine phosphorescence (*Noctiluca miliaris*) which, however, in the tropical oceans are also able to produce poisonous substances (resulting in death in the case of humans). The Chlorophyceae (green algae) are of particular significance, since several species live symbiotically with marine fungi and lichens. Zoochlorellas live symbiotically with infusaria in the tissue of sponges and corals.

Among the Phaeophyceae (brown algae), "forest forming" species (building thalli more than 50 m long) occur along the rocky coasts. Unicellular forms are the zooxanthellas, which live symbiotically in the bodies of ciliates, foraminifera, radiolaria, hydrozoa and turbellaria and, via the CO_2-balance, control the formation of the silicified skeletons of the octo- and the hexa-corals. Some of the larger species form a firm substrate for an epiphyte flora rich in species. These seaweed forests form the phytal, the biome of a diverse marine fauna. The purely marine Rhodophyceae (red algae) are found mostly in warmer waters and in shady, quiet sites at a depth of between 30 and 60 m.

This group also includes the calcareous algae (*Lithothamnion, Lithophyllum,* etc.) which often produce the calcareous algal fringes of the coral reefs which are as hard as rocks and which are exposed to the surf. The pectin substances from the cellwalls of the red algae are used, for instance, for the production of agar-agar. There are also numerous marine species among the Phycomycetes (fungi). Many sea fungi live on or in multicellular and unicellular algae (particularly diatoms) as well as on or in uni- and multicellular animals. Fungal epidemics also occur in the ocean. The lichens have conquered the littoral at least, and prefer to live on a calcareous substrate.

Of the terrestrial fauna, only the diplopods and the amphibians (exception: *Rana cancrivora*) are absent from the ocean. In addition to numerous groups with limnic and terrestrial relatives, the ocean fauna consists of numerous purely or predominantly marine orders. These include the radiolarians, whose fossil occurrences show that they have existed since the pre-Cambrian period; they occur mainly pelagically, but also in the deep sea; the Foraminifera (foraminifer sands); the Ctenophora; the cephalopods (730 Recent, 10,500 fossil species), including the Tetra- (for instance *Nautilus*) and Dibranches (among which the Decabrachia, *Sepia* and *Loligo;* the Octobrachia, including *Octopus* and the *Architeuthis* which grow to a length of 18 m; 50% of the deca- and 9% of the octobrachia are equipped with photogenic organs), the echiurida, including *Ikeda taenioides* growing to a length of 185 cm; the sipunculida (the 51 cm long *Siphonomecus multicinctus*); the Chaetognatha; the Phoronidea, solitary, utricular tentaculates living in loose secretion tubes; the Enteropneustra including the *Balkanoglossus gigas* which can be up to 2.5 m long; Brachiopodas; the Pogonophora; the Tunicata; the Acrania. Even the insects, in the forms of *Halobates* (Hemiptera), *Pelagomyia* and *Clunio* (Diptera) species have invaded the pelagic and littoral realms of the ocean. As in the undisturbed biotopes of the terra firma, 'living fossils' were also able to survive in the ocean. These include, among others, the Acrania, the monoplacophores (*Neopilina galathea, Vema ewingi*), the Crossopterygia (*Latimeria chalumnae*) and *Neoglyphea inopinata* belonging to the Mesozoic Glypheidae (Forest, Saintlaurent and Chace 1976) (Fig. 89a, 89b, 89c, 89d).

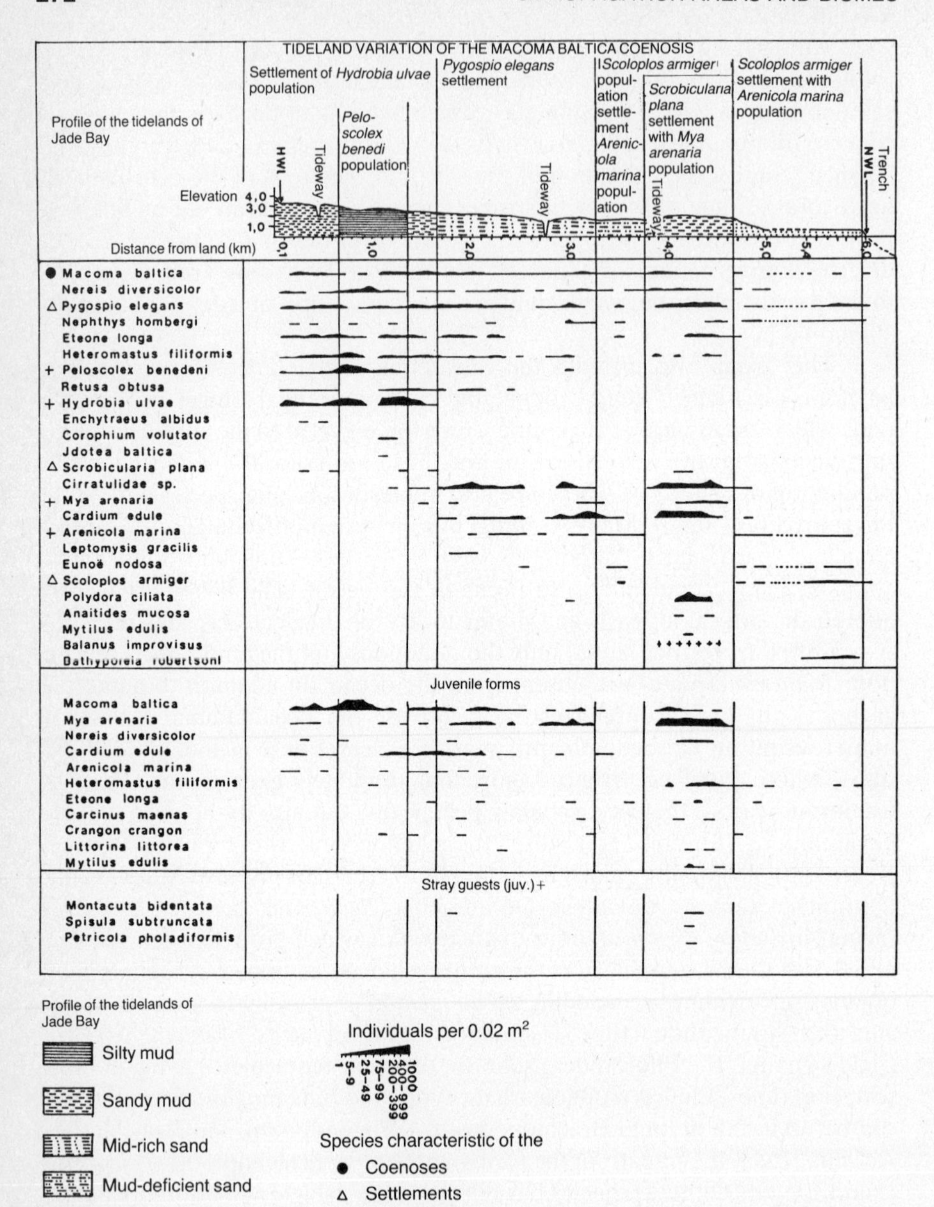

Figure 89a Profile of the tidelands of Jade Bay.

5.2.3 Ecological Macro-Structure of the Oceans

As in the case of a lake, a distinction must be made beteen two large realms of the ocean, i.e., the benthal and the pelagial. The greatest species abundance of the benthal lives close to the terra firma, i.e., in the littoral. The great

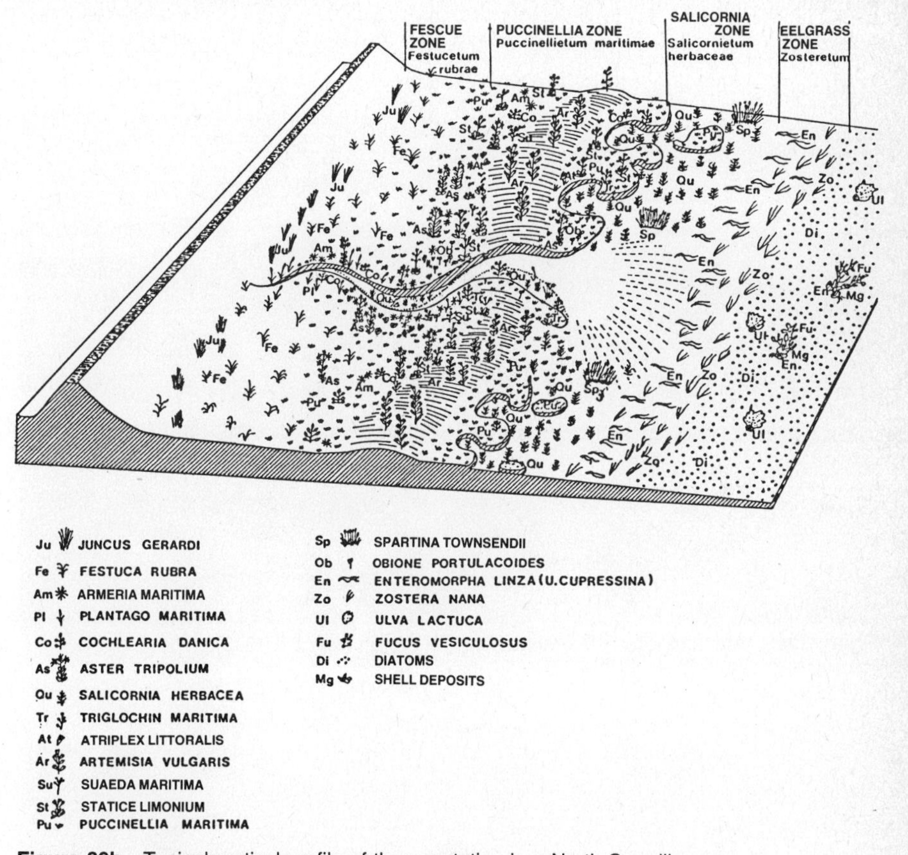

Figure 89b Typical vertical profile of the vegetation in a North Sea dike.

isolation among its biota occurs, by nature, in the form of terrestrial and oceanic barriers (East Pacific barrier, Transatlantic barrier, Euro-African barrier, American barrier). Within the littoral (neritic province), the biota of the tidal zone exhibits a vastly different composition from that of the sublittoral reaching as far as the eroded area of the shelf edge (200 m isobath). The low level of the Pleistocene glaciation is responsible for its shape.

This zone is characterized by an adequate light supply (euphotic). At 200 m depth, it changes into the archibenthal which, as the zone of dimming light, normally extends to a depth of 1,000 m. This is followed by the actual floor of the deep-sea, the abyssobenthal with its deeply incised canyons (Hadal: 10,860 m; Whitaker 1976) (the aphotic zone).

This classification, however, reflects only the ecological macro-structure. The littoral realm alone can be subdivided into a multitude of local life zones. The biota of the flat coasts (*Posidonia, Halophila, Hippocampus, Syngnathus*) change depending on the substrate (mud, sand, etc.). This also applies to the sandhole species. Rocky coasts with their seaweed forests exhibit a completely different physiognomic phenotype (Figs 90a, 90b, 91).

Figure 89c Mud tidelands in the eastern part of Amrun (July 23, 1979).

Figure 89d Female eider ducks (*Somateria mollissima*) with chicks in the tidelands of Amrun Island (July 21, 1979). Eider ducks are ocean birds skilled in diving; they are only rarely seen inland.

Figures 90a and **90b** Rocky tidelands on Runde Island (Norway, June 6, 1978). Kittiwakes (*Rissa tridactyla*) assemble during low tide in the drying pools.

Seagrasses are widely distributed in the littoral. 12 genera and 49 species are known. 7 species have a tropical dispersal pattern. In the case of the Hydrocharitaceae, *Enhalus, Thalassia,* and *Halophila,* and for the Cymodoceoiceae, the genera *Halodule, Cymodoceae, Syringodium,* and *Thalassodendron* belong to the tropical region (Fig. 92a).

The ecological border zone between land and sea is populated by a large number of semi-terrestrial animal and plant groups (Heidemann 1973, Schulte 1975, 1977).

Figure 91 Birds dominate the scenery on the Bird Mountains. Their dispersal on the islands (here: Runde Island, Norway, April 7, 1978) follows strict distribution rules which are determined by the ecological potential of the species. On the precipitous rocks close to the ocean (upper left, lower left), the kittiwakes (*Rissa tridactyla*) breed on the rocky promontories; whereas the shags (*Phalacrocorax aristotelis*) breed in the rock pebbles, the puffins (*Fratercula arctica;* right, center) in caves on the steep grass slopes; and the foolish guillemots (*Uria aalge,* upper right) and razorbills (*Alca torda,* lower right), on higher rocks with open access to the ocean.

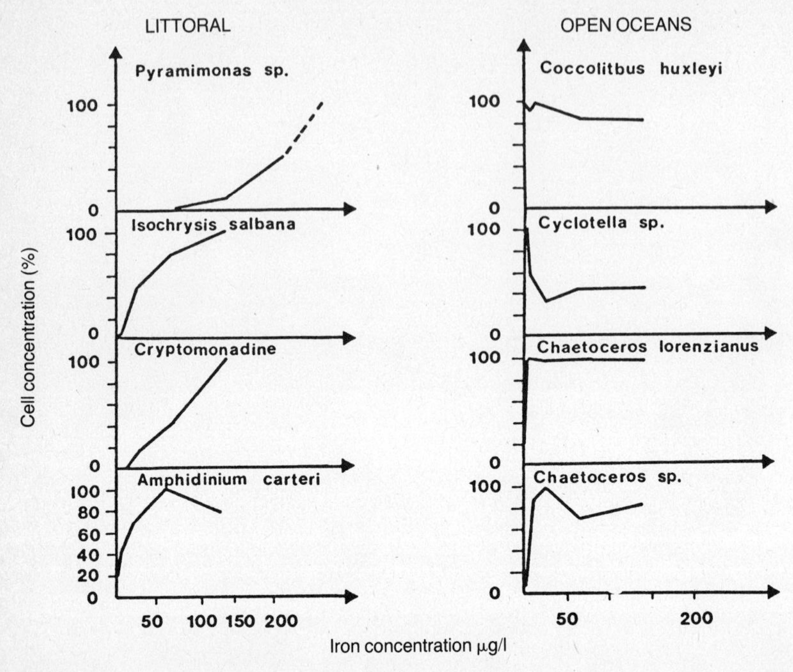

Figure 92 Varying rates of reproduction of algae in the marine littoral and pelagial (open ocean) after the addition of iron to the culture vessel (according to Ryther and Kramer, 1961).

The oceanic province, the pelagial, can also be classified according to the vertically diminishing light influence, i.e. into the epipelagial (to 200 m), the mesopelagial (200 to 1,000 m), and abyssopelagial (below 1,000 m). The

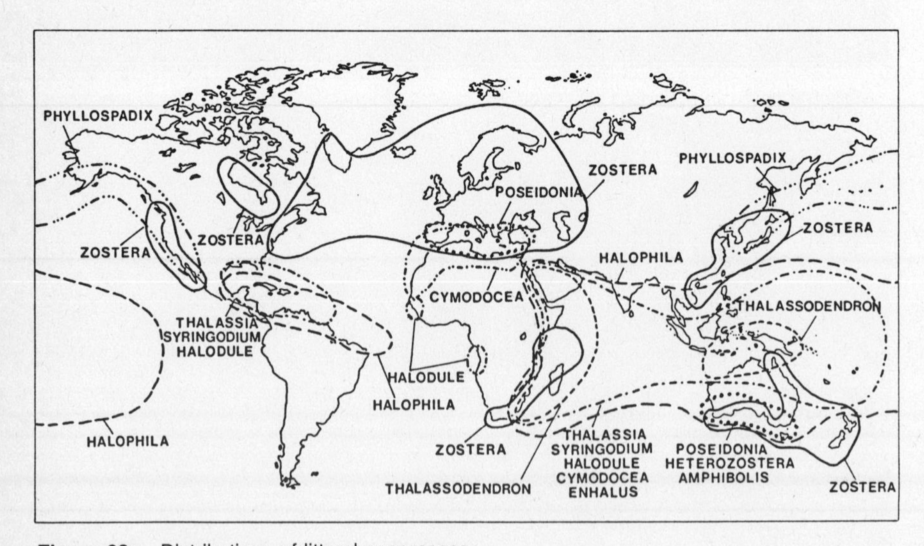

Figure 92a Distributions of littoral seagrasses.

TABLE 58 LIFE ZONES WITHIN THE REALM OF THE YUCATAN SHELF BELT

Community	Depth in m	Dominant species
1. *Acropora palmata*	0–10	*Acropora palmata; Millespora* sp.; *Palythoa mammilosa; Porolithon* sp.
2. *Diploria-Montastrea-Porites*	5–25	*Diploria* sp.; *Montastrea* sp.; *Porites astroides*
3. *Agaricia-Montastrea*	25–35	*Agaricia agaricites; Montastrea* sp.; *Solemnestrea* sp.
4. *Gypsine-Lithotamnium*	15–60	*Gypsine plana* (Foraminifera); *Lithotamnium* sp.
5. *Lithophyllum-Lithoporella*	20–60	*Lithophyllum* sp.; *Lithoporella* sp.

individual limits, however, can be shifted due to numerous factors (such as cloudiness, plant biomass, zooplankton, etc.). Along the border of the littoral, neritopelagic species occur which are characterized by special ecological capabilities. Some of them, for instance, are able to absorb much greater quantities of minerals transported in the rivers than their relatives in the open oceans.

5.2.4 Productivity

The productivity of individual ocean zones varies considerably depending on the abundance of nutrients in the ocean currents. Ryther assumes an average annual primary production of marine phytoplankton of 20×10^9 t C (Table 59).

Figures for total production (Firth 1969) which are based on the population fluctuations of the individual elements of the food chain — some of which are unknown — and on their horizontal and vertical movements are notoriously unreliable (Boney 1975, Garrod and Clayden 1972, Gulland 1977).

In 1929, Hentschel noticed — while performing plankton analyses near Ascension Island — the vertical fluctuations of the biomass. Since then,

TABLE 59 PRODUCTION OF INDIVIDUAL WATER ZONES IN THE PACIFIC (according to Nielsen 1975)

	Area (10^3 km^2)	Annual production (gC.m^{-2})
Oligotrophic water bodies	90.106	28
Mixed water zones	33.358	49
Equatorial water bodies and subpolar zone	31.319	91
Coastal region	20.423	105
Neritic zone	244	237

TABLE 60 AVERAGE BIOMASS (ZOOPLANKTON; mg/m^3) IN THE OCEANS OF THE SOUTHERN HEMISPHERE (according to Knox 1970)

	Antarctic	Subantarctic	Tropics	Subtropics
0– 50 m	55.2	55.8	33.1	40.5
0–1000 m	25.6	20.9	9.8	9.0

however, we have learned a great deal more about the distribution of albumen and chlorophyll concentrations of the oceans (Imori 1974, Sournia 1974).

Such fluctuations in populations can be particularly distinctive in shelf seas. Variable abiotic factors distinguish the North and Baltic Seas, with mean depths of 94 and 55 m respectively (Atlantic 3,300 m), at least in their edge zones. Tides in the Baltic Sea, when compared to the North Sea, are negligible. The North Sea, as an old depression (already existing as a shelf sea during the Upper Permian) with sediments of up to 6,000 m in thickness stemming from the postglacial Baltic Sea (cf. Baltic Sea Phases, section 4.1.4), exhibits very variable salt concentrations (35‰). As a result of varying abiotic factors, various plant and animal groups appear during certain seasons. There is a close connection between salinity and number of species. Distinct correlations also exist between the surface temperature and the seasonal occurrence of individual species (Fig. 92b).

TABLE 61 ESTIMATED BIOMASS (mg/m^3) AND FOOD CHAINS IN THE ANTARCTIC OCEAN (according to Holdgate 1970)

Phytoplankton	320.0	
Zooplankton		
Euphausia ("Krill")	50.0	
Others	55.0	
	105.0	
Consumers of the zooplankton		
Toothless whale	12.8	
Lobodon	1.6	
Birds	0.12	
Fishes	?	
Cephalopods	?	
Others excluding fishes and cephalopods	14.52	
End link in the marine food chain		
Toothed whale	0.5	
Hydrurga	0.026	
Other sea lions	0.15	
Birds	0.1	
	0.776	
Benthos		
Invertebrates	$400–500 \times 10^3$	ocean floor
Vertebrates (groundfish)	?	

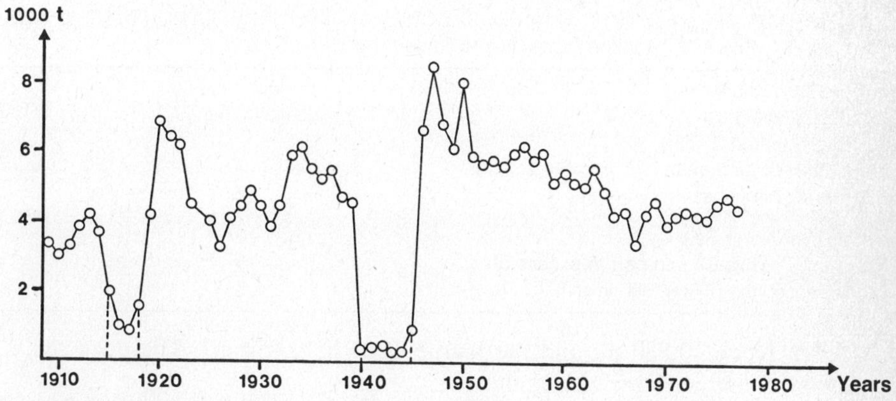

Figure 92b International landings of turbots from the North Sea (from *Bull. Stat.*).

5.2.5 Pollution

At present, the marine biota are undergoing a serious change (cf. Flowing Water Ecosystems, Chapter 4.2). They are turning into a collecting basin for radioactive substances (cf., among others, Riley and Skirrow 1975), for pesticides, oil wastes, heavy metals and numerous cumulative poisons (Gerlach 1975, 1976) (Tables 62 and 63).

In addition to the above, overfishing of the stock takes place in many parts of the oceans. As a result of this destructive exploitation, the average age of herring and cod catches in the Baltic Sea has decreased since 1923 (Magaard and Rheinheimer 1974) (Tables 64-65).

5.2.6 Abyssal Biomes

80% of the oceans belong to the deep-sea which is considerably richer than might be assumed from most graphic representations. Ocean ridges (⅓ of the ocean surface) and deep-sea mountains (with flattened peaks and terrestrial deposits — guyots) tower above the deep-sea plains; these elevations constitute a precipitous contrast to the deep-sea trenches and rifts (for instance the 2,900 km long Aleutian Rift; the Marianas Trench Deep 11,030 m; the Tonga

TABLE 62 MERCURY CONTENT IN OCEAN ANIMALS IN RELATION TO THEIR WET WEIGHT (according to Gerlach 1975)

Deep-sea edible fish	below 0.05 mg/kg
Coastal edible fish	0.1–0.2
Tuna fish	0.1–1.0
Deep-sea seals	3–19
Coastal seals	60–700
Aquatic birds	50–500
Tolerance limited for the content in edible fishes	
Mercury content of the sea water	0.03 μg/l
Toxicity possibly detectable at	0.1–0.2 μg/l

TABLE 63 DDT (AND METABOLITES) IN OCEAN ANIMALS IN RELATION TO THE WET WEIGHT (mg/kg) (according to Gerlach 1975)

Deep-sea edible fish	below 0.1
Coastal edible fish	below 0.5
Tolerance limit (FRG)	
Eel, salmon, sturgeon	3.5
Other edible fishes	2.0
Cod liver North Sea	1–3
Baltic Sea, in part in excess of	15
Tolerance limit (FRG) Fish liver	5

Trench Deep 10,880 m; and the Kurile Trench 10,050 m). The deep sea is devoid of any autotrophic organisms.

The fauna of the deep-sea (abyssal) exhibits a regional grouping which correlates with the oceanic deep-sea basins and trenches. Although the average depth of the water of the oceans is 3,800 m (average elevation of the terra firma only 840 m) and thus — in terms of volume — a large proportion of the ocean biotope pertains to the abyssal, the lack of light and high water pressure result in negligible species abundance. The species dependent on the abyssal are, however, characterized by a number of ecological and phylogenetic features. Deep-sea fishes often have photogenic organs and striking preying devices. *Chiasmodon niger* is able to dilate its stomach and skin to such a degree as to be able to accommodate fishes of its own body size. The pelican fish (*Eupharynx pelicanoides*) has a mouth which, in proportion to its eel-shaped body, is huge. The deep-sea shrimp (for instance, *Sergestes corniculum*) is equipped with antennae that measure several times the length of its body, and deep-sea crustaceans (*Cystosoma neptuni,* among others) have developed — in adaptation to the predominantly lightless biotope — excessively large eyes. As a result of the scarce food supply in the abyssal, its population density is negligible. In the case of the deep-sea monkfish (for instance, *Linophryne algibarbata*), this has led to a strong sex dimorphism. Dwarf males, which usually weigh less than 0.5% of the weight of the females, during the course of their development grow firmly onto the female under their jaw apparatus. The male is connected to the blood circulation of the female and its sole function is reproduction. Investigations into the number and structure of chromosomes have shown that the deep-sea fish, which belong to primitive Teleost families, have varying numbers of chromosomes and a varying DNA content. It is worth mentioning the tendency that

TABLE 64 AVERAGE AGE OF HERRINGS IN CATCHES FROM THE CENTRAL BALTIC SEA

Period	Average age (years)	Region
1923–1926	5.5	East Coast of Sweden
1957–1961	3.8	Bornholm Basin
1962–1966	3.1	Bornholm Basin

TABLE 65 AVERAGE AGE OF COD IN CATCHES FROM THE BALTIC SEA

Period	Western Baltic	Bornholm	Gdansk Deep	Gotland Deep
1929–1938	4.77	4.51	4.74	5.95
1939–1944	2.77	4.22	3.48	4.48
1946–1957	2.02	3.58	3.71	4.33
1960–1967	1.59	3.60	3.68	3.78

deep sea species have of often showing greater DNA content and greater numbers of chromosomes than their closest counterparts in the euphotic zone. The shallow water species *Argentina silus* and *A. sialis* (salmoniformes) have 2n = 44 chromosomes, whereas species of the closely related Bathylagidae (e.g. *Leuroglossus stilbius, Bathylagus milleri*) have 2n = 64 chromosomes (Ebeling, Atkin and Setzer 1971).

5.2.7 Coral Reef Biomes

In tropical seas there often exists a littoral symbiosis where the surrounding waters can be determined by the release of calcium from the coral reef; their dead massifs can often extend down to considerable depths. Coral reefs are essentially made up of skeletons consisting of calcium, of Madreporaria (Poritidae, Acroporidae, Astralidae), Octocorallia (Tubiporidae) and Milleporidae; Foraminifera and calcium algae (Lithothamnion) are, however, also involved. A prerequisite for optimal development of coral polyps and the zooxanthetlae that live in their cells are tropical seas with a mean annual temperature of 23.5°C, high Ca-concentrations and a high oxygen and salt content (between 30 and 40%). In areas where these conditions are not totally fulfilled the whole reef, for example, can be made up of calcium algae (algal reefs off the coast of Fernando Noronha, Brazil). Even worm snails can play a part in the reef's construction. They are able to build "vermitide" reefs (Brazilian coast), together with the bristle worms (Serpulidae, Sabellariidae). On the coral island of Funafuti (north of the Fiji Islands), a drill hole to a depth of 335 m established proof that the Foraminifera were the key participants in the construction of the reefs, whereas the corals occupied only third place.

The distribution of coral reefs is bounded by the 20°C-isochimene, a line which connects all locations with a mean winter temperature of 20°C. While this temperature condition limits the formation of Recent coral reefs to the tropical regions, fossil coral reefs have been described outside the tropics. Some well-known examples are: the Silurian reefs of Gotland and North America, the Devonian reefs in the Rhenish Slate Mountains and Ardennes, and the Triassic reefs in the Tyrolean Dolomites. Archaeocyathins, stromatopores, sponges, bryozoa, thick-shelled mussels, and calcareous algae also participate in the formation of the fossil reefs. No direct conclusion concerning individual palaeo-temperatures is possible. "The question as to whether the builders of these reefs had the same climatic requirements as today's reef-building corals cannot be answered on the basis of their biological systematic relationships, at least for for the pre-Tertiary forms. On the other hand, for different reasons — i.e., particularly because of the significant calcium deposits — it must be assumed that the large reefs of prehistoric times could indicate relatively warm water" (Schwarzbach 1974).

Many reef-building corals grow at low temperatures. Along the Norwegian coast, they reach 71°N at depths between 200 and 300 m at a temperature of 6°C. Although they do not form extensive reefs, they nevertheless build large banks. These banks, however, are very poor in species and are less light-dependent than the true reef-building corals.

One cause for the stronger growth of the reef-building corals in warmer water probably is the fact that the corals precipitate aragonite (and not calcite).

The presence of symbiotic, intra-cellular zooxanthellae within the tissue (similar to, for instance, the tropical giant tridacna, in hydrozoans, radiolarians, ascidians, poriferans, and platyhelminthians), limits the reef formation to a depth of 0 to 50 m and bonds the corals to the saltwater, leading to gaps in the reef wherever rivers empty into the ocean. The different formation of the coral reefs as fringing reefs deposited right along the coast, barrier reefs separated from the coast by channels (modified fringing reefs; for instance, the Great Barrier Reef along the north coast of Australia), or atolls encircling a lagoon can only be understood in a historical context — as Darwin pointed out in 1876 — and has contributed considerably to the knowledge of the fluctuations in sea level during the ice ages (Fairbridge 1962).

The development of the reefs is determined by a multitude of organisms. Four mechanisms must collaborate: layout of a supporting framework, cementing of individual building elements, sedimentation and cementing of the lime framework. Four principal reef types (fringing reef, barrier reef, platform reef, atoll) can be distinguished.

Fringing reefs are the most widely distributed type of reef. Deposited directly in front of the coast, their width is dependent on the structure of the ocean floor. In the case of advanced fringing reefs, a lagoon often forms in the direction of the land; the coastal fringing reef often changes into a lagoon fringing reef. The basic construction of the barrier reefs resembles that of the lagoon fringing reefs; they are, however, considerably larger. The lagoon can be many km wide and 30 to 70 m deep. A significant distinguishing feature between fringing and barrier reef is the history of their development. The barrier reef is located a long distance from the coast and does not constitute the remains of a fringing barrier slowly moving from the shore towards the ocean, but has always been in this location. Any depresssion in the substrate or an elevation of the water level furnishes the evidence for its bond to the site. The best-known barrier reefs are located in the region of the Great Barrier Reef along the Australian coast.

Platform reefs are not tied to any coast. They develop wherever an uplifting of the ocean bottom comes so close to the ocean surface (about 50

m) that reef-building corals can start their construction. Eventually, a platform reef grows evenly towards the outside. A breach in the central part of the reef may occur through erosion, making a distinction from the atoll type more difficult. This is how "pseudo-atolls" are formed.

Atolls are formed by a circular, closed reef whose interior contains a lagoon. The depth of this lagoon generally varies between 30 and 80 m, the depth being partly dependent on the diameter. The exterior slope can plunge down more than 1,000 m. The lagoon is connected with the open ocean through at least one passage (with the exception of very small atolls). The water which penetrates via the surf runs off through this passage. Often, several channels are present, particularly in the lee of the atoll. Because of the collection of sandy coral deposits, islands sometimes form on top of the reef circle. Since the reef circle is wider on the side facing the wind, most coral islands can be found here. The size of the atolls varies quite considerably; there are some with a diameter of only 1 km whereas others may measure up to 100 km. The calcium synthesis of the reef coral is controlled by the CO_2-absorption of the zooxanthellae and the CO_2-liberation of the skeleton-building cells of the polyps:

$$CO_2 + H_2O \rightarrow H_2CO_3$$

$$H_2CO_3 \rightarrow H^+ + HCO_3^-$$

Two HCO_3^- ions and Ca^{++} ions present in the seawater form $Ca(HCO_3)_2$ complexes:

$$2HCO_3^- + Ca^{++} \rightarrow Ca(HCO_3)_2$$

This $Ca(HCO_3)_2$ complex is in equilibrium with the liberated $CaCO_3$ and with the carbonic acid disintegrating into H_2O and CO_2:

$$Ca(HCO_3)_2 \rightarrow CaCO_3 + H_2O + CO_2$$

The standard expression for calculating an equilibrium constant K in this case gives:

$$K = \frac{[CaCO_3]\,[H_2O]\,[CO_2]}{[Ca(HCO_3)_2]}$$

When CO_2 is removed from this reaction, the concentration of the calcium carbonate must increase. Since the zooxanthellae absorb CO_2 through assimilations, they control the precipitation of calcium carbonate.

In the case of the polyps of reef-building corals, the above reaction process takes place in the basal section. Ectodermal cells release very fine chitin threads which fill the gaps between the polyps and complete the skeleton, and provide a structure — in the form of condensation centers and guide supports — for the calcium carbonate crystals present in the form of aragonite. Calcium carbonate synthesis is temperature-dependent (optimum 25°C to 27°C). Below 20°C and above 30°C, only small amounts of calcium carbonate are precipitated. The light intensity controls the photosynthesis of the zooxanthellae and thus also the precipitation of the $CaCO_3$ (to a depth of 50 m).

The growth rate of most coral species fluctuates between 1 and 26 cm (*Acropora cervicornis*) annually. Each reef has its own specific morphological structure. Often, roof, slope, and edge of the reef as well as the advance of the reef and lagoon are clearly developed; they are, however, often subdivided. The highest part of the reef roof is formed by the reef crown whose substrate consists of hard limestone structures.

The fauna of the coral reefs (Table 66) — in accordance with the tropical biotope — is rich in species. Among the fish species occurring here, the coral fishes (Chaetodontidae, Pomacentridae, Scaridae, Serranidae, Labridae) are especially worth mentioning.

Depending on the degree of their ecological dependence upon the coral reef, they can be divided into the following:

(a) peribionts which normally live close to the reef (coral fish), but which do not have any special morphological adaptations;
(b) parabionts which, at least part of the time, live between the branches of the coral colonies and have adapted correspondingly (crustaceans, for instance);
(c) epibionts which have grown firmly onto the surface of dead coral branches or onto calcareous algae (sponges, tunicates, among others);
(d) hypobionts living in the depths of crevices and shady reef locations (snails, sand stars);
(e) cryptobionts which are present in the interior of the reef colonies, boring themselves into them (boring clams, ship worms, for example); and
(f) endobionts which, the same as the zooanthellae, live in symbiosis with the reef corals.

Numerous consumers live from the production of the reef corals, including various fish genera (*Chaetodon, Chaetodontoplus, Oxymonacanthus*), as well as predatory snails and echinoderms.

The abundance of species is particularly great among the coral fishes, and there has been no lack of studies to establish parallels with the abundance of species in the tropical rainforests. There is no doubt that the coral reefs of

TABLE 66 NUMBER OF CORAL FISH SPECIES IN VARIOUS CORAL REEF BIOMES

Coral reef	Number of fish species
Bahamas	505
Hawaii	448
Marshall and Mariana Islands	669
Seychelles	880
Great Barrier Reef	1500
New Guinea	1700
Philippines	2177

the Indo-Pacific region surpass those of the Atlantic in this respect.

The "placard-like" designs of the coral fishes are correlated with their aggressiveness and territoriality. The design serves as a warning placard indicating the territory boundary. Morays, equipped with an excellent ability to smell, as well as the poisonous red fire fish lurk in the caves of the reef slopes; they go after their prey only at night. Out of the smaller niches, often dug by the animals themselves, wave the pinna-like arms of barnacles (Cirripedia), while slugs (Nudibranchia) and flatworms (Turbellaria), as well as sea urchins graze the surrounding coral colonies which are overgrown with algae. Crustaceans (many species of the Xanthidae, Pinnotheriidae, and Porcellanidae families), bristle worms (polychaeta, including the palolo-worm *Eunice viridis*) and gastropods creep among the sea anemones (for instance, *Discosoma, Actinodendron*), the Octocorallia (Sarcophytes, Lobophytes, Sinularia) and colonies of sponges and bryozoans (moss animalcules), while large fish schools dart by. Many predators of the high seas, such as barracudas, carangids and sharks, take advantage of the opportunity and hunt sporadically along the densely-populated reef edge and reef slopes.

Since 1962, the thorncrown starfish (*Acanthaster planci*) which can reach a diameter of 60 cm and feeds on the polyps of the corals, has been multiplying so fast within the Great Barrier Reef, on the Marshall Islands, Hawaii, and Guam that the reefs are seriously endangered. The cause of this mass development is not yet sufficiently known.

It is not only *Acanthaster,* however, which poses a problem to the coral reef biomes. Littoral sea pollution is becoming increasingly apparent and is destroying the formerly "blooming" coral gardens (cf. Johannes, 1975, for example).

Chapter 6

Evolution of Propagation Areas and History of the Landscape

The histories of landscape, climate, and organisms are interlinked. Biological, lithological and morphological indicators made important contributions to our present knowledge of the history of landscape realms. The close phylogenetic relationship between fossils and Recent forms or their ecophysiological features (for instance, flippers, drip leaf apex as an indicator of rainforest conditions; annual rings, reef formations) became indicators of the climatic developments of the realm in which they were found. We must acquire a knowledge of the structure of an area and its functional relationship to the ecological and genetic structure of the biosphere, as well as a knowledge of the ecological links between species and individuals, their population genetics and their spatial and temporal distribution. Only then are we able to understand the history of the area. Biogeography can only fulfill this task if it objectively examines the developmental history of areas and taxa, and the landscapes delineating the former. In this connection, it has primarily to deal with the history of propagation areas.

Works by Hennig (1950, 1966, 1969), Illies (1965), Brundin (1972), and Schmincke (1973, 1974) — to name just a few — have made it possible for phylogenetic systematics to become an important method for the clarification of the genesis of propagation areas. In this connection, however, it would be a mistake for biogeography to insist that "phylogenetic systematics" is its only method. It cannot start from the premise that the Recent area of a taxon (regardless of whether the taxon has different or similar characteristics to its ancestors or not) is identical to its center of origin or to one of the possible dispersal centers developed during the course of its evolution. The reconstruction of its phylogeny, i.e. the sequence through which the taxon's characteristics, and therefore the taxon itself, evolved presupposes the discovery of the groups belonging to each branch and a determination of the relative point in time of the occurrence of the branching in question.

For this method developed by Henning, recognizing the limits of its predictive possibilities, complicated taxonomic complexes and animal groups with many characteristics are of particular importance in order to distinguish between homologies and convergences. This statement also applies to biogeography in a metaphorical sense. The origin of an isolated area, of a monotypic genus, cannot generally be explained by chorological means, at least not in respect of the Tertiary development of the taxon. Only in the rarest of cases does the age of first appearance of a species at a specific site on Earth coincide with the geological age of that particular location. Apart from areas which exhibit a great spatial constancy over longer periods of time, there are those which have been characterized, until the most recent times, by remarkably great diversity of habitats. Therefore, it is important also from the point of view of biogeography to limit their chorological analysis (compare dispersal centers, section 6.5) first to such groups as are characterized by vast area diversity and by the great number of different shapes of their propagation areas. In order to reconstruct their phylogeny, fossils are unnecessary for "phylogenetic systematics", and dated fossils only serve to confirm the relative length of time sequences of the phylogenetic tree (synapomorphic scheme), of Recent forms. "Many zoologists, and probably most palaeontologists, are of the opinion that in reality, phylogeny can only be written with the aid of fossils, and therefore only for those animal groups for which an adequate number of fossils are known to us. This opinion, stated simply, is erroneous, even if we have to concede that it contains a grain of truth" (Hennig 1969 p. 13).

To explain area genesis, biogeography must make use of phylogenetic systematics as well as of palaeontology. In addition, however, it has its own chorological methods which, starting from the present, penetrate into the past. In the following, we shall illustrate in more detail how far we are able to explain history by means of chorological-biogeographical methods.

6.1 PALAEONTOLOGY AND HISTORY OF CLIMATE

The importance of *fossils* for the understanding of climatic history was already known in the 17th century. In 1686, the English physicist Robert Hooke inferred a formerly warm climate from the existence of fossil turtles and giant ammonites in Portland, and Lyell, in his *Principles of Geology*, related fossil Mesozoic reef corals to former tropical conditions in northern oceans. The indicator value of the fossils, however, is closely linked to the assumption of the approximate constancy of their ecological potential. Today, we know of the existence of numerous taxa for which we cannot make this assumption (Hennig 1969, Müller 1974, 1976).

Similar life forms of organisms were and are discussed whenever there is a problem in explaining the ecological relationships of areas. Since "similarity" does not necessarily mean relationship, here, too, caution must be exercized when undertaking historical reconstructions (Hennig 1949, 1960, 1969, Brundin 1972, Schmincke 1974, Zwick 1974).

Pollen analyses were first performed by Steenstrup in 1841. He recognized the connection between pollen succession and the climatic and vegetational history of an area. Today, pollen spectra and pollen diagrams are used to explain climatic successions in large areas (Frenzel 1960, 1968, etc.) (Fig. 93). The relatively great mobility of pollen (Stix 1975), the differential productivity of the different plant species (high pollen production in the case of *Pinus, Alnus, Betula;* low pollen production for *Quercus, Fagus,* and *Castanea*), and the differential preservation of the pollen make it clear that, in most cases, pollen diagrams only allow general statements to be made.

Without verification by means of pollen profiles, however, many of the theories which have been developed with regard to Pleistocene vegetation displacements within the Neotropics (compare, among others, van der Hammen 1974), or in New Zealand (cf. Harris and Norris 1972) would not be meaningful. Together with tree ring studies, they are excellent indicators of the postglacial history of the vegetation and thus also of the landscape (Schwarzback 1974, cf. also Olsson 1971).

While the significance of palaeontology for the understanding of geological history was recognized a long time ago, the biogeographic method based on the analysis of Recent propagation areas was generally ignored.

6.2 LAND BRIDGES AND THEORIES OF LAND DISPLACEMENT

If the existence of closely-related groups of organisms in various areas or continents is attributed simply to the presence of former land or water

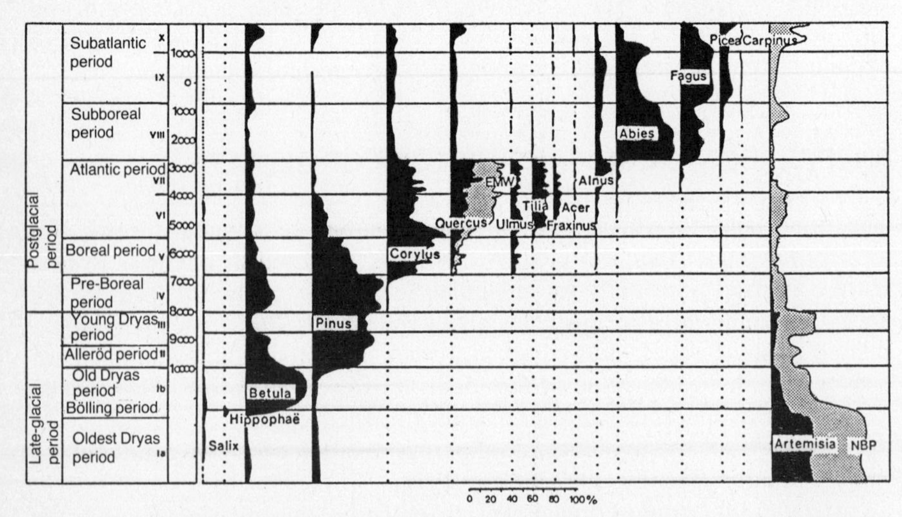

Figure 93 Postglacial vegetation history on the basis of pollen profiles from a bog in southern Germany (according to Lang, from: Müller 1977).

bridges, then there is not a spot on Earth which would not need to be spanned by a bridge. It is important to mention in this context that many of these bridges actually did exist. Their existence, however, cannot simply be proven by the existence of such groups. This often superficial "bridge-building" which has been particularly damaging to the science of biogeography (Hesse 1924). "The facts of geographic distribution, even the paths of migratory birds, constantly lead to new hypotheses of lost continents, sunken islands, or bridges across former oceans" (Thomson 1973).

A short description of some of the bridges "constructed" by biogeographers follows.

6.2.1 Land Bridges

According to Schuchert and Ihering (1927), *Schuchert Land* was a North American–Pacific mountain range which, during the Cretaceous period, connected North and South America, flanked by the Pacific Coast in the west and by the Missouri Basin (which was occupied by the Tethys Sea) in the east.

Sclater and Ihering claimed that *Lemuria* was a Cretaceous or Eocene land bridge between Madagascar and India. It was "created" in order to be able to explain today's disjunct distribution of the Madagascan and Indian lemurs. More recent biogeographic examinations have revealed the independence of the Recent Madagascan fauna (Günther 1970) and were able to furnish proof that most of it immigrated across the ocean (Peake 1971).

At the time of Buffon, *Beringia* had already been proposed as the land bridge between Asia and North America across the Bering Strait. According to more recent palaeontological, geological, and biogeographic findings, Beringia existed several times during the Tertiary and the cold periods of the Pleistocene. The exchange of fauna which took place *via* Beringia was of great importance to both the Old and the New World. Successive marine and land deposits on the Bering–Chukchi platform containing *Sequoia* and other taxa, show that land connections via the Bering Strait existed repeatedly during the early Tertiary. At the beginning of the Quarternary, however, Asia and North America were separated by marine flooding, the extent of which can be ascertained by means of an old marine cliff line (from the Arctic Coast of Alaska to the Yukon River). Because of a lowering of sea level during the cold periods of the Quarternary, however, Beringia was formed again; pollen-analytical findings, drill cores and radiometric dating show that the bridge was flooded during interglacial periods and that by the beginning of the postglacial climate optimum — after man had immigrated into America via the bridge — it had finally disappeared (Table 67).

Archiplata, according to Ihering (1927) is a land bridge which developed during the Cretaceous period between a South Pacific land bridge (Archinotis) and the Schuchert Land in today's Andes in South America. A faunal exchange took place during the Tertiary between North and South America via Archiplata.

Archiguiana represents an assumed Cretaceous island in the region of

TABLE 67 INTERGLACIAL AND LATER OCEAN TERRACES IN THE BERING STRAIT (according to Hopkins 1967, 1972)

Transgression	Elevation of the terrace	Climate in comparison with the present	Age in years	Stratigraphic classification
Krusenstern	About the same as today	Same as today	10,000–<5,000	Late Wisconsin to early Postglacial
Woronzof (incl. Bootlegger Cove Clay)	Probably a few meters below mean sea level	Colder	14,000–15,000?	Late Wisconsin
Peluk	+7 to 10 m	Warmer	ca. 100,000	Sangamon
Kotzebue	+20 m	Same as today	120,000	Pre-Illinoian
Einahnuhtan	ca. +20 m	Same as today	<300,000–>100,000	
Anvillian	<+100 m >+20 m	Warmer	Presumably <1,900,000–>700,000	
Beringian	?	Considerably warmer	Last phase before 2,200,000	Plio-Pleistocene boundary

Venezuela and the Guyanas. The distributions of numerous species of monotypic genera, for instance, the parakeet *Gypopsitta vulturina,* the cotingids *Haematoderus militaris* and *Perissocephalus tricolor,* and the tyrant flycatcher *Microcochlearis josephinae* are evidence for the existence of this "island". Modern biogeographical investigations showed, however, that the most recent shifts in the rainforests also influenced the distribution patterns.

The Mesozoic *Archhelenis* is supposed to have been a land bridge connecting the eastern part of South America with Southwest Africa via Tristan da Cunha. It was used in order to explain the similarities between the old African and South American faunas. Kossig (1944) showed that the African relatives of the South American fauna could be explained satisfactorily by means of other considerations. This applies to the spirostreptides (millipedes, Kraus 1964), the platannas, clawed toads, cichlids, Characidae, numerous parasites (*Nesoelecithus;* cf. Manter 1963), ostracods, Mutelidae, and the *Mesosaurus* species only known in fossil form (Kurtén 1967), whose areas include South America as well as considerable parts of Africa.

Archatlantis (Ihering 1927) was assumed to have been a Cretaceous land bridge connecting the Antilles and Florida with North Africa and Southern Spain (including the Azores, the Canaries and the Cape Verde Islands). The existence of Archatlantis is not necessary to explain the occurrence of sea cows on both sides of the Atlantic. They are, in any case, unknown as fossils in pre-Pleistocene deposits.

A *Canaries Bridge* has been claimed in recent times on the basis of geological, palaeontological (cf. Sauer and Rothe 1972), and phylogenetic relationships between the eastern Canary Islands and the African continental block. Discussions are still being conducted at present as to whether we are dealing with a bridge or a "mutual divergence". Separation of the eastern Canaries from Africa might have occurred through a rift, and a land connection might still have existed in the lower Pliocene (Sauer and Rothe 1972).

The *Thyrrhenian Bridge* was presumed to be a land bridge during the penultimate glaciation between Tuscany (Italy) and Corso-Sardinia (Corsica and Sardinia) and was necessary for the biogeographers of the past century and of the present to explain the close correlations of the herpetofauna (cf. Schneider 1971). Additional land bridges of varying age (for instance, between Corso-Sardinia and Africa = Galita Bridge; between Corso-Sardinia-Balearics and Spain = Balearics Bridge; between Corso-Sardinia and Provence = Provence Bridge) were assumed in this region in order to interpret the various faunal correlations. So far, in-depth analyses of the potential for passive dispersal of taxa of this region are missing. Pleistocene faunas in North America and Sicily argue against a land bridge between Sicily and the African continent, at least during the last glaciation.

Possible Tertiary or Pleistocene land connections from Crete have not been clarified so far. The Pleistocene mammal faunas of Crete show an older phase with *Kritimys* and a younger phase with *Mus minotaurus* (Kuss 1970). The *Kritimys* phase encompasses an older section including *Elephas antiquus* and the moderately dwarfed *Hippopotamus creutzburgi* and a younger section including an extremely dwarfed hippopotamus. The *Mus minotaurus* phase, too, is divisible into two periods. At the beginning of this phase, *Elephas creutzburgi, Mus minotaurus,* and probably a hominid, immigrated after *Kritimys, Hippopotamus* and *Elephas creticus* had become extinct.

A Cretaceous and early Tertiary land bridge postulated by biogeographers of the past century is *Archinotis*. It connected the southern part of South America with New Zealand and Australia via Antarctica and the South Pacific islands. This land connection is still being claimed by today's biogeographers in order meaningfully to explain the close correlation between many animal groups distributed on both sides of the South Pacific (Brundin 1966, Illies 1965, Mertens 1958, Müller and Schmidthüsen 1970, Noodt 1977, Schmincke 1974). This applies, for instance, to the southern beech (Nothofagus), the freshwater crab family Parastacidae, the stone flies of the family Eustheniidae, the chironomids, land snails (Bulimulidae) and the snake-necked turtles (Chelidae).

6.2.2 Continental Drift

Geologists have recently unearthed some decisive facts supporting Wegener's (1912) theory of continental drift. Gondwana thus obtained new prominence (Hallam 1973, Plumstead 1973, Reyment 1972, Vandel 1972, Gosline 1972, Cracraft 1972, Colbert 1972, Patterson 1972, Paulian 1972, Axelrod 1972, Tarling and Runcorn 1973).

In 1912, when A. Wegener presented his idea about a former connection between Africa and South America and about the formation of the Atlantic, the facts and foundations upon which such a hypothesis could be based were rather sparse; hence the lively discussions. "Of great importance, however, is (and this is to Wegener's lasting credit) the realization that every geotectonic hypothesis which does not include the vast oceanic areas in its conceptual framework but instead limits itself to the analysis of the Sialic continental blocks and which extrapolates from there, must be incomplete and insufficient" (Beurlen 1974).

The Atlantic which today gives an impression of such uniformity has only grown since the Upper Cretaceous period out of four different old parts — South, Central, North Atlantic, Scandic — each with a different history. North Atlantic and Scandic both have a Palaeozoic prehistory; both, however, are a new formation, independent of the Palaeozoic North Atlantic (Beurlen 1974). The disintegration of East Gondwana, which started towards the end of the Palaeozoic Period, and the movement in a westerly direction of West Gondwana resulting from it, had a decisive effect on the formation of the Atlantic Ocean.

Gondwana was a Palaeozoic continent in the southern hemisphere, encompassing South America, Africa, Madagascar, India, and Australia. In Wegener's opinion, it was torn apart by continental drift. Its positions provide a key for explaining the distribution of numerous Mesozoic animal and plant species (e.g. for the Gondwana flora with *Glossopteris* and for the *Mesosaurus* group). The primitive continent Laurasia (North America, Greenland, Scandinavia and parts of Siberia) was located northwest of Gondwana. According to Wegener, until the Mesozoic period, Laurasia consisted of two continental blocks closed off by the Tethys Ocean, one of which included the northern parts of North America (Laurentia), whereas the other was a forerunner of Eurasia (Angaria). The history of evolution of the Permian and Mesozoic reptiles can be correlated with the assumed degrees of separation of the continents during these eras. While the chelonians (*Eunotosaurus,* for instance), the Mesosauria, Eosuchia, Rhynchocephalia, and Ornithischia underwent their development and segregation in Gondwana, the Sauropterygia and Therapsida had their origin in Laurasia.

The significance of the Laurasian and Gondwanan landmasses for biogeography was disputed for a long time. Nelson (1969) pointed out that the value of historical biogeography depends on the "reconstructability" of its results. The results of geophysical analyses performed in the past few years have been confirmed in fundamental work by Hennig (1960), Brundin (1966) and Illies (1965).

Hennig was responsible for turning the evolutionary history of species into an indicator of the development of land areas and their habitats. His "phylogenetic systematics" has a deeper meaning for the science of biogeography (Brundin 1972, Peters 1972), although, by itself, it is not able satisfactorily to clarify chorological problems (for a critique, see Darlington 1970, Nelson 1969).

6.3 EFFECTS OF THE LAST GLACIATION ON PROPAGATION AREAS

When considering the temporal development of propagation areas as discussed in connection with land bridges, it becomes clear that our knowledge decreases gradually as we pass from the Holocene through the Pleistocene all the way to the Archaean period (Table 68).

While it is difficult for us to estimate the effects of, for instance, climatic fluctuations on propagation areas during the Tertiary (Glenie *et al.* 1968, Goodell *et al.* 1968, Rutford *et al.* 1968, Tanner 1968), the Pleistocene glaciations and the drop in temperature and sea level had a lasting influence on propagation areas (Lamb 1971, DeLattin 1967, Frenzel 1967, Matsch 1976). Even tropical ecosystems were not spared, although in those regions the effect was less a change in temperature than in humidity.

The reafforestation and resettlement of glaciated areas did not proceed in such a manner that the zones which had retreated pushed the vegetation and its communities as an entity northwards into the formerly glaciated areas; rather, new commuities developed from species which either again immigrated or which arrived for the first time in the ice-free area (Table 69) (Knapp 1974, Chaine 1973, Hoffman and Jones 1970, Schultz 1972, Martin 1970). This is also true for animals, which even during the Pleistocene greatly affected the composition of the vegetation. Thus, for instance, the high grass pollen proportions in interglacial deposits of the lower Thames Valley can be correlated with the occurrence of a graminivorous mammal fauna including *Hippopotamus* (Turner 1975).

Eustatic fluctuations in sea level occurred, partially correlated with the climatic fluctuations (Milliman and Emergy 1968, Milliman and Summerhayes 1975). During the last glaciation, the sea level along the South American East Coast — as was the case in other oceans — was at least 80 m below today's level (Alt 1968, Stevenson and Cheng 1969, Griggs and Kulm 1969) (Fig. 93b and Table 70).

Considerable shifts of river courses and enormous fluctuations in the sea level took place (compare Lakes, Flowing Water Bodies, sections 4.1 and 4.2). During the Midle Miocene, the Nile Delta of today was still a wide bay of the Mediterranean. The Pleistocene is characterized by tectonic

TABLE 68 GEOLOGICAL PERIODS (in million years)

Era	Period	Epoch/Series	Million years
Cenozoic	Quaternary	Holocene	0.01
		Pleistocene	2–2.5
	Tertiary	Pliocene	
		Miocene	25
		Oligocene	
		Eocene	
		Paleocene	65
Mesozoic	Cretaceous	Senonian	
		Turonian	
		Cenomanian	
		Gault	
		Neocomian	135
	Jurassic	Malm	
		Dogger	
		Lias	180
	Triassic	Keuper	
		Bunter	220–225
Palaeozoic	Permian	Upper Permian	
		Lower strata of new red sandstone	280
	Carboniferous	Upper Carboniferous	
		Lower Carboniferous	340–355
	Devonian	Upper Devonian	
		Middle Devonian	
		Lower Devonian	400
	Silurian		440
	Ordovician		500
	Cambrian		570–600
Precambrian	Algonkian		2000
	Archaean		>3000

movements in the area of the Red Sea. Its present appearance developed only during the past 12,000 years (Rzoska 1976). Lake Victoria is well known palaeolimnologically. Although many authors "give" it a Miocene age, ^{14}C-data show that it was not formed until the Pleistocene (Kendall 1969, Butzer *et al.* 1972).

7,500 to 5,500 B.C., the water level of Lake Rudolf was still 70 m higher than today's level. Fluctuations in the vegetation occurred in line with this development (Bonnefille 1972).

While propagation areas were greatly influenced by the Pleistocene, its evolutionary significance was mainly at the intrageneric level. Fundamental, phylogenetic innovations are usually related to expansion into totally new biomes. The biomes existing today, however, were already present during the

TABLE 69 GLACIAL TEMPERATURE DECLINE IN MIDDLE LATITUDES (according to Schwarzback 1974)

Climatic evidence	Temperature decline in °C Year	Summer	Author
Dryas flora in Central Europe	10		Gabel 1923
Coleoptera (England)	13	7	Voope 1971
Decline of the snow line in the Alps	6		A. Penck 1938
Tundra polygons in England	13.5		Shotton 1960
Ice wedges in Central Germany	11		Soergel 1936
Pinus koraiensis in Japan	7.5		Miki 1956
Decline of the snow line in Japan		4.5–6.5	Hoshiai 1957
Picea glauca and *P. mariana* in Texas		8	Potzger and Tharp 1947
Glacial advance over living forests in Ohio	15	11	Goldthwait 1959
Decline of the snow line in Colorado	5.5		Antevs 1954
Frost crevasses in Montana	8		Schäfer 1949
Marine coastal fauna in Massachusetts	6		Gustavson 1973
Periglacial phenomena in Lesotho (Africa)	5.5–9		Harper 1969
Decline of the solifluction limit in Australia and Tasmania	9	5	Galloway 1965
Decline of the snow limit in New Zealand	5 –7		Willett 1950 Gage 1966

Pleistocene. Brodkorb (1971) even assumes that the number of bird species (in accordance with the number of families) during the Pleistocene and the Middle Tertiary was greater than it is today. In this connection the Quaternary, encompassing 2 million years, is considered to be a single unit which, however, it was not in reality.

	Families	Bird Species
Recent	148 (95 passeriforms + 53 nonpasseriforms	8,656
Pleistocene	153 (100 + 53)	10,653
Pliocene	154 (101 + 53)	10,705
Miocene	155 (102 + 53)	10,753
Oligocene	130 (82 + 48)	8,157
Eocene	94 (70 + 24)	5,164

Of the 848 Recent nonpasseriform genera, 10 are known from the Oligocene, 42 from the Miocene, and 34 from the Pliocene. 245 Recent

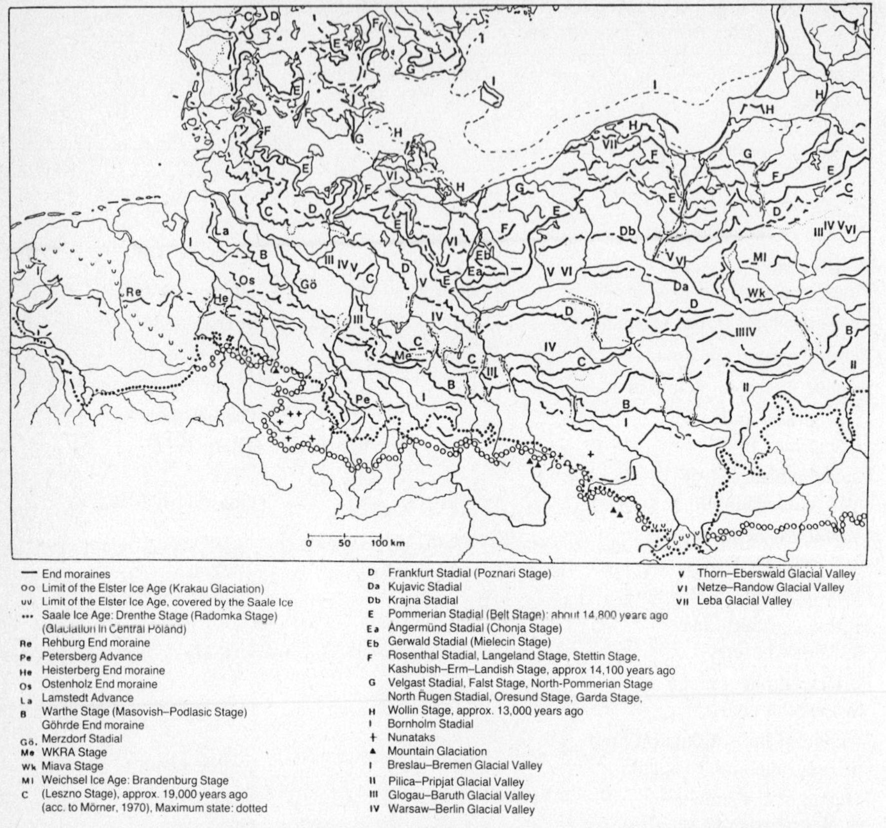

Figure 93a Major locations of Moraine and glacial valleys during the North European Glaciation in the Central European lowlands (according to Liedke, 1979).

genera have been established as also occurring in the Pleistocene. The passeri-
forms are less well documented by fossils. Of the 1,420 genera living today, two are known from the Miocene, two from the Pliocene, and 144 from the Quaternary. Of the 207 bird species from the Pleistocene in the U.S.S.R., 11 species are extinct, whereas the others are identical with recent species (Burčak-Abramovič 1975).

A large number of recent mammals was present in North America during the Pleistocene, including the following groups which in the meantime have become extinct there:

1. Elephants (*Mammoteus*/*Mastodon*)
2. Bovids (*Bison, Euceratherium*)
3. Reindeer (*Ranifer*)
4. Moose (*Cervalces*)
5. Ground sloths (*Parmylodon*)
6. Giant armadillos (*Chlamydotherium*)
7. Water hogs (*Neochoerus*)

TABLE 70 SIZE OF RECENT AND PLEISTOCENE ICE MASSES (according to Flint 1971)

Present	Pleistocene	10^6 km^2
Antarctic		12.6
Greenland		1.7
	Antarctic	13.8
	Greenland	2.3
	Laurentian Ice (N. America)	13.4
	Cordilleran Ice (N. America)	1.6
	Scandinavian Ice (+ England)	6.7
	Alps	0.04
	Asia	4.0
	Southern South America	0.7
	Australia and New Zealand	0.03
Whole Earth		15.0
	Whole Earth	44.4

8. Short-tailed bears (*Arctodus,* comparable to the polar bear)
9. Sabre-toothed tigers (*Smilodon*)
10. Canids (*Canis*)
11. Giant beavers (*Castoroides*)
12. Tapirs (*Tapirus*)
13. Perissodactylians (*Equus*)
14. Camels (*Camelops*)
15. Pronghorns (*Breamerys*)
16. Peccaries (*Platygonus*)
17. Musk-oxen (*Bootherium*)

During the same period, the genera *Castor, Rangifer, Ovibos, Coelodenta, Alces, Macaca, Hemitragus, Cervus, Canis, Saiga, Equus, Mammoteus, Megaceros, Hippotamus, Panthera* and *Tapirus* occurred in Eurasia.

The history of the Central American biomes is of particular significance for an understanding of faunal exchange between North and South America. Numerous pollen analyses have shown that the last glacial and postglacial periods of Central America are characterized by changes in vegetation and climate (Bartlett and Barghoorn 1973, Müller 1973, among others).

Around 14,000 years ago the temperatures in the Canal Zone of Panama were about 2.5°C lower than they are today. This has been concluded from the occurrence in the lowlands of types of pollen which occur only in the mountains during the recent period. Around 9,300 years ago the climatic conditions were drier. "Pollen from the interval from about 9,300 B.P. to 6,200 B.P. suggests a drier, more seasonal, and perhaps cooler climate during this period" (Bartlett and Barghoorn 1973). During this period, the first cultigens appeared in the Canal Zone. Dated *Zea* pollen is 7,300 to 6,200 years old.

The precipitation conditions changed world-wide during the last glaciation (Lamb 1971). How this manifested itself in small areas is still not sufficiently well known (compare Krolopp 1969). The idea that during the Pleistocene the tropics were climatically stable is wrong. We know today that vast dry and rainy periods alternated during the past 2 million years in South America, Africa, New Guinea, and Australia. Strong climate and vegetation shifts have also been shown for Madagascar (Battistini 1972), although by no means all of the associated problems connected therewith have been satisfactorily explained so far. During the Quaternary of Madagascar (named the Aepyornian after the occurrence of *Aepyornis* and other large ratites) obviously three pluvial periods occurred. ^{14}C-dates from the last 'pluvial period' (the Lavanonian Period, which corresponds to the last glaciation) on continental mollusc strata (primarily *Tropidophora*) located at a depth of 2.5 m, indicate that they are 32,600 years old. The top stratum, abundant with fossil *Aepyornis* eggs, is 6760 ±100 years old. During the Quaternary, huge dune formations took place in Madagascar (Brenon 1972). Subfossil giant

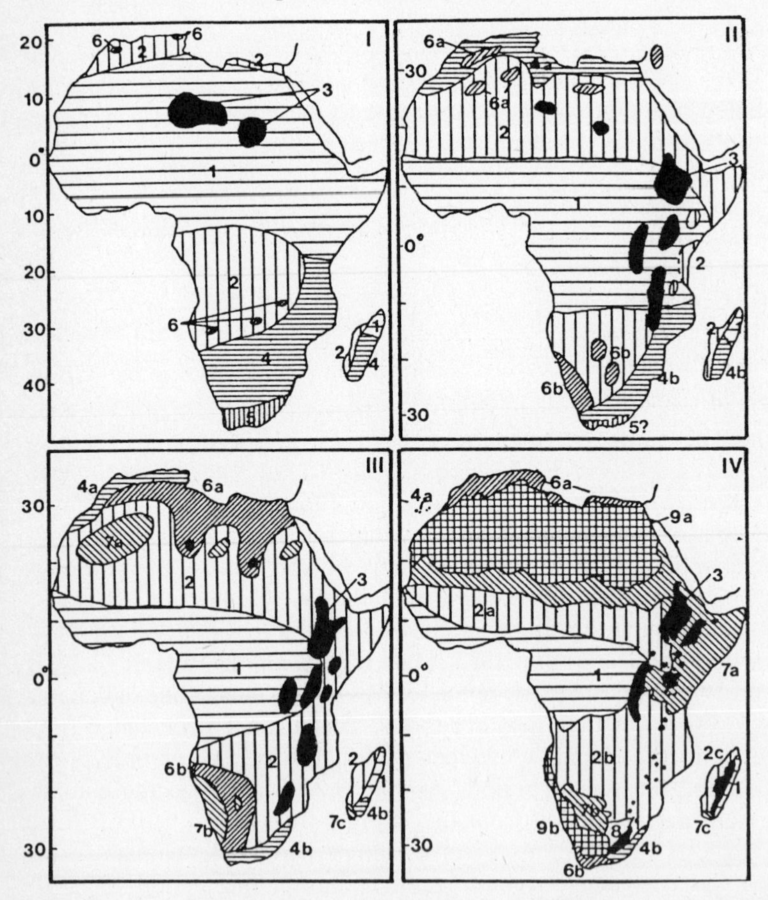

Figure 94 Schematized dynamics of the distribition of Ethiopian biomes since the beginning of the Tertiary (according to Greenway, 1970). I = Upper Cretaceous and Paleocene; II = Oligocene to Miocene; III = End of Miocene; IV = Present; 1 = Tropical forests of the lowlands; 2 = Savannas and dry forests; 3 = Montane forests; 4 = Subtropical forests; 5 = Non-tropical forest types; 6 = Sclerophyllous vegetation.

lemurs, hippopotamus and giant birds (*Aepyornis, Müllerornis*) were still alive during the first colonization by man who started early on with the destruction of the forests and triggered immense erosion processes.

On the basis of examinations made in bird and mammal habitats, Eisentraut (1968, 1970) and Moreau (1963, 1966, 1969) postulated far-reaching vegetation shifts in tropical Africa which were of decisive importance for the speciation of the African fauna (compare the distribution patterns with Bernardi 1969, 1974) (Fig. 94).

At the beginning of our century, fluctuations in African vegetation were already being claimed (Lönnberg 1918, 1926, 1929; Moreau 1931, Chapin 1932, Braestrup 1935) in order to be able satisfactorily to explain the speciation of the savanna and rainforest taxa. Isolated rainforest islands played an important role in this process. "The faunas of these isolated patches show very close affinities to that of the great western rainforest, having many forms in common (for instance, *Anomalurus, Perodicticus*) which could not possibly have crossed the vast intervening steppes. From this Lönnberg concluded that the said forest islands must have once been continuous with the main western forest owing to a moister climate" (Braestrup 1935).

Gentilli (1949) and Keast (1959, 1961, 1968) also suspected that a large part of the subspecies and species formation of Australian taxa was due to the isolation of populations during "drier" climatic conditions along the edge of the continent (compare also Horton 1972), and Pianka (1972) required for the evolution of Australian reptiles "habitats fluctuating in space and time." The establishment of a chronology for this drought period determining the distribution centres of Australia is still awaited.

In 1954, Rambo recognized the varying spatial and historical influences which give the flora of the Rio Grande do Sul its special charm. In his *Análise histórica da Flora de Pôrto Alegre* (Rambo 1954), he enumerated 1288 phanerogams (about 28% of the phanerogam flora of Rio Grande do Sul), the majority of which he subdivided into 5 'centres of origin':

1. Flora of the rain forest	176 species
2. Mountain flora	25 species
3. Flora of the Northeast (Chaco)	36 species
4. Flora of the Central and South Brazilian Campos	677 species
5. 'Insular Flora' (between Buenos Aires and Porto Alegre)	234 species
	1,148 species

The 'Insular flora', characterized by a considerable proportion of endemic species, is of special interest. Rambo (1954) subdivided it into three groups:

1. Endemic species of Andean affinity	74 species
2. Species of the Southern and Central Andes	23 species
3. Species with affinities to the Central and South Brazilian campos	90 species
	(uncertain 47 species)

He interpreted this composition of the flora by assuming:

1. The existence of a former Tertiary island landscape between Buenos Aires and Porto Alegre, within which the evolution of the endemic "insular" flora took place. Its original settlement took place "from the Andes, from the north coast of the Chaco Sea, and from the South Brazilian highlands."
2. "After the fusion of the islands to the mainland, a stronger immigration took place, in which the Brazilian campos supplied the majority of the total; the rainforest was the last unit which today is in the process of advancing slowly."

Of interest in his analysis of the insular flora is the statement that rainforest elements are completely missing from the endemics (Rambo 1954). Sehnem (1977), who analyzed the migration paths of the South Brazilian ferns arrived at similar conclusions.

6.4 ISLAND BIOGEOGRAPHY

Islands are isolated ecosystems. The fact that they can be viewed as a unit permits us to use them to test the dynamic processes taking place in propagation areas. Fundamental biogeographic processes, such as the history of colonization, dispersal, the effect of competitive factors, adaptation problems, displacement rates and extinction rates can be followed more closely on islands.

6.4.1 Island Theory

Independently of whether islands are of oceanic or continental origin, MacArthur and Wilson (1963, 1967) based on Preston (1962) developed their "theory of island biogeography", which suggests implies that an equilibrium exists between the number of newly immigrating species on an island and that of species becoming extinct. Even assuming that there is only approximate equilibrium between the rates of immigration and extinction, this assumption can be very useful. It permits prediction and controlled experiments. The success of the original colonization of an island is dependent on the size of the island and its distance from the source biota. The gradient of the colonization curve can consequently vary very considerably. The level it finally reaches depends on the size of the island and its ecological variability. Additional species can only appear once the extinction of the species already present has started. The equilibrium theory was experimentally tested by Wilson and Simberloff (1969). (Compare also Schoener (1974, 1976), and Simberloff (1972, 1976), and see Table 71.)

Close correlations exist between the size of the island, species diversity, and population density (Diamond 1970, Krebs, Keller and Tamarin 1969, MacArthur *et al.* 1973), and also between size of the island and number of species. The latter can be expressed by the formula $S = CA^Z$ (S = number of species, A = area of the island, C = population-dependent factor, Z =

TABLE 71 NUMBER OF ARTHROPOD SPECIES AS A FUNCTION OF THE SIZE OF THE ISLAND AND OF TIME ON SMALL MANGROVE ISLANDS (The Island Faunas were exterminated by methyl bromide at the start of the Test) (Sugarloaf Islands, Florida; according to Simberloff 1976)

	Number of species (in parentheses = days after gassing)			
Year	Island I	Island II	Island III	Island IV
0	20 (0)	35 (0)	22 (0)	25 (0)
1	13 (360)	32 (371)	28 (379)	28 (322)
2	17 (726)	37 (730)	27 (725)	28 (691)
3	19 (1235)	48 (1256)	33 (1239)	30 (1167)

empirically determined parameter in the range between 0.2 and 0.35). See Tables 72 and 73.

The island biota can, depending on the knowledge of the fundamental, ecological relationships between surface area and number of species, be used as indicators for the distributional dynamics of related populations on the continent (Müller 1969, 1972). This, however, can only be done via a clarification of the genetic structures of the island and mainland populations.

6.4.2 Oceanic Islands

Oceanic islands never had land connection with the continents during their evolutionary history (Wallace 1880).

Surtsey A very good example of this is the Island of Surtsey (63°18′N, 20°36′30″W), with an area of 2.7km^2, located 30 km south of Iceland; it came into being on November 14, 1963, as a result of a submarine volcanic eruption. The natural colonization on the island, which was originally devoid of any life, has been studied since 1964 by numerous biologists and since 1965 by the Surtsey Research Society (Schwabe 1970). The results of their work have been published since 1964 in the Surtsey Research Progress Reports in Reykjavik. The island closest to Surtsey, which is somewhat smaller, is Geirfuglasker, 5 km away. The first insect discovered on the island on May 14, 1964, was a chironomid, *Diamesa zernyi.* In the autumn of 1964, one noctid, *Agrotis ypsilon,* was captured. By 1968, 70 living arthropods were collected on the island, among which the Diptera dominated with 43 species.

TABLE 72 CORRELATION BETWEEN SIZE OF ISLAND AND NUMBER OF SPECIES OF THE MOLLUSC GENUS *PARTULA* ON THE SOCIETY ISLANDS (according to Purchon 1977)

Island	Area in km^2	Number of *Partula* species
Bora Bora	8	1
Huahine	19	5
Tahaa	32	5
Moorea	40	10
Raiatea	60	21
Tahiti	350	8

TABLE 73 SURFACE AREA AND FLOWERING PLANTS ON WEST AFRICAN ISLANDS (according to Williams 1964)

Island	Area in km^2	Number of species
Fernando Po	2000	826
San Thomé	1000	556
Principe	126	276
Annobon	16	115

In contrast with the arthropods which were carried onto the island primarily by the wind, the five higher plants identified by 1968 arrived on Surtsey via the ocean. These include *Cakile edentula, C. maritima, Elymus arenarius, Honckenya peploides,* and *Mertensia maritima.*

No lichen species had reached Surtsey by 1968. On the other hand, numerous land algae had been identified on this small island. Surtsey is an excellent example of how a new ecosystem can develop accidentally on a sterile island.

Krakatau Another example (compare Dammerman 1922, 1948) is the volcanic island of Krakatau (elevation 832 m), situated 40 km west of Java. On August 26/27, 1883, a volcanic eruption destroyed the plant and animal life (the island had been covered with forest up to the summit), and the original surface area covering 32.5 km^2 shrank to 10.67 km^2. In 1886, 27 higher plants were identified on the remaining island, whereas in 1897 the number had already increased to 62 and by 1906 to 114, the majority of which had dispersed through the air. The nearest land is Sibesia island which is located at a distance of 18.5 km from Krakatau. In 1889, 40 arthropods, 2 reptiles and 16 bird species were present, and by 1923 almost 500 arthropods, 7 land gastropods, 3 reptiles (among them one snake), 26 breeding birds, and 3 mammals (2 bats, 1 rat) were identified. Surtsey, Krakatau and many other similar cases (Diamond 1974, among others) illustrate with their species composition that organisms possess highly variable dispersal capabilities during their individual developmental phases. At the same time, they also illustrate how the "ecological potential" decisively determines the dispersal process of a given species. Although many species would be capable of making use of their dispersal capabilities (flight, for instance), they don't, whereas others apparently occur regularly in the aerial plankton. Such islands contribute to our understanding of the development of other island biota whose chronological classification often presents difficulties (Günther 1969).

Hawaii In addition to Hawaii (10,438 km^2), the Hawaiian islands include the following larger islands: Maui (1,885 km^2), Oahu (1,564 km^2), Kauai (1,437 km^2), Molokai (673 km^2), Lanai (365 km^2), Niihau (186 km^2), and Kahoolawe (116 km^2). The Hawaiian group is separated from North America, the Aleutians, and Japan by 3,900 to 4,200 km.

The 2,800 km long volcanic Hawaiian ridge, with an elevation of up to 8 km, developed over the course of the past 100 million years in a fault zone in the North Pacific running from west-northwest to east-southeast; its location in relation to the sea level repeatedly underwent important changes. Due to a tectonically-dependent drop in the level of the ocean bed of a total of 2 to 3 km, a number of volcanic island groups has already ceased to exist, and in the case of others, only the highly eroded remains still jut out of the ocean. It is only the volcanoes whose primary building phase took place in the Quaternary and, in some instances, still takes place today, which form significant landmasses. Thus, Mauna Kea and Mauna Loa rise 4,205 and 4,170 meters above the sea level respectively, and, together with the 3,056 m high Haleakala, form the group of high volcanoes which stand out from all the other massifs by their many habitat zones.

Hennig (1974) was of the opinion that a small group of the endemic vascular plants had already immigrated during the Upper Cretaceous — Tertiary Period via a large number of volcano islands functioning as stepping stones.

Erythrina sandwicensis (Leguminosae), *Osmanthus sandwicensis* (Oleaceae), two characteristic trees of the leeward dry forests, and the screw palm liana *Freycinetea arborea* (Pandanaceae) of the humid forests are represented by the same species on all of the islands. Each of the five large islands had its own *Gunnera* species which all originated from a single species which immigrated from the Neotropics (Table 74).

The lobelias possess fibrous, narrow, often very colorful blooms to which the beaks of the honeyeaters of the family Drepaniidae have adapted perfectly. The 2,000 newly imported plant species (the secondary flora) completely changed the original vegetation pattern, particularly in the arid zones.

Noteworthy endemic animal families are the snail families Achatinellidae and Amastridae, and the bird family Drepaniidae (11 genera, 21 species), whose evolution has been discussed by Mayr (1943) and Amadon (1950).

Amerson (1975) examined the avifauna of smaller atolls in the northwest of the Hawaiian island group. It could be demonstrated in all the cases that the abundance of species depends greatly on the structural diversity of the vegetation (Table 75).

Zimmerman (1948) provided an overview of the Hawaiian invertebrates. Freshwater fishes, amphibians, and numerous invertebrates (including the snail *Achatina fulica*) were introduced by man.

TABLE 74 ORIGIN OF THE ENDEMIC FLORA ON THE HAWAIIAN ISLANDS (according to Hennig 1974, in %)

Origin	Angiosperms	Pteridophytes
Indo-Pacific	40.1	48.1
Australian	16.5	3.7
American	18.3	11.9
Boreal	2.6	4.4
Pantropical, cosmopolitan	12.5	20.8
Unknown	10.0	11.1

TABLE 75 PROPORTION OF ENDEMICS AMONG CERTAIN TAXA OF THE BIOTA OF THE HAWAIIAN ISLANDS (according to Carlquist 1970)

	No. of species and varietes	% Endemic	Number of genera	% Endemic
Insects	3750	99	377	53.5
Land molluscs	1064	99	37	51
Birds	71	98.6	40	37.5
Seed plants	1729	94.4	216	13
Fern plants	168	64.9	37	8.1

Endemicity of Avian Taxa on the Hawaiian Islands

Families		Genera		Species		Subspecies	
Total	% Endemic	Total	% Endemic	Total	% Endemic	Total	% Endemic
12	8	25	60	35	86	68	99

Galapagos Islands The Galapagos Islands, whose fauna is one of the best documented island faunas, have geological preconditions similar to those existing in Hawaii. They provided Darwin (1833) with the first impetus for his theory of evolution. It was here that he recognized the significance of spatial isolation for the formation of species. There are no endemic plant families. The abundance of endemic species increases from the ferns and Monocotyledonae to the Dicotyledonae (Table 76).

According to Wiggins and Porter (1971), although there are ten fern families with 39 genera and 89 species on Galapagos, they contain only 10 endemic species (Table 76).

The well-examined vertebrate fauna of the Galapagos is characterized by the lack of amphibians and primary freshwater fishes. The blind brotulid *Caecogilba galapagosensis,* described by Poll and Leleup (1965) lives in caves of varying pH values (5.8 to 6.2) and highly fluctuating salt contents (2.97 to 8.71 ‰). The other species exhibit very close affinities to the Andean-Pacific area, to Central America, and, to a much lesser degree, to the Antilles.

Among the herpetofauna, the endemic iguana *Amblyrhynchus cristatus* occurs in closely related subspecies on the islands of Narborough, Albemarle,

TABLE 76 FLORA OF THE GALAPAGOS ISLANDS (according to Wiggins and Porter 1971)

Taxa	Ferns and relatives	Apetalae	Gamopetalae	Polypetalae	Monocotyledonae	Total
Families	10	15	23	44	15	107
Genera	39	32	107	114	56	348
Species	89	68	186	185	114	642
Endemics	10	43	88	70	17	228
Recent, introduced or accidentally carried onto the island	0	6	23	39	9	77

and Indefatigable, which are all located within the 200 meter isobath, whereas James is inhabited by the more strongly differentiated *mertensi* subspecies.

Apart from *Amblyrhynchus,* there exists a further endemic iguana species (*Conolophus* sp.) on the Galapagos islands. The strong differentiation of *Conolophus* and *Amblyrhynchus* suggests an early immigration (Plio-Pleistocene). The closest relatives of both species (*Ctenosaurus, Cyclurus, Iguana*) occur in Central and South America. The remaining Galapagos reptiles cannot be separated as special genera from the mainland groups; their closest relatives are Andean-Pacific faunal elements. These include the sand lizard *Tropidurus* and the snake genus *Dromicus* which can be divided into three morphological groups (*Dromicus biserialis, D. dorsalis, D. slevini*). These may be derived from *Dromicus chamissonis,* dispersed in the Andean-Pacific region, which is equipped with scale cavities characteristic of the Galapagos snakes. The distribution of the *Tropidurus* species, the degree of their morphological differentiation and their behavioural pattern would indicate — similarly to the *Amblyrhynchus* — a close affinity to the central group of islands (Narborough, Albermarle, James, Indefatigable). Barrington (also within the 200 meter isobath) still — although it is at a distance from Indefatigable — exhibits very close relationships with the central group of islands; e.g., a form of *T. albemarlensis* which occurs here is only subspecifically differentiated. All of the islands beyond the 200 meters isobath, however, contain populations with a very distinctly differentiated morphology and which, in spite of their allopatric distribution, are still classified as species. The existence of *Tropidurus* on the various islands is no indication of a former connection of the archipelago to the mainland as was assumed by Denburgh and Slevin (1913), but it can be explained by passive drifting.

The distribution of the Galapagos geckos is consistent with that of the other reptiles; however, the lack of gecko species on Narborough is remarkable. The Galapagos giant tortoises (*Geochelone elephantopus*), too, can be traced back to their ancestors on the South American mainland. In this connection, the occurrence of a giant tortoise (*Testudo cubensis*) during the Pleistocene in Cuba is worth mentioning; it bears a very close resemblance to the Galapagos tortoises. *Testudo praestans* is known from the Argentinean Pleistocene and possesses characteristics of the species living today, as well as of *T. cubensis,* and might be considered the possible point of origin. There are many other indications that the giant tortoises of Galapagos only reached the group of islands — by swimming — at the beginning of the Pleistocene at the earliest.

The 'swimming inability' of *Geochelone elephantopus,* however, induced Denbrugh (1914), Beebe (1924) and Vinton (1951) to 'construct' a Miocene continental land bridge between Central America and the Galapagos Islands via the Cocos Islands, in order to interpret the existence of the Galapagos giant tortoise. Since this tortoise exhibits such an obviously rapid rate of differentiation, it is surprising that it has not deviated further from its ancestral group. A more plausible explanation of the biogeography is that the ancestors of *G. elephantopus* were able to swim, just as *G. carbonaria* can in its present habitat in the Campos of South America.

The 89 breeding birds of the Galapagos Islands also immigrated over the ocean. This applies to the flightless cormorant *Nannopterum harrisi* which is present only in the most western islands with about 800 breeding pairs, as well as to the penguin *Spheniscus mendiculus* and to the endemic gull species *Craegrus furcatus* and *Larus fuliginosus,* the mockingbirds of the genus *Nesomimus* (*N. parvulus, N. Trifasciatus, N. macdonaldi, N. melanotis*), and Darwin's finches which populate the islands (including the Cocos Islands) in 14 species (Lack 1947, Bowman 1961, Harris 1974). The Galapagos finches probably share their ancestors with those of the mainland genus *Tiaris.* Apart from these endemics, there are widely-dispersed species which immigrated only in historical times, such as the cattle egret (*Bubulcus ibis*) or which, being cosmopolites, belonging to the original fauna, such as the osprey (*Pandion haliaetus*), the peregrine falcon (*Falco peregrinus*), the night heron (*Nycticorax nycticorax*), the gallinule (*Gallinula chloropus*), the barn owl (*Tyto alba*), and the short-eared owl (*Asio flammeus*). The lack of families which are widely spread on the American continent, however, is remarkable; this includes the hummingbirds (Trochilidae), the antbirds (Formicariidae), the cotingas (Cotingidae), and the true finches (Fringillidae). The similarities between the avifauna of the individual islands of the Galapagos are dependent more on the vegetation structure and less on their isolated location (Power 1975).

Apart from the fur seal, *Arctocephalus galapagoensis,* the mammals of the Galapagos consist only of small species (*Lasiurus brachyotis, L. cinereus, Rattus rattus, Mus musculus, Megalomys curioi, Nesoryzomys indefessus, N. darwini, N. swarthi, N. narboroiughi, Oryzomys galapagoensis, O. bauri*). Patton (1975) analyzed the chromosome sets of *N. narboroughi* and *O. bauri* and was able to demonstrate that *Nosorzyomys* karyologically deviates distinctly from the *Oryzomys*-related rodents, while *O. bauri* exhibits a set of chromosomes identical to that of *O. xantheolus* which occurs in the Peruvian coastal zone.

The *Rattus rattus,* which was introduced, has, incidentally, the chromosome number of European populations (2n = 38) and not that of Asian ones (2n = 42). Electrophoretic analyses by Patton *et al.* (1975) confirmed that the Galapagos populations had been carried into the area at various times. The occurrence of *Megalomys,* discovered in Galapagos only in 1964 by J. Niethammer, is interesting. The species was found in the pellets of *Tyto alba punctatissima* and *Asio flammeus galapagoensis* on Indefatigable. It is possible that until then, it had been known only on the Antilles Islands of Martinique, Barbados and Saint Lucia (where it is probably extinct today). This is a typical relict distribution which suggests a formerly wide dispersal of the genus.

It is not necessary to assume a firm land connection to the continent in order to interpret the fauna of Galapagos. Two groups of immigrants (based on their age) can be distinguished according to the degree of their differentiation:

1. A group morphologically strongly differentiated from Recent mainland populations (*Amblyrhynchus, Conolophus, Nesoryzomys,* and others), which probably reached the Galapagos from the north during the early Pleistocene (presuming that *Cyclura* and *Ctenosaura,* the probable original forms, were limited to Central America during the Pliocene, as assumed by Savage 1966). In that case, the conditions of the marine currents within the area in question must have been different from today. Possibly, the marine transgression during the Pliocene within the Panama area was the cause of these different current conditions.
2. A group whose morphological differentiations are only slight (as compared to (1) above) from the mainland populations (*Tropidurus, Dromicus, Phyllodactylus, Gonatodes, Oryzomys*) which most probably reached the Galapagos only during the Pleistocene from the direction of the Peruvian coast.

Canaries Many groups of islands do not have uniform geological history. The Canary Islands (Fuerteventura, Lanzarote, Gran Canaria, Tenerife, Gomera, Hierro, La Palma) provide a good example of such a group. Their geological origin was questioned for a long time. The question was, were the Canaries oceanic or continental? While their geological "basement" (compare Hansen 1966) has much in common with parts of the African continent, which might make a large "Canaries Land" of Eocene age conceivable, other geological and palaeontological findings only speak for a Mio-Pliocene connection of the eastern islands to Africa, whereas a purely volcanic origin is assumed for the western islands rich in endemics. Miocene deposits, Pleistocene ocean straits, a rich fossil and subfossil fauna and flora, numerous island endemics, phylogenetically old relict forms and taxa with strongly disjunct areas clearly point to the fact that a distinction must be made between at least two types of settlement: Plio-Pleistocene and glacial-postglacial type. This Plio-Pleistocene type certainly includes the fossil ostriches (Sauer and Rothe 1972) of the eastern islands, the dermapteran *Anatella canariensis* of Tenerife, called a 'living fossil', the mosquito *Protoculex arbieeni,* the giant tortoise *Testudo burchardii* growing to a length of 80 cm and now extinct, possibly also the giant lizards *Lacerta goliath* and *Lacerta maxima* (which grew to a length of 1.5 m) but are also extinct (Bravo 1966), the Canarian giant mouse *Canariomys bravoi* which probably still existed during historical times, and numerous plant species in the laurel forests of the western islands (Figs. 95 and 96). Man is among the most recent immigrants. The aboriginal population of the Canaries, the Guanches, who were destroyed after the conquest by the Spanish (1402, Juan de Bethencourt) exhibited distinct traits of North African population groups (Fuste 1966, Schwidetzky 1963). The earliest fossil findings of man date back to the 2nd Century B.C.

It is not only the individual islands which differ in their macro-ecological structure and permit at best a differentiation between a dry eastern (Lanzarote and Fuerteventura) and a moister western group of islands, but each

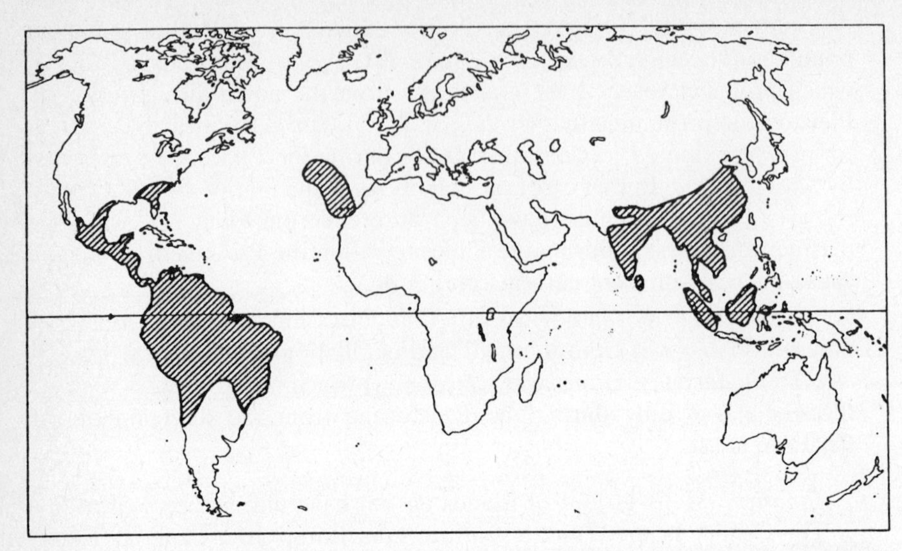

Figure 95 Disjunct distribution of the plant genus *Persea* (Lauraceae; according to Bramwell, 1976).

island in itself is composed of markedly varying biomes.

The course of the cool Canaries Current, the range of the Sahara winds (the Harmattan), the topographic location, the orography and particularly the elevation result in shifts of the vegetation formations. Thus, the rainy northern slopes of Tenerife (the Miocene Sierra de Anaga, north of Santa Cruz de Tenerife) and of Gomera differ fundamentally from the south side where the rainfall is sparse all year long (for instance, El Medano). The northern side of the western islands is persistently covered by clouds even in summer, whereas the sun shines brilliantly on the southern side. Parts of

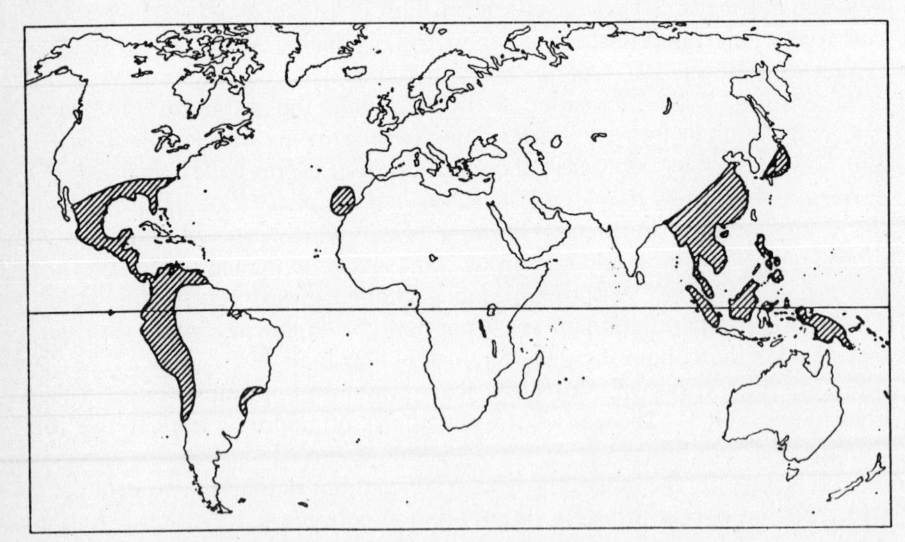

Figure 96 Disjunct distribution of the plant family Clethraceae (according to Bramwell, 1976).

Fuerteventura, Lanzarote and the south of Gran Canaria are characterized by semideserts and deserts. Only the highest elevations of Fuerteventura (Jandia in the south) and Lanzarote (Famara, near Haria) receive the moisture during the summer months which is necessary for a richer vegetation. Four zones characterized by varying plant formations can clearly be distinguished on the more strongly differentiated western islands:

(a) *Xerophyte and succulent zone (0 to 700 m).* This is the lowest altitude vegetation zone on all of the islands. Important plants include the man-high columnar *Euphorbia canariensis,* which is on all the islands to an elevation of approx. 900 m, and which is unmistakable by its brownish-red capsules. It is the forage plant of the tenebrionid beetle *Pelleas crotchi* and its milky juice was used by the aborigines, the Guanches, as a purgative and as a fish poison. Also present are the bushy *Euphorbia bourgaeana* (Tenerife, 100 to 600 m), columnar *Euphorbia handiensis* which can be as high as 80 cm and is reminiscent of cacti (very rare, Fuerteventura), and the leafless *Euphorbia aphylla,* characterized by succulent green stems. The wide Tertiary dispersal of the Canarian taxa indicates that the Recent Canaries area of many species should most probably be understood as a relict area. The dragon trees (*Dracaena draco*) have posed a puzzle ever since Alexander von Humboldt's *Voyage in the Equinoctial Areas of the New Continent,* during the course of which he stayed in Tenerife from June 19 to June 24, 1799, and climbed the Pico de Tide from the Orotava Valley. He estimated the age of the *Dracaena* in the garden of the Marquez de Sanzal of Orotave, shown in his *Atlas Pittoresque du Voyage* (Table 58), and which was still seen in 1866 by Ernst Haeckel and was then destroyed by a storm on January 2, 1868, to be between 5,000 and 6,000 years. The *Dracaena* at Icod de los Vinos, about 22 m high and often visited by tourists, is also accorded a great age (1000 years).

Pütter (1925), Simon (1974) and Mägdefrau (1975) were able to prove — by means of growth measurements and a determination of flowering rhythms — that the surviving dracaenas are considerably younger than assumed by von Humboldt. Since the inflorescence of *Dracaena* is in a terminal position, the vegetative bud ceases to exist at the time of the formation of blooms. New vegetative buds form below the inflorescence which grow into limbs and do not branch out until the next flowering period (about every ten years — greatly fluctuating). By establishing the interval between the blooming periods, the age can be determined from the number of branchings. In 1971, the age of the tree at Icod de los Vinos was estimated by this method to be at most 365 years.

Deep canyons, lava fields varying in age, and re-deposited material in this region exhibit atypical vegetation. Raised beaches have formed in many locations on the island, and on some of them roads have been constructed; for instance, in the capital of Tenerife, Santa Cruz de Tenerife, which is heavily polluted by the petrochemical industry. The proportion of cliff endemics is remarkably high; some of the endemics probably evolved on their present site which had survived the recent eruptions and which became highly isolated (for instance, the endemic composite genera *Vieraea, Heywoodiella,* or *Sonchus bornmuelleri,* and *S. gummifer).*

The lowlands of all of the islands are heavily used for agricultural purposes (potatoes, tomatoes, bananas and, behind walls built as wind shelters, figs, corn, wheat and strawberries. Up to an elevation of 1,000 m, terrace cropping and irrigation canals — some of which are several hundred years old and in many areas in decayed condition — can be encountered. In spite of polluted seas and beaches the coastal area is classified as a tourist zone. There is a lack of sewage treatment plants and considerable pollution caused by oil and power stations; for instance, near Las Caletillas there is a power station without effluent desulfuration, resulting in serious damage to the fig trees. Prickly pear (*Opuntia*) was deliberately imported to the lowlands and settled with cochineal insects. These insects furnish a brilliant red dye which is used for, among other things, Persian carpets and cosmetics. About 700 different plant species (for instance, *Agave americana, A. sisalana, A. fourcroydes, Hibiscus, Bougainvillea, Ficus carica, Pelargonium, Philodendron, Aloë, Acacia, Pheonix dactylifera, Castanea sativa, Mangifera indica, Carica papaya, Cirus, Eucalyptus globulus, Pinus radiata, P. halepensis*) were introduced into the Canaries by man (Kunkel 1976).

(b) *The Laurel Forest Zone.* The laurel forests are the areas most influenced by humans. About 1% of the original *Laurus* forest surface still exists on Gran Canaria, and on Tenerife it amounts to 10% at best. Pine afforestations have, to a large extent, taken place on former *Laurus* forest soil.

In accordance with their climatic differentiation, the best preserved forest stands are found on the northern slopes of the western islands; isolated occurrences of laurel forest species on the south sides (for instance, La Ladera de Guimar on Tenerife) show that a wider occurrence of this forest type would be possible under present conditions. Characteristic tree species of these western island forests are *Laurus azorica; Persea indica; Picconia exelsa,* which is related to the olive tree; *Apollonias barbusana; Ocotea foetens,* growing to a height of up to 40 m; the Canarian strawberry tree *Arbutus canariensis; Pleiomeris canariensis,* and *Heberdenia bahamensis,* both of which belong to the Myrsinaceae; *Prunus lusitanica;* the two holly trees *Ilex platyphylla* (Tenerife and Gomera) and *I. canariensis; Visnea mocanera,* growing to a height of 15 m and belonging to the Ternstroemiaceae; and the Canarian willow *Salix canariensis.* Degraded stages of the former laurel forests were able to survive on some sites on the western islands; and although *Laurus* is generally missing in these locations, other laurel forest indicators usually exclude any doubts about the growth potential of these sites. Such a site is located in the Ladera de Guimar, a steeply rising mountain face, breeding place of rock doves and kestrels, northwest of Guimar, to which one can easily hike starting from the small village of Medida, first through vinegrowing terraces, then along goat paths (up to about 600 m elevation). Two aqueducts, of which the upper is partially in ruins (8/8/1977), have been artistically laid out to run in parallel through the forest where one can recognize from afar the deep green color of isolated *Arbutus canariensis.* In the higher elevations, the bushy *Isoplexis canariensis* and the man-high branched *Echium virescens* are still in bloom in August. The pink-flowered *Cistus symphytifolius* and the red berries of the nearly one meter high relative of the arum, *Dracunculus canariensis,* shine between the *Erica arborea,* which can be up to 8 m high.

(c) *The Pinus canariensis Zone.* This zone is located between 800 and 1,900 m, particularly on the south slopes and above the summer cloud level of the western islands of Tenerife, La Palma, Gran Canaria, and Hierro. The forest stands consist of the endemic *Pinus canariensis,* which has also been planted on Gomera, where it was not a native originally. Its habitat on the other western islands has in some instances been pushed far into the original laurel forests as a result of an afforestation program started in 1950. Its needles, growing to a length of 30 cm, form the only ground litter in some areas. This suggests an insufficient supply of consumers. A postglacial *Pinus canariensis* stand (3,075 ±50 years) has been described in the Caldera de los Arenas on Gran Canaria (Nogales and Schminke 1969).

(d) *The Canarian High Mountain Zone.* The recent upper forest limit is formed by *Pinus canariensis* and *Juniperus cedrus* above the summer cloud cover between 1900 and 2250 m. The High Mountain flora is most varied in the Las Canadas National Park on the Pico de Teide (Tenerife). It can be characterized by numerous endemics, including, among others, the well-known, yellow-blooming Teide violet *Viola cheiranthifolia* occurring up to an altitude of 2,800 m, which is becoming increasingly endangered by mass tourism and whose closest relative occurs near the peak of La Palma (*V. palmensis*). While both of these species are inconspicuous because of their small size, others leave a clear mark on the landscape. The red-bloomed *Echium wildprettii,* which can reach 2 m in height, must be counted among them. It belongs to a genus rich in species in the Canaries, including giant forms in the lowlands (for instance, the 2.5 m high *E. giganteum* on the north coast, the 2 m high *E. virescens* in the forests on Tenerife). The blue-flowering *E. auberianum* occurs considerably more rarely in Las Canadas.

The oceanic islands located within the Palaeotropic-holartic transition area in front of the West African coast can in most cases be clearly assigned to a specific animal or plant realm. While Cape Verde belongs to the Ethiopian Realm, the Canaries can be assigned to the Palaearctic Realm in spite of their remarkable local endemic (470 of the 1700 plant species are endemic) and the plant and animal species which they have in common with the Ethiopian Realm and the Macaronesian Islands (Macaronesian endemics = 110 plant species from the Canaries).

The composites are dominant among the endemic plant genera. When examining the dispersal of the endemic genera and their species, it is noteworthy that Fuerteventura and Lanzarote are reached only by *Schizogyne sericea, Drusa glandulosa,* and *Plocama pendula,* which also occur on most of the western islands. The remaining endemic genera are limited to the western islands.

From the cytogenetic point of view, the small proportion of polyploids among the Canarian plants is surprising. It indicates that the flora is relict (Bramwell 1976).

The abundance of endemics within the herpetofauna is also greater on the western islands than on Lanzarote and Fuerteventura. There are no endemic families. Of the two frog species it is certain that the Spanish laughing frog (*Rana perezi*) has been carried onto the island. The absence of snakes and agamids is noteworthy.

The reptiles *Tarentola mauritanica* and *Chalcides ocellatus,* living on Lanzarote and Fuerteventura, occur in other subspecies on the North African and the Spanish mainland, while *Lacerta atlantica* is an endemic whose closest relation, *L. galloti,* lives on the western islands. The gecko *Tarentola delalandii* has a 'Macaronesian dispersal type' (Cape Verde, Salvages, Madeira, Canaries), while the skinks *Chalcides sexlineatus* and *viridanus* are endemics of the Canaries. Klemmer (1976) attributes the Canaries lizards whose closest continental relative is represented by *L. lepida,* to two groups of immigrants (*L. galloti atlantica, L. simonyi stehlinii*). Fossil and subfossil lizards and giant tortoises of Tenerife (Burchard and Ahl 1927, Mertens 1942) lead us to believe that the ancestors of the Canarian *Lacertas* reached the islands only during the last glaciation. We know little at this point about the rate of evolution of the Canarian taxa although it may be assumed that the differentiation processes were accelerated not only by their isolated location, but particularly also by the marked ecological differences between and on the individual islands.

The avifauna of the Canaries confirms this assumption. Although here, too, there exist remarkable endemic species (*Fringilla teydea polatzeki* on Gran Canaria; *F. teydea teydea* on Tenerife which is ecologically dependent on *Pinus canariensis*), most faunal elements can be ecologically interpreted. The dependence of the number of species on the size of the area is decisively modified by the vegetation formation present in the corresponding region.

Within the avifauna, noteworthy species are those which, like *Columba jonuniae* or *C. trocaz bollii,* are dependent on the laurel forest and which are relicts of a formerly vast forestland as indicated by distributional pattern. Apart from these, there are species which, ecologically speaking, are strictly adapted to the open country.

When comparing the avifauna of the Canaries with that of Madeira and the Azores, it is clear that Madeira and the Azores were reached only by those species which are present on all of the islands or only on the western islands. The Firecrest *Regulus ignicapillus* is an exception; on the Canaries and on the Azores, it is replaced by *R. regulus* (Bannerman and Bannerman 1965).

The lepidoptera confirm the Canarian pattern developed in the case of the birds. 26 butterfly species are present on the islands (Fernandez 1970), predominantly populating the western islands. With the exception of *Pieris rapae, Colias croceus,* and *Vanessa cardui,* all are missing on Lanzarote.

6.4.3 Continental Islands

At least since the Pleistocene, as a result of eustatic sea level fluctuations, the history of the continental islands has been closely related to that of the continent and its biota (Cromwell 1971). These islands can often be used to study the problems of the rate of differentiation of taxa and changes in their continental distributional area, Mertens 1934, Müller 1970).

Excellent examples of such island types can be found close to continental coasts. Although their biota generaly reflect the conditions on the continent located close by, they are often characterized by a different composition and differentiation, depending on the size of the island, the ecology, the period of isolation and the type of isolation barriers. Endemic species whose speciation is not always due to the conditions prevailing on the island (i.e. relicts), are able to survive here.

Examples of this are provided by the islands within the 60 m isobath off the Brazilian coast. One of them, the 350 km^2 island of São Sebastião, which is located close to the mainland, is isolated from the mainland only by a narrow (3.3 km wide) channel. Because of the size of postglacial eustatic sea level fluctuations, it must be assumed that it has been isolated for about 7,000 years (Fairbridge 1958, 1960, 1961, 1962; Fray and Ewing 1963). Two factors, however, may mean that we cannot rely on this estimate of how long its isolation has lasted. First of all, the ocean depths in the channel — at practically the same sea level — during the year are subject to certain fluctuations, which are influenced above all by the Brazil Current often depositing large quantities of sand in this area; secondly, it has also been observed, that in parallel with the eustatic sea level fluctuations of the Pleistocene, a tectonic rise of Eastern Brazil took place, as can be determined on the basis of the variation in height of the shoreline (Machatschek 1955). Moreover, the erosion activity of some rivers of the Serra do Mar, mentioned by Franca (1954) and Freitas (1944, 1947) as the cause of the island's separation, is surely of only secondary significance. The morphology of the 1379 m high island (Pico de São Sabastião) shows that 86.6% of the land is located at over 100 m above sea level (Franca 1954); of this 6.9% lies above 900 m. A flat coastal strip of Pleistocene or Holocene age, has formed only on the west coast between the settlements of Ilhabela and Perequê and also near Castelhanos. The elevated part of the island consists of alkaline volcanic rock and the highest peaks — the Pico de São Sebastião, the Pico de Papgeio, the Morro da Serraria, the Morro do Eixo, and the Morro Ramalho — are all formed out of this rock. The peripheral zone of the island and the low hills (Morro de Cantagalo, Morro des Enxovas, for instance) consist of granite and gneiss: rocks which we also find in the coastal zone of the Serra do Mar of Brazil (Costa cristalina).

The flora of the island also exhibits this direct connection to the Serra do Mar (Hoehne 1929, Edwall 1929, Luederwaldt 1929). Luederwaldt (1929), however, points out that it is poorer in species than that of the continent.

The flora of the island also exhibits this direct connection to the Serra do Mar (Hoehne 1929, Edwall 1929, Luederwaldt 1929). Luederwaldt (1929), however, points out that it is poorer in species than that of the continent.
These findings can be definitely confirmed only for the west side of the island where the rainforest starts only above 500 m, while the zone between the rainforest and the coast is covered — due to anthropogenic influences (fires, etc.) — only by grassland and sparse shrubbery. In some spots, for instance along the river courses, the rainforest penetrates far into the grass zone; but even here, there are clear indications of yearly grass fires. It is obvious that as a result of these fires, no classical floral classification can be demonstrated for the west side such as has been described by Paffen (1955) and Hueck (1966) for the Serra do Mar. The vegetation has generally survived on the east side of the island, and the original floral gradient there is still largely unaltered. In the northern part of the island signs can be found indicating that during the past century several attempts have been made at utilizing the land for agriculture, but this has been abandoned as evidenced by the extremely young dense forest growth. Parts of the south of the island, exhibiting habitat conditions similar to those prevailing in the north, bear a silty deposit in the coastal zone on which mangroves have formed. The vegetation here is composed solely of *Rhizophora mangle* and *Laguncularia racemosa*. The *Rhizophora* stands which penetrate the furthest into the ocean, attain a height of only 5 m (maximum) on the island. 'Tall forest formations,' such as they have been described by Gerlach (1958), do not exist.

Among the herpetofauna (Müller 1968), only the Gymnophionid *Siphonops insulanus* occurs as an endemic. All the other species are also known on the continent (Table 77). Most representatives of the island fauna of São Sebastião belong to the native rainforest fauna of the Serra do Mar. The absence of native savanna elements proves that this area was also covered by forest during the last glacial period.

6.4.4 Island Dispersal of Cave Biota

From a biogeographical point of view, all well-isolated populations must be assigned comparable importance to that attributed to oceanic island biota. The island theory established by MacArthur and Wilson (1963) is applicable even to the continents and to their interspersed habitat islands. A very good comparison to the oceanic "island dispersal patterns" is provided by the dispersal areas of Alpine (compare High Mountain Biomes, section 5.1.9) and cave populations. Surprising progress has been made in research in the

TABLE 77 HERPETOFAUNA ON THE ISLAND OF SÃO SEBASTIÃO AND THE NEARBY CONTINENTAL MAINLAND

Island fauna	Continent	
Caeciliidae		
1. *Siphonops insulanus*	–	
Leptodactylidae		
2. *Basanitia lactea*	+	
3. *Eleutherodactylus binotatus*	+	
4. *Eleutherodactylus guentheri*	+	
5. *Eleutherodactylus parvus*	+	
6. *Elosia aspera*	+	
7. *Elosia lateristrigata*	+	
8. *Elosia nasus*	+	
9. *Eupsophus miliaris*	+	
10. *Leptodactylus marmoratus*	+	
11. *Leptodactylus ocellatus ocellatus*	+	
12. *Leptodactylus pentadactylus flavopictus*	+	
13. *Physalaemus biligonigerus*	+	
14. *Physalaemus olfersi*	+	
15. *Physalaemus signiferus*	+	
16. *Cycloramphus asper*	+	
Bufonidae		
17. *Bufo crucifer crucifer*	+	
Brachycephalidae		
18. *Dendrophryniscus brevipollicatus*	+	
Hylidae		
19. *Hyla faber*	+	
20. *Hyla albopunctata*	+	
21. *Hyla goughi goughi*	+	
22. *Hyla hayii*	+	
23. *Hyla albomarginata*	+	
24. *Hyla microps*	+	
Cheloniidae		
1. *Chelonia mydas mydas*	+	
2. *Eretmochelys imbricata imbracata*	+	
3. *Caretta caretta*	+	
4. *Dermochelys coriacea*	+	
Chelidae		
5. *Hydromedusa maximiliani*	+	
Gekkonidae		
6. *Hemidactylus mabouia*	+	*(introduced)*
7. *Gymnodactylus geckoides darwinii*	+	
Iguanidae		
8. *Enyalius iheringi*	+	
Teiidae		
9. *Tupinambis teguixin*	+	
10. *Placosoma cordylinum champsonotus* (cf. Uzzell, Th.M. 1962: Additional notes on teiid lizards of the genus *Placosoma* Copeia 1962 (4): 833–835)	+	

TABLE 77 (*contd*)

Scincidae	
12. *Mabuya*	+
Amphisbaenidae	
13. *Leposternom microcephalum*	+
Colubridae	
14. *Chironius bicarinatus*	+
15. *Chironius pyrrhopogon*	+
16. *Spilotes pullatus anomalepis*	+
17. *Leimadophis melanostigma*	+
18. *Liophis miliaris miliaris*	+
19. *Rhadinaea affinis*	+
20. *Simophis rhinostoma rhinostoma*	+
21. *Clelia clelia clelia*	+
22. *Pseudoboa doliata*	+
23. *Thamnodynastes*	+
24. *Dipsas albifrons*	+
Elapidae	
25. *Micrurus corallinus*	+
Crotalidae	
26. *Bothrops jajaraca*	+
27. *Bothrops jararacussu*	+

area of cave biota, particularly during the past few years (Ford and Cullingford 1976). In lightless caves, the vegetation consists only of fungi and bacteria (Cubbon 1976). Wherever diffuse light penetrates, liverworts (*Conocephalum*), mosses (*Eucladium verticillatum*) and a few vascular plants (*Adoxa, Phyllitis*) are able to survive. Where caves are artificially illuminated, an actual 'lamp flora' may develop. Bacteria and fungi can provide the necessary energy for a cave-specific invertebrate fauna which is often marked by numerous 'degenerate' characters (for instance, loss of pigments, loss of eyes) (Jefferson 1976, Wilkings 1973). Regarding their "cave preference", cave animals can be classified into the following ecological groups:

Troglobites = Obligately dependent on lightless caves; not viable in above surface habitats.

Troglophiles = Optionally cavernicole species which live in caves but can also be found outside of the caves.

Trogloxenes = Cave dwellers which spend only part of their life cycle in caves. This group can be divided into species which "accidentally" arrived in the caves and others which visit the caves only at certain times (for instance, bats and the butterfly *Scolioptery libatrix*).

When compared with surface biomes, caves are subject to limited environmental fluctuations (slight temperature fluctuations, practically constant air humidity, absence of strong air currents, constant darkness, etc.). Like other isolated populations, cave populations provide a special opportunity for historical interpretations (Gueorguiev 1973). Still, Recent ecological factors must always be kept in mind when attempting such interpretations. Thus, an analysis of 48 caves in the Tessin shows that there is a direct correlation between the quality and duration of isolation and the habitat diversity, or the species diversity and the number of endemics. The size of water body present in the caves and its nutrient supply (for instance, the guano of bats) regulate the number and density of the species of cave populations (Vuilleumier 1973). Numerous organisms living on the surface have cave relatives lacking pigmentation (for instance, *Proasellus cavaticus, Niphargellus glenniei, Niphargus fontanus,* numerous fishes and urodelans). The distribution of cave species generally is dependent on areas which were free of ice during the last glaciation. Examples are the English cave crayfishes *Niphargus fontanus, N. keshignus,* and *Niphargellus glenniei.* The occurrence of permafrost soils and of fossil coleoptera from the last glacial period in England fit this pattern (Williams 1969, Coope 1970, Coope *et al.* 1971).

Other species such as the collembolan *Onvchiurus schoetri,* however, show that a survival in the caves was possible even under the glacial icecap. "The controls on evolution of organisms in caves and subterranean water bodies are no different from those of surface habitats. In caves also, therefore, selection and isolation are the only factors causing speciation. They act on the material provided by natural genetic variations of the organisms, which originate without exception through random mutations. In this connection, it should not be surprising that the morphological diversity in caves is of a different composition than above ground. In caves certain forms of degeneration (loss of eyes, pigmentation and wings) are able to exist as well as normal ones; on the surface, such forms are nearly always eliminated through the process of selection before any propagation can take place. Thus, this isolated group of organisms is particular valuable — because of its peculiarities — in providing a clear and impressive contribution to our knowledge of the macroevolutionary processes to which all living beings, without exception, owe their creation" (DeLattin 1941, p. 297).

6.5 ANALYSIS OF DISPERSAL CENTERS AND HISTORY OF LANDSCAPES

Numerous examples show that fundamental phylogenetic innovations usually arise accompanied by significant habitat changes and thus, an alteration in the ecological potential of populations or parts of them. In the case of vertebrates, particularly of poikilothermic reptiles and homoiothermic birds, allopatrically dispersed sub- or semispecies within a species or closely-related species group usually belong to the same or similar biomes (Keast 1961, Moreau 1966, Müller 1973, 1974). At the genus and family levels, however, widely different major ecosystems are often utilized. Since 'similarity' does not necessarily mean phylogenetic relationship, generally speaking only groups with complicated structural complexes and many characteristics are suitable for the clarification and phylogenetic reconstruction of the relationships of higher taxa in order to distinguish between homologies and convergences (Hennig 1949, 1960, 1969, Illies 1965, Brundin 1972, Schmincke

1973, 1974, Zwick 1974).

The circumstances in the case of propagation areas are analogous. While a monotypic genus can be used for the reconstruction of phylogenesis, i.e., the *sequence* of the development of characteristics and thus the formation of the taxa, the geographical origin of the (possibly totally isolated) distribution area of this genus can be clarified neither by means of chorological nor phylogenetic explanations. Only in the rarest of cases is the age or the presence of a species at a certain site on Earth tied to the geological age of that site. The hypothesis that the Recent area of a taxon — regardless of the presence of different or similar characteristics — is identical to its center of origin or to one of the many possible dispersal centers created during the course of its evolution must confirmed by careful analyses, in each individual case.

The theoretical point of departure, therefore, is the Recent propagation area of a taxon with all of its differentiations, and the assumption that populations are adapted to a specific environment and that a clarification of their area history and phylogeny can shed light on the landscapes and their development. A method which has proved to be particularly effective is the analysis of *dispersal centers* (Müller 1973, 1974, 1976).

For the sake of a better understanding of the statements that follow, I feel it is important at this point to go into more detail about the problems

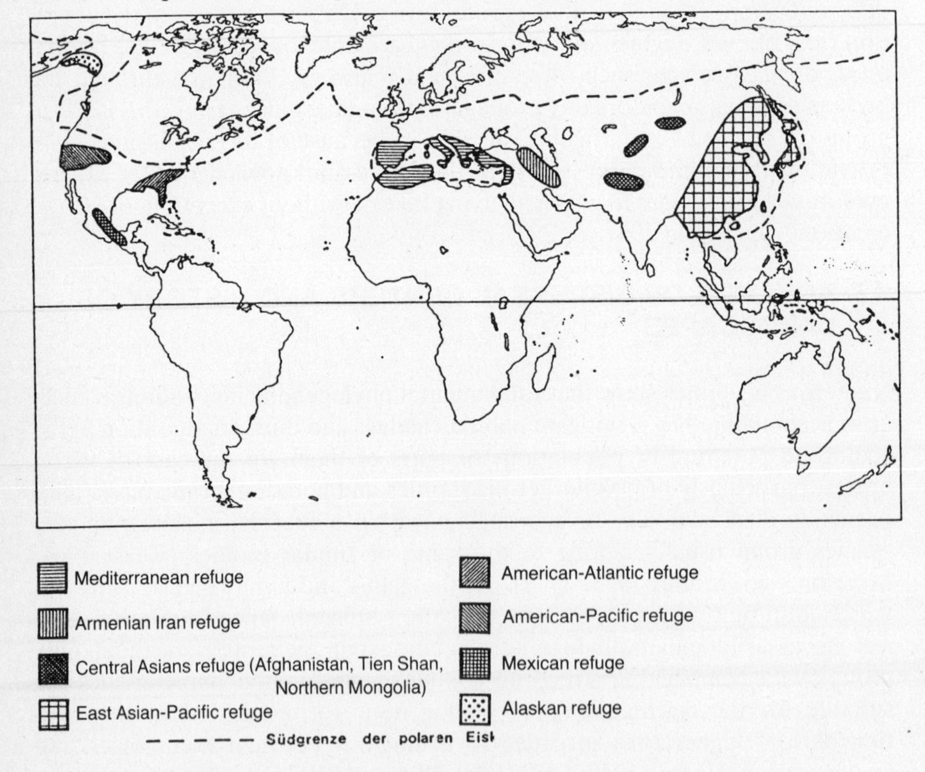

Figure 97 Reinig's (1937) proposed glacial refuges for the forest flora and fauna.

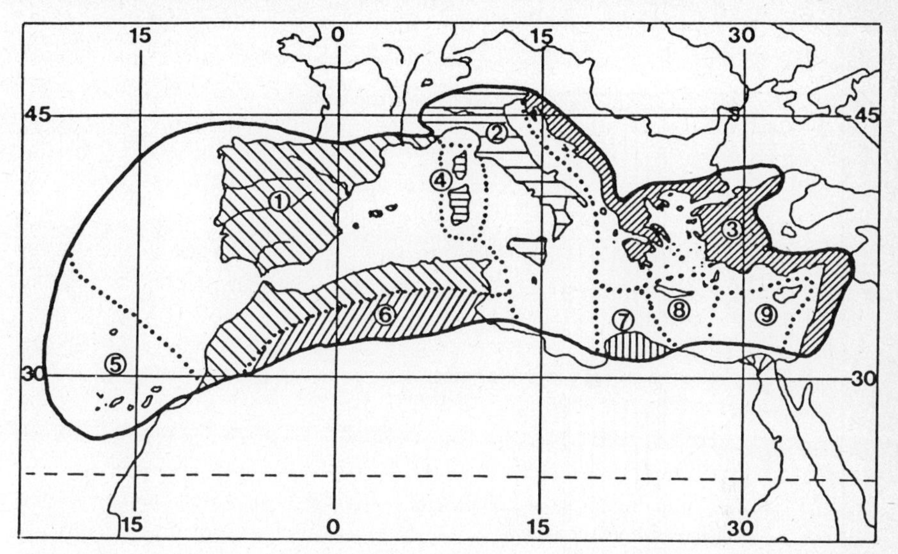

Figure 98 Subdivisions of the Mediterranean dispersal center, according to De Lattin (1967).

connected with the analysis of dispersal centers. The results obtained from their examination and explanation can contribute towards a better understanding of the recent history of evolution of organisms and towards a clarification of geological and climatological facts, and thus to a better knowledge of the present conditions of the landscape. The importance of dispersal centers, therefore, is as great for research on evolution as it is for geography.

Dispersal centers are those geographic areas in which populations survived under — for them — unfavorable environmental conditions (Figs. 97 and 98). By its nature, a geographic area can only function as a dispersal center if the environmental conditions did not cause the extinction of the biotic communities present there. The populations remaining in the dispersal centers for the duration of adverse environmental conditions are isolated from other regions and other populations. This enables a force important to the formation of species and subspecies, i.e. geographic isolation to be effective. To avoid misunderstandings, it must be pointed out that dispersal centers are not necessarily identical to centers of origin (Cook 1969, Croizat, Nelsun and Rosen 1974). In order to assert that they are identical it is necessary to examine the distribution areas to determine whether they are located in the vicinity of the center of origin (plesiochorous) or whether, during the course of the history of evolution of a species, they moved very far from the center of origin (apochorous, Müller 1972, 1974, 1975) (Figs. 99a and 99b). Since for higher systematic units (genera, families) an answer to these questions is dependent on a complete understanding of the history of their evolution — which so far has been satisfactorily analyzed only for a few animal species — it seems appropriate to limit the analysis of dispersal centers to species, superspecies and subspecies. In the case of sub- or superspecies, the establishment of proof of the apo- or plesiochory of their distribution area

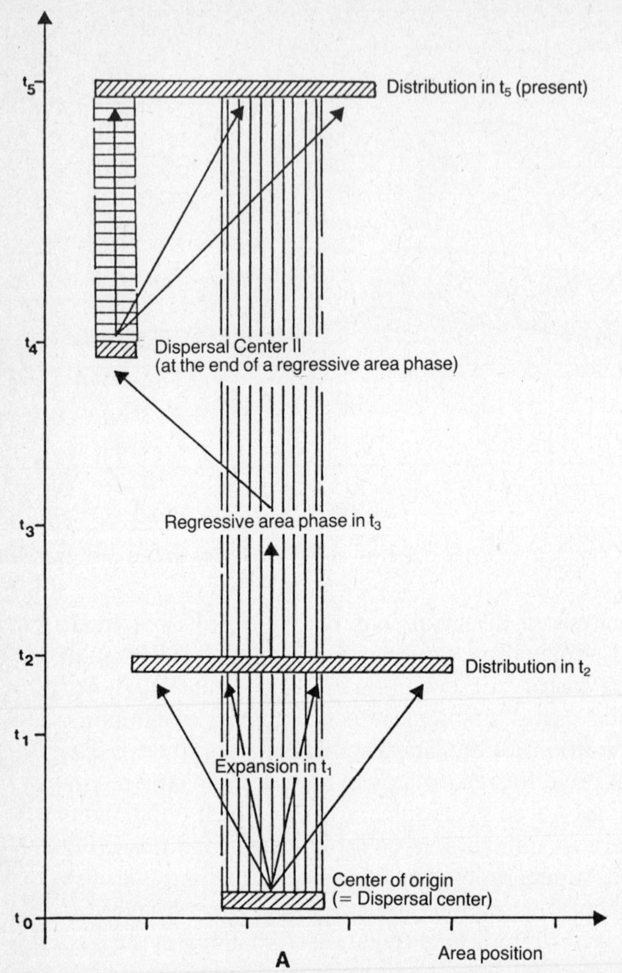

Figures 99a and **99b** Spatial and temporal connections between centers of origin and dispersal centers of a taxon. Case A assumes the Recent area to be plesiochorous in relation to the center of origin and to the various expansion centers. In case B, the Recent area is apochorous in relation to the last dispersal center (t_4) as well as to the center of origin. In reality, apochory and plesiochory, respectively, must be proven by a number of processes of analysis. In most cases, proof can only be furnished for propagation areas rich in characteristics.

to the last functioning dispersal center is much easier (allopatric distribution, for instance) than in the case of species. It must be emphasized, however, that subspeciation is not exclusively tied to the geographical isolation of originally uniform populations (Fig. 99c).

The relationships between refugial area phases, ecological changes in those areas, migrations, and genetic differentiations make a distinction between three types of subspeciation appropriate (Fig. 100):

1. Refugial subspeciation,
2. Extra-refugial subspeciation, and

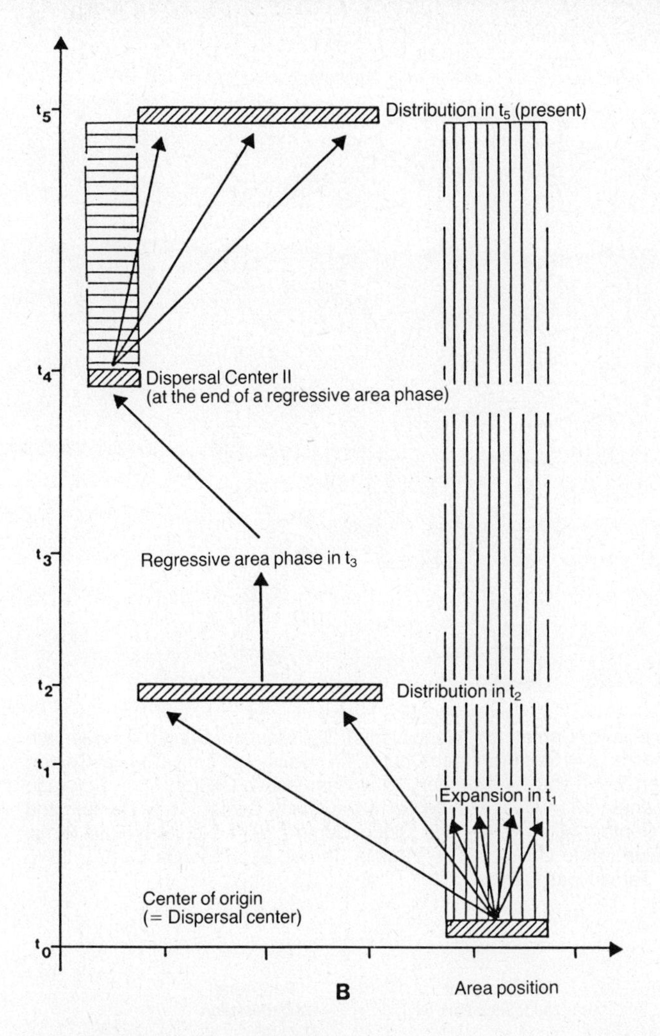

3. Peripheral subspeciation.

Peripheral subspeciation can be the result of fluctuations in area boundaries. It can also, however, as pointed out by Reinig (1970), be the result of suppression which he interprets to mean the suppression of phylogenetically older subspecies by younger ones possessing dominant alleles, without the effect of selection. According to this theory, phylogenetically older peripheral subspecies would then be relict populations of originally more widely dispersed subspecies which were "overrun" in the core of the distribution area by younger ones. These older types, however, do not have to be exclusively amassed along the geographical edge of the range but may also have survived in the region of ecological barriers (for instance, mountains) within the distribution range.

These limiting points are a necessary prerequisite for the correct interpretation of the importance of a dispersal center. The analysis of a dispersal

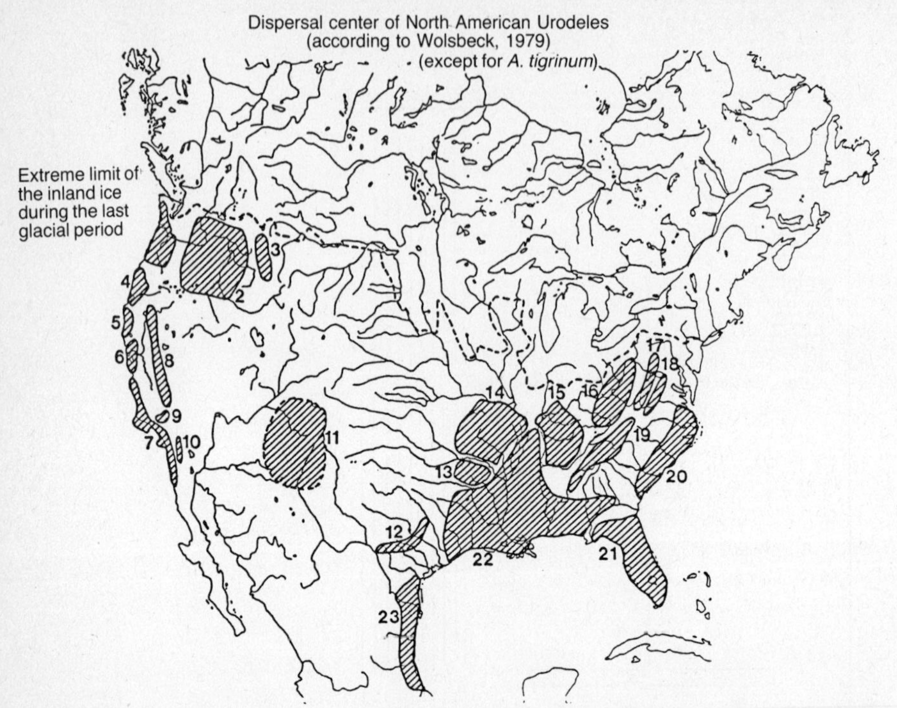

Figure 99c 1 = Oregon Coastal Center; 2 = Idaho Center; 3 = Montana Center; 4 = Klamath Center; 5 = Mendocine Center; 6 = San Francisco Center; 7 = South California Coastal Center; 8 = Sierra Nevada Center; 9 = Tehachapi Center; 10 = San Jacinto Center; 11 = Colorado Center; 12 = Balcones Center; 13 = Ouachita Center; 14 = Ozark Center; 15 = Cumberland Center; 16 = Kanawha Center; 17 = Valley and Ridge Center; 18 = Northern Blue Ridge Center; 19 = Southern Blue Ridge Center; 20 = Virginia Center; 21 = Florida Center; 22 = Mississippi Center; 23 = Tamaulipas Center.

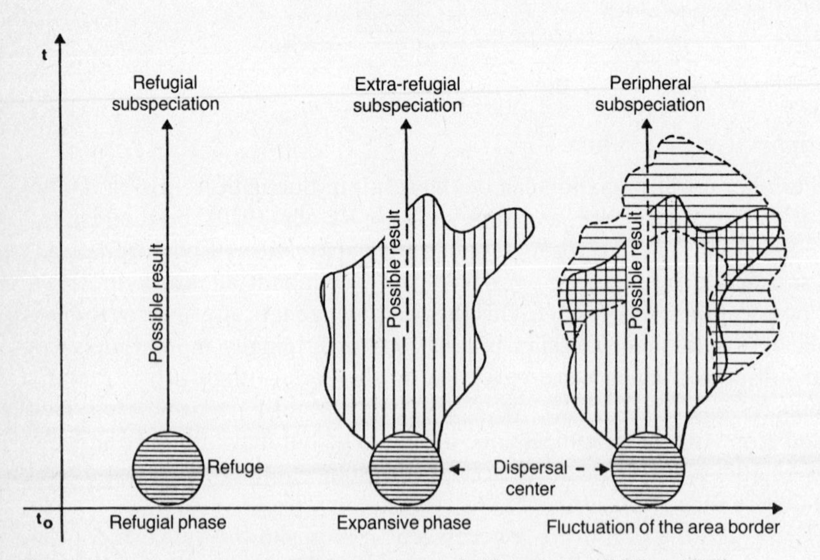

Figure 100 Connections between isolation and differentiation (Müller, 1974).

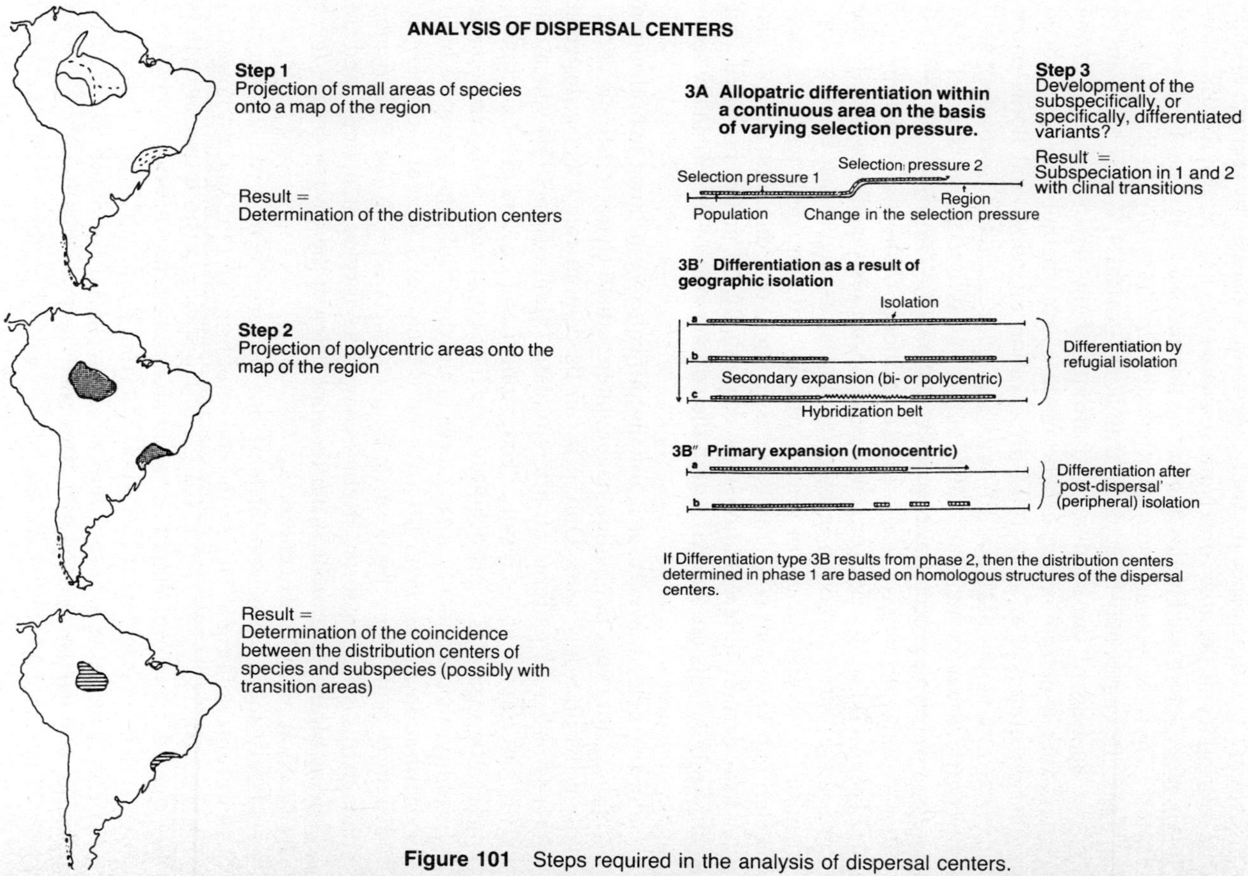

ANALYSIS OF DISPERSAL CENTERS

Step 1
Projection of small areas of species onto a map of the region

Result =
Determination of the distribution centers

Step 2
Projection of polycentric areas onto the map of the region

Result =
Determination of the coincidence between the distribution centers of species and subspecies (possibly with transition areas)

Step 3
Development of the subspecifically, or specifically, differentiated variants?

3A Allopatric differentiation within a continuous area on the basis of varying selection pressure.

Selection pressure 1
Selection pressure 2
Region
Population
Change in the selection pressure

Result =
Subspeciation in 1 and 2 with clinal transitions

3B′ Differentiation as a result of geographic isolation

Isolation
a
b
Secondary expansion (bi- or polycentric)
c
Hybridization belt

Differentiation by refugial isolation

3B″ Primary expansion (monocentric)

a
b

Differentiation after 'post-dispersal' (peripheral) isolation

If Differentiation type 3B results from phase 2, then the distribution centers determined in phase 1 are based on homologous structures of the dispersal centers.

Figure 101 Steps required in the analysis of dispersal centers.

center requires three steps (Fig. 101). The first step consists of projecting the smallest part = areas of species, semispecies and subspecies onto a map of a continent or zoogeographical region. Although boundaries of individual areas only very rarely coincidence precisely, they always overlap to a certain extent; the overlap forms the *area core.* Numerous analyses have shown that distribution centers determined in this manner do not necessarily have to be dispersal centers. Those centers are regions of greatest *area diversity.* In this phase of analysis, the causes may be ecological and/or historical. It is only by means of further analysis of the conditions of affinity of the faunas assigned to the centers that a determination can be made as to whether they are indeed expansion centers, i.e., centers of preservation of faunas and floras under adverse environmental conditions. Therefore, in the second phase, polycentric areas (widespread distribution areas containing several area cores) of polytypic species must be projected onto the same region. The result often — but by no means always — is a coincidence of the dispersal centers of the species and subspecies (or semi-species) with restricted area size. In the third phase, often forgotten in biogeographical examinations, the development of the subspecifically or semispecifically differentiated vicariants must be clarified. Decisions concerning the assignment of a differentiated population to a certain differentiation type can usually be made, provided that hybridization belts are formed within the contact area of the originally separated populations.

In the field, however, it is usually difficult to pin down hybridization belts since not every population having intermediate characteristics is necessarily a hybrid population. Only if it can be demonstrated that a dispersal center is based on patterns of differentiation (Montanucci 1974) which can only be interpreted as the result of geographic isolation, and that the populations can be regarded as pleiochorous, can it be concluded, with a reasonable probability, that the identical structures of the dispersal centers are the basis for distribution centers (Fig. 101). As far as their development is concerned, distribution centers are by no means only associated with the Pleistocene, although this is often asserted. Refuges have been formed at all times, and are still being formed at present. After the conclusion of the adverse phase, these can function as dispersal centers.

Every species possesses at least one dispersal center which coincides with its center of origin. During the course of history of the species, however, the two centers can be separated by vast distances. The dispersal centers determined, therefore, simply represent areas where populations survived the last adverse environmental conditions affecting them.

Dispersal centers have recently been analyzed for South and Central America (Müller 1973) (Fig. 102). This analysis shows that there are at least 40 centers whose location was affected by glacial or postglacial fluctuations in climate and by vegetation fluctuations. The latest rate of differentiation of the faunal elements (at subspecific level, among others) assigned to these centers is to be considered as an indicator of the time of the last time the centers

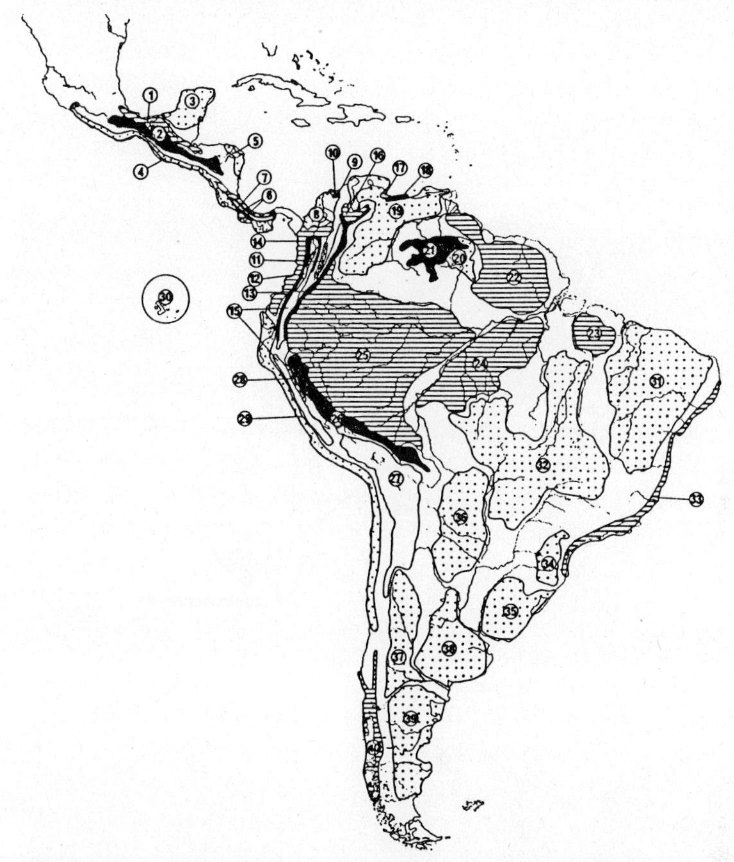

Figure 102 The dispersal centers of the Neotropic terrestrial fauna (according to Müller, 1972, 1973). Black = montane forest centers; hatched = rainforest centers of high mountain areas; dotted = nonforest centers. The numbers in the center signify the following: 1 = Central American Rainforest Center; 2 = Central American Montane Forest Center; 3 = Yucatan Center; 4 = Central American Pacific Center; 5 = Cocos Center; 6 = Costa Rica Center; 7 = Talamanca-Paramo Center; 8 = Barranquilla Center; 9 = Santa Marta Center; 10 = Sierra Nevada Center; 11 = Magdalena Center; 12 = Cauca Center; 13 = Colombian Montane Forest Center; 14 = Colombian Pacific Center; 15 = North Andean Center; 16 = Catatumbo Center; 17 = Venezuelan Coast Forest Center; 18 = Venezuelan Montane Forest Center; 19 = Caribbean Center; 20 = Roraima Center; 21 = Pantepui Center; 22 = Guyana Center; 23 = Para Center; 24 = Madeira Center; 25 = Amazon Center; 26 = Yungas Center; 27 = Puna Center; 28 = Maranon Center; 29 = Andean Pacific Center; 30 = Galapagos Center; 31 = Caatinga Center; 32 = Campo Cerrado Center; 33 = Serra do Mar Center; 34 = Parana Center; 35 = Uruguay Center; 36 = Chaco Center; 37 = Monte Center; 38 = Pampa Center; 39 = Patagonian Center; 40 = Nothofagus Center.

functioned as preservation centers of faunas and floras during unfavourable environmental phases. When attempting to classify the 40 centers, one has to realize that the affinity between two or several centers is a function of the phylogeny of the faunal elements assigned to the centers.

By this means, polytypic and polycentric species whose ecological potential is known, are given special consideration. Supraspecific units, on

Figure 103 Distribution centers of the semispecies of the South American birds of the *Crax rubra* group. The phylogenetic relationships are indicated by arrows.

the other hand, are not considered since the often great phylogenetic age of such groups provided the individual taxa with enough time to adapt to very different habitats.

The ecological potential indicating the amplitude of environmental conditions within which a species is able to grow, can often change considerably during the course of evolution of a taxon, and even from population to population. In the case of vertebrates, the distributional area of specifically or subspecifically differentiated populations is in most cases strictly correlated with vegetation formations or climatic zones, etc. The problem of the size of the habitat island is relevant here.

Ecologically, the polytypic cortalid *Lachesis mutus* is strictly tied to the rainforests of the lowlands. Its disjunctly distributed monocentric subspecies are faunal elements of the Serra do Mar Center (*Lachesis mutus noctivagus*), of the Amazon Center (*L. m. mutus*), and of the Costa Rica Center (*L. m.*

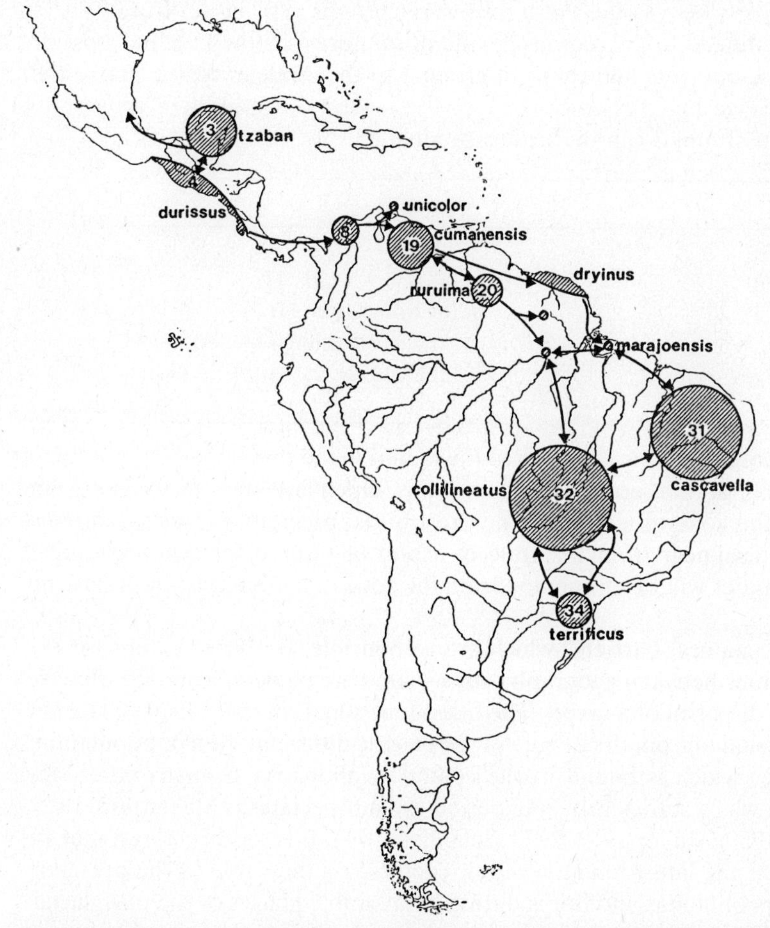

Figure 104 Distribution centers of the subspecies of the South American savanna rattlesnake *Crotalus durissus* and the presumed migration paths of Recent isolated populations. The numbers in the circles correspond to those of the dispersal centers (according to Müller, 1973).

stenophrys). *Crax rubra* is a polycentric group of birds which is strictly adapted to the forest. It forms a superspecies complex encompassing seven species which are monocentric for centers 14, 17, 22, 24, 25, and 33 (area numbers as in Fig. 102; Fig. 103).

Crotalus durissus (Fig. 104) is a polytypic rattlesnake species which, in contrast to *Lachesis mutus,* strictly avoids the rainforest. Looking at the monocentric subspecies areas of this crotalid, one can see that, with the exception of the two island subspecies on Marajo (*C.d. marajoensis*) and Aruba (*C.d. unicolor*), as well as of the subspecies *dryinus,* endemic in the coastal savannas of Guyana, the other subspecies must be assigned, as faunal elements, to centers 3, 4, 8, 19, 20, 31, 32 and 34.

In accordance with the degree of taxonomic and phylogenetic relationship of their faunal elements, the 40 centers can be defined as dispersal centers and can be classified into three large groups. The dispersal centers of group 1 are linked to the regions devoid of rainforest below 1500 m, those of group 2 to rainforests and those of group 3 to the Andean region above the forest limit (see Fig. 102).

Groups 1 and 2 can be further subdivided:

Group 1:	Group 2:
Subgroups:	Subgroups:
(a) 3, 4, 5	(a) 1, 6, 9, 14, 16, 17
(b) 8, 11, 12, 19, 20	(b) 2, 10, 13, 18, 21, 26
(c) 31, 32, 34, 36	(= montane forest species)
(d) 35, 37, 38, 39	(c) 22, 23, 24, 25, 33
(e) 28, 29, 30	(d) 40

The potential zone of action of the faunas in the three groups is modified by the degree of their ecological adaptations. Open landscapes will simply not appear in the zone of action of rainforest birds, precipitation-rich rainforest areas will disappear from the zone of action of Campo species, and cooler Alpine climates will be eliminated from the zone of action of tropical flatland species.

Those natural barriers, which are responsible to a large extent for the discontinuities between geographically isolated areas, determine or influence the rate of dispersal of a taxon. The latest functional role of a dispersal center can be concluded from the degree of subspecific differentiation of populations of a species which is bound to the center by allopatry. In many cases, the postglacial was a sufficiently long period for subspeciation (Mayr 1967, 1975, de Lattin 1959, Müller 1969, 1970, Selander 1976). It is for this reason that we surmise that the latest function of our centers was their role as the preservation centers of faunas and floras during unfavorable phases of the postglacial. We must therefore start out with the assumption that the unfavorable phases for the forest faunas were caused by the expansion phases of nonforest faunas and vice-versa, and that the location of dispersal centers was primarily determined by glacial and postglacial climatic fluctuations and vegetation fluctuations. The significance of these for the differentiation of taxa has already been determined for Africa by Chapin (1932), Moreau (1933, 1963, 1966, 1969), and Eisentraut (1968, 1970); for Australia by Gentilli (1949) and Keast (1959, 1961, 1968); for South America by Vanzolini (1963, 1970), Haffer (1967, 1969, 1970), Müller (1968, 1970, 1971), and Müller and Schmithüsen (1970).

During the Submergência Ilha do Mel (Quaternary period; Bigarella 1965), Campo expansions occurred in the course of which Campo cerrado faunal elements, among others, immigrated in to Amazonia (Vanzolini 1963, Haffer 1967, Müller 1968, 1970).

South American island faunas can be used as indicators of the date of these shifts in major ecosystems within the central and eastern parts of South America (Müller 1970, Müller and Schmithüsen 1970). The Campo expansion phase can be determined as occurring in the period from 6,000 to 2,000 B.C.

Figure 105 Migration routes of savanna species through the rainforests under more arid conditions during the Pleistocene and Holocene (according to Müller 1973).

(Ab'Saber 1962, 1965, Bigarella 1965, Fränzle 1976, Goosen 1964, Pimienta 1958, Tricart, Vogt and Gomes 1960); it led to a widening of the Restinga from Cabo Frio to Rio Grande do Sul (Delancy 1963, 1966, Hurt 1964, Vanzolini and Ab'Saber 1968, Vuillemier 1971, 1975) and to an arid period in the northeast of Brazil (Tricart, Santos, Silva and Silva 1958). The migration routes of non-forest species were interrupted by a new forest expansion phase starting around 4,400 B.C. which, with a few exceptions, is still continuing today (Fig. 105). The Campo islands within the Amazonian rainforest, which exhibit a specific Campo fauna and flora (Reinke 1962), must be considered as relicts of these postglacial arid phases which greatly influenced the location of the forest centers and the rates of differentiation of the forest fauna. Recent intermixing zones of subspecifically differentiated forest populations (Haffer 1969, Vanzolini and Williams 1970) are located within the area of these Campo migration routes which have been reclaimed by the forest over the past 4,000 years.

The beginning and the end of the postglacial arid phase are characterized by a pluvial phase with rainforest expansion. This led to a regression and isolation of the nonforest biomes. The Campo Cerrado, since the beginning of the postglacial arid phase until the present, made gene flow exchange between the rainforest populations of the Serra do Mar and Amazonia difficult or prevented such gene flow completely. This is indicated by the subspecific differentiation of numerous disjunct distributed forest popula-

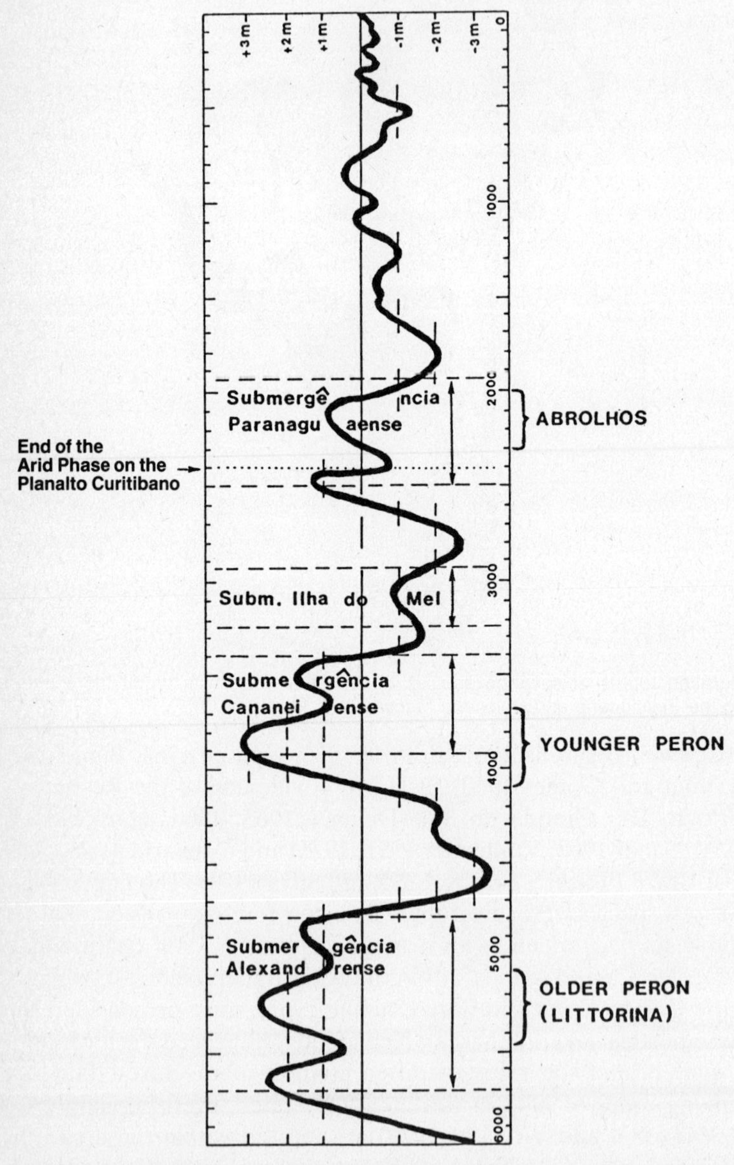

Figure 106 Sequence of sea level fluctuations during the postglacial period along the Brazilian Atlantic coast (according to Müller 1973).

tions. The Camp Cerrado must have contained considerably more trees in certain locations around 7,000 B.C. than it does today (Müller 1968). Regrettably, there are insufficient pollen-analytical studies with respect to the rainforest of Amazonia. This is in part due to the fact that even the pollen of recent species have not been sufficiently analyzed. Work on the pollen of Amazonian plants, published by Batista and Andrade (1975) is therefore of great importance (compare also Carvalo 1971, Absi 1979).

As the climate grew warmer starting with the late glacial period (Gonzales, van der Hammen and Flint 1965, van der Hammen 1974), an increased isolation of the montane forest and high montane faunas took place due to a vertical shift; these faunas, around 11,000 B.C., still preferred foothills biomes (Haffer 1970, 1974). It is not yet certain what effect this vertical shift (Heine 1974) had on the rainforest fauna of the lowlands (below 1,500 m). A large part of the subspeciation among montane forest faunas is attributable to this isolation of the montane forests which started around 8,000 B.C.

The number of Neotropic dispersal centers can be correlated with the species richness of the South and Central American biomes. The dispersal centers must also be viewed as differentiation centers, whose location was greatly influenced by Quaternary changes in major ecosystems and by climatic fluctuations (see Fig. 106). We consider this supports our assumption that most species of the Neotropic forest fauna developed in forest refuges during the arid phases (Müller 1968, 1970, Müller and Schmithüsen 1970, Haffer 1967, 1969, 1970, Vanzolini 1970, Vanzolini and Williams 1970, Vuilleumier 1975). In this connection, it is remarkable that forest refuges, independently analyzed by Haffer (1969) for the Amazonian forest birds, by Vanzolini and Williams (1970) for the *Anolis chrysolepis* group, by Brown (1976, 1977), Brown *et al.* (1974), and Turner (1971, 1972) for the Heliconidae, by Spasski *et al.* (1972) and Winge (1973) for the *Drosophila* species, and by Prance (1973) for plants, are located within our dispersal centers (Müller 1972, 1973, 1975). For critical observations, see Endler (1979).

No doubt, major problems remain in the delineation of the centers, as has been highlighted by Vuilleumier (1977). It has become obvious, however, that habitats located between two centers which are delineated according to our interpretation generally exhibit a mixed fauna.

References

Abbot, M. and van Ness, H.: Thermodynamik, Theorie und Anwendung. MacGraw-Hill Book Comp., New York, London, Düsseldorf, 1976.

Ab'Saber, A.N. Revisão dos conhecimentos sobre o horizonte subsuperficial de cascalhos inhumados do Brasil Oriental. Bol. Univ. Parana 2. 1962.

Ab'Saber, A. N.: A Evolução geomorfologica. In: A baixada Santista. São Paulo, 1965.

Almquist, E.: Floristic notes from the railways, Sv. Bot. Tidskr. *51:* 223–264. 1967.

Alt, D.: Pattern of Post-Miocene eustatic fluctuations of Sea level. Palaeogr., Palaeodim., Palaeoecol. *5:* 87–94. 1968.

Amadon, D., Stuart, E.B., Gal-Or, B. and Brainard, A.J.: The Hawaiian honey-creepers (Aves, Drepaniidae). Bull. Amer. Mus. Nat. Hist. *95:* 151–262. 1950.

Amerson, A.B.: Species richness on the nondisturbed northwestern Hawaiian Islands, Ecology *56:* 435–444. 1975.

Anderson, S.: Patterns of Faunal Evolution. Quart. Rev. Biol. *49:* 311–322. 1974.

Andrade, M.A.B. de: Contribuição ao conhecimento da ecologia das plantas das dunas do litoral do Estado de São Paulo. Faculdade de Filosofia, Ciências e Letras, Bot. *22:* 3–170. 1967.

Andrade, M.A.B., Rachid-Edwards, M. and Ferri, M.G.: Informações sobre a transpiração de duas gramineas frequentes no cerrado. Rev. Brazil, Biol. *17* (3): 317–324. 1957.

Andrewartha, H.G.: Introduction to the study of Animal Populations. Univ. Chicago Press, Chicago. 1961.

Andriashev, A.P.: A general review of the antarctic fish fauna. In: Biogeography and Ecology in Antarctica. 491–550. 1965.

Ant, H.: Die malakologische Gliederung einiger Buchenwaldtypen in Nordwestdeutschland. Vegetatio *18:* 374–386. 1969.

Archer, T.E.: Stability of DDT in foods and feeds, transformation in cooking and food processing, removal during food and feed processing. Residue Rev. *61:* 29–36, Springer Verl., New York, Heidelberg, Berlin. 1976.

Arens, K.: O cerrado como vegetação oliogotrófica. Bol. Fac. Fil. Ciênc. e Letr. USP. 224, Botanica *15:* 59–77. 1958.

Arens, K.: As plantas lenhosas dos campos cerrados como vegetação adaptada a deficiências minerais do solo. Simpósio sobre o cerrado, 249–266, Ed. da Univ. São Paulo. 1963.

Arndt, U.: Langfristige Immissionswirkungen an ungeschütztem Nutzholz. Staub-Reinhaltung der Luft *34:* 225–227. 1974.

Ashley, H., Rudman, R.L. and Whipple, C.: Energy and the Environment: A Risk Benefit Approach, Pergamon Press, New York. 1976.

Ashton, P.S.: The quaternary geomorphological history of western Malasia and lowland forest phytogeography. Trans. Aberdeen-Hull Symp. Malasian ecology, Univ. Hull, Dep. Geogr. Misc. Ser. *13:* 35–49. 1972.

Askew, R.P., Cook, L.M. and Bishop, J.A.: Atmospheric pollution and melanic moth in Manchester and its environs. J. applied ecology *8* (I): 247–256. 1971.

Aurand, K.: Die natürliche und künstliche Strahlenexposition des Menschen. In: Kernenergie und Umwelt, 179–190, E. Schmidt Verl., Berlin. 1976.

Axelrod, D.J.: Evolution of the Madro-Tertiary Geofloras. Bot. Rev. *24:* 433–509. 1958.

Bach, H., Beck, P. and Goettling, D.: Energie und Abwärme. Verl. E. Schmidt, Berlin. 1973.

Bach, W.: Changes in the composition of the Atmosphere and their impact upon climatic variability – an overview. Bonner Meteorol. Abh. *24:* 1–51. 1976.

Bailey, R.H.: Ecological Aspects of Dispersal and Establishment in Lichens. In: Lichenology: Progress and Problems, Acad. Press, London. 1976.

Baker, R.H.: Mammals of the Guadina Lava Field, Durango, Mexico. Mich. State Univ. Biol. Ser. *I:* 305–327. 1960.

Balgooyen, Th. and Moe, L.M.: Dispersal of Grass Fruits. An Example of Endoornithochory. Americ. Midl. Naturalist *90* (2): 454–455. 1973.

Banarescu, P.: Principusi probleme de zoogeografie. Acad. Rep. Soc. Rom, Bucuresti. 1970.

Banarescu, P.: Competition and its Bearing on the Fresh-Water Faunas. Rev. Roum. Biol. *16* (3): 153–164. 1971.

Bannerman, D.A. and Bannerman, M.N.: Birds of the Atlantic Islands. A History of the Birds of Madeira, the Deserts and the Porto Santo Islands. Oliver & Boyd, Edinburgh and London. 1965.

Bartlett, A.S. and Barghoorn, E.S.: Phytogeographic History of the Isthmus of Panama during the last 12,000 years (A history of Vegetation, Climate and Sealevel-Change). In Vegetation and Vegetational history of Northern Latin-America. Amsterdam-London-New York. 1973.

Batista, H.P. and Andrade, T.A.P. de: O pólen em plantas da Amazônia. V. Contribuição ao estudo da Familiam Icacinaceae. Bol. Mus. Paraense Emilio Goeldi, Botânica *47:* 1–14. 1975.

Battistini, R.: Madagascar relief and main types of Landscape. In: Biogeography and Ecology in Madagascar. The Hague. 1972.

Baumgartner, A.: Wald als Austauschfaktor in der Grenzschicht Erde/Atmosphäre. Forstwiss. Centralbl. *90* (3): 174–182. 1971.

Baumgartner, A. and Reichel, E.: Die Weltwasserbilanz. Oldenbourg Verl., München, Wien. 1975.

Beadle, L.C.: The Inland Waters of Tropical Africa. Longman, New York. 1974.

Beadle, N.C.W.: Soil temperatures during forest fires and their effects on the survival of vegetation. J. Ecology *28* (1): 180–192. 1940.

Beavington, F.: Heavey Metal Contamination of Vegetables and Soil in Domestic Gardens Around a Smelting Complex. Environm. Pollut. *9:* 211–217. 1975.

Beck, A.M.: The ecology of stray dogs: a study of free-ranging urban animals. York Press, Baltimore. 1973.

Becker, J.: Art und Ursachen der Habitatbindung von Bodenarthropoden (Carabidae, Diplopoda, Coleoptera, Isopoda) xerothermer Standorte in der Eifel. Diss. Univ. Köln. 1972.

Becker, J.: Die Carabiden des Flughafens Köln/Bonn als Bioindikatoren für die Belastung eines anthropogenen Ökosystems, Decheniana *20:* 1–9. 1977.

Benjamin, Y.H.L.: Fine Particles, Aerosol Generation, Measurement, Sampling and Analysis. Acad. Press, New York and London. 1976.

Benson, S.B.: Concealing coloration among desert rodents of southwestern United States. Univ. Calif. Pub. Zoolo. *40:* 1–70. 1933.

Berg, L.S.: Die bipolare Verbreitung der Organismen und die Eiszeit. Zoogeographica. *1* (4): 449–484. 1933.

Berhausen, E.H.: Human pathogene Helminthen aus Fäkalien des Haushundes von Kinderspielplätzen im Stadtgebiet von Mainz. Mz. Naturwiss. Archiv. *12:* 23–41. 1973.

Bernardi, G.: Aréatypes et Chorologie de l'Ouest Africain principalement d'après les Pieridae (Insect. Lépid.). J. West African Science Assoc. *11:* 49–67. 1966.

Bernardi, G.: Polymorphisme et mimétisme chez les Lépidoptères Rhopalocères, Mém. Société Zool. France *37:* 129–165. 1974.

Bernatzky, A.: Großstadtklima und Schutzpflanzungen. Nat. u. Mus. *102* (11): 425–431. 1972.

Beschel, R.: Flechtenvereine der Städte; Stadtflechten und ihr Wachstum. Ber. Naturiwss.-Med. Ver. Innsbruck *52:* 1–158. 1958.

Besuchet, C.: Répartition des insectes en Suisse, influence des glaciations. Mitt. Schweiz. Entom. Ges. *41:* 337–340. 1968.

Beurlen, K.: Die geologische Entwicklung des Atlantischen Ozeans. Geotekt. Forsch. *46:* 1–69. 1974.

Bevenue, A.: The "bioconcentration" aspects of DDT in the environment. Residue Rev. *61:* 37–112, Springer Verl., New York, Heidelberg, Berlin. 1976.

Bick, H. and Kunze, S.: Eine Zusammenstellung von autökologischen und saprobiologischen Befunden an Süßwasserciliaten. Int. Revue ges. Hydrobiol. *56* (3): 337–384. 1971.

Bigarella, J.J.: Sand-ridge structures from Paraná Coastal Plain, Marine Geology *3:* 269–278. 1965.

Bigarella, J.J.: Subsidios para o estudo das variações de nivel oceanico no quaternario brasileiro. A. Acad. Brasil, Ci. *37:* 263–278. 1965.

Bishop, J.A.: An experimental study of the cline of industrial melanism in Biston betularia (L.) (Lepidoptera) between urban Liverpool and rural North Wales. J. animal ecology *41:* 209–243. 1972.

Blair, W.F.: Ecological distribution of mammals in the Tularossa Basin, New Mexico. Contrib. Lab. Vert. Biol. Univ. Mich. *20:* 1–24. 1943.

Blanchard, R.: Sur quelques variétés françaises du Lézard des murailles. Mém. Soc. zool. France, *4:* 502–508. 1891.

Blau, G.E. and Neely, W.B.: Mathematical Model Building with an Application to Determine the Distribution of Dursban Insecticide added to a Simulated Ecoystems. Adv. Ecol. Research *9:* 133–163. 1975.

Böcher, T., Holmen, K. and Jakobsen, K.: The Flora of Greenland. Haase and Son. Publ., Copenhagen. 1968.

Boesch, E.: Psychopathologie des Alltags. Zur Ökopsychologie des Handelns und seiner Störungen. Verl. Huber, Bern, Stuttgart, Wien. 1976.

Boney, A.D.: Phytoplankton, Arnold Publ. Studies in Biology *52.* London. 1975.

Bonnefille, R.: Associations polliniques actuelles et quaternaires en Ethiopie. Thèse, Univ. de Paris VI, CNRS A 07229 T., 1,513 p. 1972.

Bonte, L.: Beiträge zur Adventivflora des rheinisch-westfälischen Industriegebietes 1913–1927. Decheniana *86:* 141–255. 1930.

Bornkamm, R.: Vegetation und Vegetations-Entwicklung auf Kiesdächern. Vegetatio *10:* 1–24. 1961.

Bornkamm, R.: Die Unkrautvegetation im Bereich der Stadt Köln. Decheniana *126* (1/2): 267–306, 307–332. 1974.

Bourgat, R.M.: Cytogénétique des Caméléons de Madagascar. Incidences taxónomiques, Biogéographiques et phylogénétiques. Bull. Soc. Zool. France *98* (1): 81–90. 1973.

Bovden, J. and Johnson, C.G.: Migrating and other terrestrial insects at sea. In: Marine insects, North-Holland Publ. Comp., Amsterdam, Oxford, New York. 1976.

Bowman, R.: Morphological differentation and adaption in the Galapagos finches, Univ. California Publ. Zool. *58.* 1961.

Braestrup, F.W.: Remarks on climatic change and Faunal Evolution in Africa. Zoogeographica *2* (4): 484–494. 1935.

Bramwell, D. and Bramwell Z.Y.: Wild flowers of the Canary Islands. Pitman Press, London. 1974.

Braun-Blanquet, J.: Pflanzensoziologie. Grundzüger der Vegetationskunde. Wien. 1964.

Brauns, A.: Praktische Bodenbiologie, Fischer Verl. Stuttgart. 1968.

Bridges, E.M.: World Soils, Cambridge Univ. Press. 1970.

Briggs, J.C.: Marine Zoogeography. McGraw-Hill, New York. 1974.

Brisbin, I.L. and Smith, M.H.: Radiocesium concentrations in Whole-Body Homogenates and several Body Compartments of naturally contaminated White-Tailed Deer. Mineral Cycling in Southeastern Ecosystems, Springfield, Virginia. 1975.

Brock, T.D.: Microbial growth under extreme environments. Symp. Soc. gen. Microbiol. *19:* 15–41. 1969.

Brocksieper, R.: Der Einfluß des Mikroklimas auf die Verbreitung der Laubheuschrecken, Grillen und Heldheuschrecken im Siebengebirge und auf dem Rodderberg bei Bonn (Othoptera: Saltatoria). Decheniana, Beiheft *21:* 1–141, Bonn. 1978.

Brodkorb, P.: Origin and Evolution of Birds. In: Avian Biology, Acad. Press, New York, London. 1971.

Brown K.: Centros de evolução, refuios quaternarios e conservação, de patrimônios genéticos na regiás neotropical: padrões de diferenciação em Ithomiinae (Lepidoptera: Nymphalidae). Acta Amazonica *7* (1): 75–137. 1976.

Brown, K.: Geographical patterns of evolution in neotropical forest Lepidoptera (Nymphalidae: Ithomiinae and Nymphalinae-Heliconiini). In: Biogéographie et Evolution en Amérique tropical, Publ. Lab. Zool. de l'Ecole Normale Superieure *9:* 118–160. 1977.

Brown K.S., Sheppard, P.M. and Turner, J.R.: Quaternary refugia in tropical America: Evidence from race formation in Heliconius butterflies. Proc. R. Soc. London *187:* 369–378. 1974.

Brüll, H.: Das Leben europäischer Greifvögel. Ihre Bedeutung in den Landschaften. 3. Aufl. Verl. Fischer, Stuttgart. 1977.

Brulotte, R.: Study of Atmospheric Pollution in the Thetford Mines Area, Cradle of Quebec's Asbestos Industry. In: Benarie, Atmospheric Pollution, 447–469. Elsevier Publ., Oxford. 1976.

Brundin, L.: Transantarctic relationships and their significance, as evidenced by chironomid midges, with a monograph of the subfamilies Podonominae and Aphroteniiae and the austral Heptagyiae. Kungl. Svenska Vetenskapsakademeins Handlinger, Fj. Ser. *11.* 1966.

Brundin, L.: Circum-Antarctic distribution patterns and continental drift, XVII Congrès internat. Zool. Biogéographie et liaisons intercontinentales au cours du Mésozoique. Monte Carlo (Manuscript). 1972.

Buchwald, K.: Der Ländliche Raum als ökologischer Ausgleichsraum für die Verdichtungsgebiete. Umwelt und Gesellschaft, 51–61. 1974.

Buchwald, K. et al.: Gutachten für einen Landschaftsrahmenplan Bodensee Baden-Württemberg. Stuttgart. 1973.

Buckup, L. and Rossi A.: O gênero Aegla no Rio Grande do Sul, Brasil (Crustacea, Decapoda, Anomura, Aeglidae). Iheringia. 1977.

Bull, P.C. and Whitaker, A.H.: The Amphibians, Reptiles, Birds and Mammals. In: Biogeography and ecology in New Zealand, Junk, The Hague. 1975.

Bullock, St.H.: Consequences of Limited Seed Dispersal within Simulated Annual Populations. Oecologia *24:* 247–256. 1976.

Bunting, B.T.: The Geography of Soil. Hutchinson. 1965.

Burchard O. and Ahl, E.: Neue Funde von Riesen-Landschildkröten aus Teneriffe. Z. deutsch. geol. Ges. *79:* 439–447, Berlin. 1927.

Burrage, S.W.: Aerial Microclimate around plant surfaces. In: Dickinson and Preece, Microbiology of Aerial Plant Surfaces, 173–184, Acad. Press, London, New York, San Francisco. 1976.

Burton, J.D. and Liss, P.S.: Estuarine Chemistry. Acad. Press, London, New York, San Francisco. 1976.

Butzer, K.W., Issac, G.L., Richardson, J.L. and Washbourn-Kamau, C.: Radiocarbon dating of East African lake levels. Science *175:* 1069–1076. 1972.

Cabrera, A. .: Ecologia vegetal de la puna. Geoecologia de las regiones montanosas de las Americas tropicales. Proc. Unesco Mexico Symposium, *1966:* 91–116. 1980.

Carpenter, J.R.: The Biome. Amer. Midl. Natural. *21* (1). 1938.

Carvalho, M.J.M.: O polen em plantas da Amazônia. Gênero Poraqueiba Aubl. e Emmotum. Bol. Mus. Para. Emilio Goeldi, Botânica *42*. 1971.

Cernusca, A.: Ökophysik: Neue Wege zur quantitativen Ökologie. Umschau in Wissenschaft und Technik, *18:* 663–668. 1971.

Chaline, J.: Biogéographie et fluctuations climatiques au Quaternaire d'après les faunes de rongeurs. Acta Zoologica Cracoviensia *18* (7):141–165. 1973.

Chandler, T.R.: Selected Bibliography on Urban Climate, World Meteorological Organization 276, T.P. 155, Geneva 1970.

Changnon, S.A.: Recent studies of urban effects on precipitation in the United States, World Meteorological Organization *108:* 325–341. 1970.

Chapin, J.P.: Birds of the Belgian Congo. Amer. Mus. Nat. Hist. *65*. 1932.

Chapman, V.J.: Mangrove Vegetation. Pergamon Press, Oxford. 1974.

Chapman, V.J.: Wet Coastal Ecosystems. Elsevier Scient. Publ. Amsterdam. 1977.

Cheng, L.: Marine insects, North-Holland Publ., Amsterdam, Oxford, New York. 1976.

Clark, A.H.: The invasion of New Zealand by people, plants and animals. New Brunswick. 1949.

Claussen, T.: Die Reaktionen der Pflanzen auf Wirkungen des photochemischen Smogs. Acta Phytomedia 3, Verl. P. Patey, Berlin and Hamburg. 1975.

Cleef, A.M.: Characteristics of Neotropical Paramo Vegetation and its Subantarctic Relations. Erdwiss. Forschung. 1977.

Clements, F.E. and Shelford, V.E.: Bio-ecology. New York. 1939.

Cleve, K.: Die Erforschungen der Ursachen für das Auftreten melanistischer Schmetterlingsformen im Laufe der letzten hundert Jahre. Z. angew. Ent. *65:* 371–387. 1970.

Cloudsley-Thompson, J.L.: Man and the Biology of Arid Zones. Arnold, London. 1977.

Cody, M.L. and Walter, H.: Habitat selection and interspecific interactions among

Meditteranean sylviid warblers. Oikos. 27. 1977.
Coker, R.E.: Das Meer — der größte Lebensraum. Verl. Parey, Hamburg and Berlin. 1956.
Cole, M.M.: Recent developments in biogeography. In: Phillips, A. and Turton B.: Environment, Man and Economic Change. Longman, London, New York. 1975.
Cook, R.E.: Variation in species density of North American birds. Syst. Zoology. 1969.
Coope, G.R.: Interpretations of quaternary fossil insects. Ann. Rev. Entomol. *15:* 97–120. 1970.
Coope, G.R., Morgan A. and Osborne, P.J.: Fossil Coleoptera as indicators of climatic fluctuations during the last glaciation in Britain. Palaeogeography, Palaeoclimatology, Palaeoecology *10:* 87–101. 1971.
Cornaby, B.W., Gist, C.S. and Crossley, D.A.: Resource partitioning in Leaf-Litter faunas from Hardwood and Hardwood-Converted-To-Pine forests. Mineral Cycling in Southeastern Ecosystems, Springfield. 1975.
Cour, P. and Duzer, D.: Persistance d'un climat hyperaride en Sahara Centrale méridional au cours de l'holocene. Rev. de Géog. phys. et de Géol. dyn. *18:* 175–198. 1976.
Crawford, M.D., Gardner, M.J. and Morris, J.N.: Changes in water hardness and local death-rates. Lancet *2:* 327–329. 1971.
Crocker, R.L.: Fast climatic fluctuations and their influence upon Australian vegetation. In: Keast, A. et al. (eds.), Biogeography and ecology in Australia, The Hague. 1959.
Croizat, L., Nelson, G. and Rosen, D.E.: Centers of origin and related concepts. System. Zool. *23:* 265–287. 1974.
Cromwell, J.E.: Barrier coast distribution. A world-wide survey. Second Natl. Coastal Shallow Water Res. Conf. 1971.
Cronin, L.E.: Estuarine Research. Chemistry, Biology and the Estuarine System. Acad. Press, Inc., New York, San Francisco, London. 1975.
Cuatrecasas, J.: Aspectos de la vegetación natural de Colombia. Rev. Acad. Col. Cienc. E. F. Nat. *10* (40): 1958.
Cuatrecasas, J.: Paramo vegetation and its life forms. Geo-ecology of the mountainous regions of the tropical Americas. Coll. Geogr. *9:* 163–186. 1968.
Dahl, E.: Flora and Plant Sociology in Fennoscandian Tundra Areas. In: Ecol. Studies *16:* 62–67, Verl. Springer, Heidelberg, New York. 1975.
Dammermann, K.W.: The fauna of Krakatau. Amsterdam. 1948.
Dansereau, P.: Zonation et succéssion sur la restinga de Rio de Janeiro. Rev. Canad. Biol. *6* (3): 448–477. 1947.
Dansereau, P.: Biogeography, an Ecological Perspective. Ronald Press. Comp., New York. 1957.
Darlington, P.: Area, climate and evolution. Evolution *13:* 488–510. 1959.
Darlington, P.: Zoogeography. New York. 1957.
Darwin, C.: Structure and distribution of Coral reefs, 2nd (revised) edition, 1874. Stuttgart. 1876.
Delaney. P.J.V.: Quaternary Geologic History of the Coastal Plain of Rio Grande do Sul, Brazil, Louisiana State University Press, Baton Rouge. 1963.
Delaney, P.J.V.: Geology and Geomorphology of the Coastal Plain of Rio Grande do Sul, Brazil and Northern Uruguay, Louisiana State University Press, Baton Rouge. 1966.

Diamond, J.M.: Colonization of exploded volcanic islands by birds: the supertramp strategy. Science *184:* 803–806. 1974.

Diamond, J.M.: Ecological Consequences of Island Colonization by Southwest Pacific Birds, II. The Effect of Species Diversity on Total Population Density. Proc. Nat. Acad. Sciences *67* (4): 1715–1721. 1970.

Diamond, J.M.: The island dilemma: Lessons of modern biogeographic studies for the design of natural reserves. Biol. Conser., *7:* 129–146, Applied Science Publ., Great Britain. 1975.

Dierl, W.: Grundzüge einer ökologischen Tiergeographie der Schwärmer Ostnepals (Lepidoptera-Sphingidae). Khumbu Himal *3:* 313–360. 1970.

Domrös, M.: Möglichkeiten einer klimaökologischen Raumgliederung der Insel Ceylon. Geogr. Z., Beiheft *27:* 205–232. 1971.

Dose, K. and Rauchfuss, H.: Chemische Evolution und der Ursprung lebender Systeme. Wissensch. Verlagsgesellsch., Stuttgart. 1975.

Drozdowicz, A.: Equilibrio Microbiologico dos solos de cerrados.4. Simpos. Cerrado, 233–245, Ed. Univ. São Paulo, São Paulo. 1976.

Düll, R.: Neuere Untersuchungen über MOose als abgestufte ökologische Indikatoren für die SO_2-Immissionene im Industriegebiet zwischen Rhein and Ruhr bei Duisburg. VDI-Kommission Reinhaltung der Luft, Düsseldorf. 1974.

Duvigneaud, P. et al.: L'écosystème. L'écologie, science moderne de synthèse. Brussels. 1962.

Ebeling, A.W., Atkin, N.B. and Setzer, P.Y.: Genome sizes of Teleostean Fishes: Increases in some Deep-Sea Species. American Naturalist *105:* 549–561. 1971.

Edney, E.B.: Water Balance in Land Arthropods. Verl. Springer, Berlin, Heidelberg, New York. 1977.

Edwards, R.W. and Garrod, D.J.: Conservation and Productivity of Natural Waters. Acad. Press, New York. 1972.

Eggers, J.: Zur Siedlungdichte der Hamburger Vogelwelt. Hamb. Avifauna. Beitr. *13:* 13–72. 1975.

Eggers, J.: Zur Vogelwelt einer großstädtischen Brachfläche (City Nord, Hamburg). Hamb. Avifaun. Beitr. *14:* 47–53. 1976.

Ehrendorfer, F.: Cytotaxonomische Beiträge zur Genese der mittel-europäischen Flora and Vegetation. Ber. Deutsch. Bot. Ges. *75:* 137–152. 1962.

Eigen, E.: Quart. Reviews of Biophys. *4:* 149. 1971.

Eisentraut, M.: Die tiergeographische Bedeutung des Oku-Gebirges im Bamenda-Banso-Hochland (Westkamerun). Bonn. Zool. Beitr. *19:* 170–175. 1968.

Eisentraut, M.: Eiszeitklima und heutige Tierverbreitung im tropischen Westafrika. Umschau *3:* 70–75. 1970.

Eisentraut, M.: Die Wirbeltierfauna von Fernando Poo und Westkamerun. Bonner Zool. Monogr. *3.* 1973.

Ekman, S.: Tiergeographie des Meeres. Leipzig. 1935.

Ekman, S.: Biologische Geschichte der Nord- und Ostsee. In: Grimpe, Tierwelt der Nord- u. Ostsee, *I.* Leipzig. 1940.

Ellenberg, H.: Belastung und Belastbarkeit von Ökosystemen. In: Belastung und Belastbarkeit von Ökosystemen, 19–26, Blasaditsch Verl., Augsburg. 1973.

Ellenberg, H.: Die Ökosysteme der Erde, Versuch einer Klassifikation der Ökosysteme nach funktionalen Gesichtspunkten. In: Ökosystemforschung, Verl. Springer, Berlin, Heidelberg, New York. 1973.

Ellenberg, H.: Das Reh in der Landschaft. Jb. des Vereins zum Schutz der Bergwelt, *42·* 225–240. 1977.

Emmel, T.: Population Biology. Harper and Row, New York. 1976.

Endler, J.: Evolution in the Tropics. Symp. Venezuela, (Manuscript). 1979.

Engler, A.: Die Entwicklung der Pflanzengeographie in den letzten 100 Jahren. Humboldt-Centenarschrift d. Ges. Erdkunde, Berlin. 1879.

Eriksen, W.: Probleme der Stadt-und Geländeklimatologie. Wiss. Buchgesellsch., Darmstadt. 1975.

Erz, W.: Populationsökologische Untersuchungen an der Avifauna zweier norddeutscher Großstäde. Z. Wiss. Zool. *170:* 1–111. 1964.

Fairbridge, R.W.: World Sea-Level and climatic changes. Quaternaria *6:* 111–134. 1962.

Farnworth, E.G. and Golley, F.B.: Fragile Ecosystems, Evaluation of Research and Applications in the Neotropis. Verl. Springer, Berlin, Heidelberg, New York. 1974.

Feldhaus, G. and Hansel, G.: Bundes-Immissionsschutzgesetz mit Durchführungsverordnung sowie TA Luft und TA Lärm. Deutsch. Fachschr.-Verl., Mainz und Wiesbaden. 1975.

Fernandez, J.M.: Los Lepidopteros diurnos de las Islas Canarias. Enciclop. Canaria, Santa Cruz de Tenerife. 1970.

Ferri, M.G.: Transpiração de plantas permanentes dos cerrados. Bol. Fac. Fil. Ciênc. e Letr. USP *41,* Botanica *4:* 159–224. 1944.

Ferri, M.G.: Ecologia dos Cerrados. IV. Simp. sobre o Cerrado, 15–36, São Paulo. 1976.

Firth, F.E.: Encyclopedia of marine resources. Van Nostrand Reinhold, New York. 1969.

Fittkau, E.: Ökologische Gliederung des Amazonas-Gebietes auf geochemischer Grundlage. Münster. Forsch. Geol. Paläont. *20/21:* 35–50. 1971.

Fittkau, E.: Artenmannigfaltigkeit amazonischer Lebensräume aus ökologischer Sicht, Amazoniana *4* (3): 321–430. 1973.

Fjerdingstad, E.: Taxonomy and saprobic valency of benthic phytomicro-organisms. Int. Revue Ges. Hydrobiol. *50:* 475–604. 1965.

Fleming, C.A.: The Geological History of New Zealand and its Biota. In: Biogeography and Ecology in New Zealand, Junk, The Hague. 1975.

Flint, P.S. and Gersper, P.L.: Nitrogen Nutrient Levels in Arctic Tundra Soils. In: Soil Organisms and Decomposition in Tundra, 375–387. Stockholm. 1974.

Ford, T.D. and Cullingford, C.H.D.: The Science of Speleology. Acad. Press, London, New York. 1976.

Forest, J., Saintlaurent, M. de and Chace, F.A.: Neoglyphen inopinata: A Crustacean "Living Fossil" from the Philippines. Science *192*, 884. 1976.

Förstner, U. and Müller, G.: Schwermetalle in Flüssen und Seen. Springer-Verl., Berlin, Heidelberg, New York. 1974.

Förstner, W. and Hübl, E.: Ruderal-, Segetal- und Adventivflora von Wien. Vienna. 1971.

Fox, R.: Naturwissenschaften, *60:* 359. 1973.

Franke, Parasitologie. Verl. Ulmer, Stuttgart. 1976.

Franken, E.: Der Beginn der Forsythienblüte in Hamburg 1955, ein Beitrag zur Phänologie der Großstadt. Meterol. Rdsch. *8:* 113–114. 1955.

Franz, H.: Die Bodenfauna der Erde in biozönotischer Betrachtung. 2. Bd. Steiner Verl., Wiesbaden. 1975.

Franz, H.: Die Rolle der Böden in den hochalpinen Ökosystemen. Verhdl. Ges. Ökologie, Wien, *1975:* 41–48, Junk, The Hague. 1976.

Fränzle, O.: Die Schwankungen des pleistozänen Hygroklimas in Südost-Brasilien und Südost-Afrika. Biogeographica *7:* 143–162, Junk, The Hague. 1976.

Fränzle, O.: Der Wasserhaushalt des amazonischen Regenwaldes und seine Beeinflussung durch den Menschen. Amazoniana *6* (1): 21–46, Kiel. 1976.

Fränzle, O.: Biophysical aspects of species diversity in tropical rain forest ecosystems. Biogeographica *8:* 69–83. 1977.

Freitas, A.S.W. et al.: Factors in whole body retention of methyl mercury in fish. Richland, Wash. 1975.

Freitas, F.G. de and Silveira, C.O. de: Principais solos sob vegetacão de Cerrado e sua aptidao agricola. 4. Simp. Cerrado, 155–194, Ed. Univ. São Paulo, São Paulo. 1976.

Frenzel, B.: Die Vegetations- und Landschaftszonen Nord-Eurasiens während der letzten Eiszeit und während der postglazialen Wärmezeit. Akad. Wiss. Lit. Mainz, Abhd. Math.-Nat. Kl. *13.* 1960.

Frenzel, B.: Grundzüge der pleistozänen Vegetationsgeschichte Nord-Eurasiens. Wiesbaden. 1968.

Frenzel, B.: Vegetationsgeschichte der Alpen. Studien zur Entwicklung von Klima und Vegetation im Postglazial. Verl. Fischer, Stuttgart. 1972.

Friedlander, S.K.: Smoke, Dust and Haze. Fundamentals of Aerosol Behavior. John Wiley, New York. 1977.

Friedrich, G.: Ökologische Untersuchungen an einem thermisch anomalen Fließgewässer (Erft/Niederrhein). Schriftenr. Landesanst. Gewässerkd. und Gewässerschutz NW *33:* 1–125. 1973.

Froebe, H. and Oesau, A.: Zur Soziologie und Propagation von Iva xanthifolia im Stadtgebiet von Mainz. Decheniana *122* (1): 147–157. 1969.

Funke, W.: Das zoologische Forschungsprogramm im Sollinprojekt. Verhdl. Ges. Ökologie *1976:* 49–58, Junk, The Hague. 1977.

Furch, K.: Haupt- und Spurenmetallgehalte zentralamazonischer Gewässertypen (erste Ergebnisse). Biogeographica *7:* 27–43. 1976.

Fuste, M.: Aperçu sur l'anthropologie des populations préhistoriques des îles canaries. Acta del V Congr. Panafricano de Prehistoria y de estudio del cuaternario, 69–80, Santa Cruz de Tenerife. 1966.

Gansen, R.: Grundsätz der Bodenbildung. B.I. Hochschultaschenbücher, Bibl. Inst., Mannheim, Vienna and Zürich. 1965.

Gardner, A.L. and Patton, J.L.: Karyotypic variation in Oryzomyine Rodents (Cricentinae) with comments on chromosomal evolution in the neotropical Cricetine complex. Occasional Pa. of the Mus. of Zool., Louisiana State University, Baton Rouge. 1976.

Gardner, W.S. et al.: Concentrations of total Mercury and Methyl Mercury in fish and other coastal organisms: implications for Mercury Cycling. Mineral Cycling in Southeastern Ecosystems, Springfield, Virginia. 1975.

Garrod, D.J. and Clayden, A.D.: Current biological problems in the conservation of Deep-Sea Fishery resources. In: Conservation and Productivity of Natural Waters. Acad. Press, New York. 1972.

Gentilli, J.: Foundations of Australian bird geography. Emu *49:* 85–129. 1949.

Gerassimov, I.P. and Glazovskaja, M.A.: Grundlagen der Bodenkunde und Bodengeographie. Moscow. 1960.

Gerlach, S.A.: Die Mangroveregion tropischer Küstren als Lebensform. Z. Morph.-Ökol. Tiere *40.* 1958.

Gerlach, S.A.: Über das Ausmaß der Meeresverschmutzung. Verhdl. Ges. Ökologie, Erlangen, 201–208, Junk, The Hague. 1975.

Gerlach, S.A.: Meeresverschmutzung. Diagnose und Therapie. Springger-Verl., Heidelberg. 1976.

Gibbs, R.J.: The geochemistry of the Amazon River system: Part I. The factors that control the salinity and the composition and concentration of the suspended solids. Geol. Soc. America Bull. *78:* 1203–1232. 1967.

Giebel, J.: Untersuchungen zur Simulation der Immissionsbelastung durch Schwefeldioxid in der Umgebung von Ballungsräumen. Schriftnr. LIS, Nordrhein-Westfalen *42:* 17–31. 1977.

Gilbert, O.L.: Further studies on the effect of sulphur dioxide on lichens and bryophytes, New Phytol. *69:* 605–627. 1970.

Gilbert, O.L.: Urban Bryophyte communities in North-East England. Trans. B.B.S. *6* (2): 3066316. 1971.

Gill, D. and Bonnet, P.A.: Nature in the urban landscape: a study of city ecosystems. York Press, Baltimore. 1973.

Gish, C.D. and Christensen, R.E.: Cadmium, Nickel, Lead and Zinc in Earthworms from Roadside Soil. Environ. Sci. Technol. *7:* 1060–1062. 1973.

Glenie, R.C., Schofield, J.C. and Ward, W.T.: Tertiary Sea Levels in Australia and New Zealand. Palaeogeogr., Palaeoclimat., Palaeoecol. *5:* 141–163. 1968.

Gliesch, R.: A Fauna de Torres, Escola de Engenharia, *5-74,* Porto Alegre. 1925.

Godley, E.J.: Flora and Vegetation. In: Biogeography and Ecology in New Zealand, Junk, The Hague. 1975.

Goebbels, R.: Die Ruderalflora der Trümmer Kölns. Dissertation, MS. Cologne. 1947.

Goeze, J.A.E.: Europäische Fauna oder Naturgeschichte der europäischen Thiere in angenehmen Geschichten und Erzählunger für allerley Leser vorzüglich für die Jugend. Leipzig. 1795.

Goldman, C.: Antarctic Freshwater Ecosystems. In: Antarctic Ecology, Acad. Press, London and New York. 1970.

Golterman, H.L.: Physiological Limnology. An Appraoch to the Physiology of Lake Ecosystems. Elsevier Scient. Publ., Amsterdam. 1975.

Good, R.: The Geography of the Flowering Plants. Longmans, London. 1964.

Goodell, H.C., Watkins, N.D., Mather, T.T. and Koster, S.: The Antarctic glacial history recorded in sediments of the southern Ocean. Palaegeogr., Palaeoclimat., Palaeoecol. *5:* 41–62. 1968.

Goodland, R.: An ecological study of the cerrado vegetation of South Brazil. McGill University, 224, Montreal. 1969.

Goodland, R.: The Savanna controversy: Background Information on the Brazilian Cerrado Vegetation, Savanna Res. Ser. *15:* McGill University, Montreal. 1970.

Goodland, R.: A physiognomic analysis of the cerrado vegetation of Central Brazil. J. Ecol. *59:* 411–419. 1971.

Goodland, R. and Irwin, H.S.: Amazon Jungle: Green hell or red desert? Elsevier Publ. Amsterdam, Oxford, New York. 1975.

Goodland R. and Pollard R.: The Brazilian cerrado vegetation: a fertility gradient. J. Ecol. *61:* 411–419. 1973.

Goosen, D.: Geomorfologia de los Llanos orientales. Rev. Acad. Col. Cien. Ex., Fis. y Nat. *12:* 129–139. 1964.

Gosline, W.A.: A reexamination of the similarities between the fresh-water fishes of

Africa and South America. Manuscript, Monte Carlo. 1972.
Gottschalk, W.: Die Bedeutung der Polyploidie für die Evolution der Pflanzen, Stuttgart. 1976.
Gray, D.: Soil and the city. In: Urbanization and Environment. Duxbury Press, 135–168, Belmont. 1972.
Greenberg, B.: Flies and Disease. Princeton, New Jersey. 1971.
Gressit, J.L.: Pacific Basin Biogeography. Bishop Mus. Press, Honolulu, Hawaii. 1963.
Griggs, G.B. and Kulm, L.D.: Glacial Marine Sediments from the Northeast Pacific. J. Sedimentary Petrology *39* (3): 1142–1148. 1969.
Grindon, L.H.: The Manchester flora. London. 1859.
Guderian, R.: Air Pollution. Ecological Studies 22. Verl. Springer, Berlin, Heidelberg, New York. 1977.
Guderian, R., Krause, H.M. and Kaiser, H.: Untersuchungen zur Kombinationswirkung von Schwerfeldioxid und schwemetallhaltigen Stäuben auf Pflanzen. Schriftenr. LIS *37,* Essen. 1977.
Guilland, J.A.: Fish Population Dynamics. John Wiley & Sons, London. 1977.
Günther, K.: Die Tierwelt Madagaskars und die zoogeographische Frage nach dem Gondwana-Land. Sitz.-ber. Ges. naturf. Freunde, Berlin (N.F.) *10:* 79–92. 1970.
Haeckel, E.: Generelle Morphologie der Organismen. G. Reimer Verl. Berlin. 1866.
Haeger-Aronson, B.: Studies on the urinary excretion of delta amino-levulinic acid and other haem precursors in lead workers and in lead intoxicated rabbits. Scand. J. clin. Lab. Invest. *12:* 1–128. 1960.
Haeupler, H. and Schönfelder, P.: Bericht über die Arbeiten zur floristischen Kartierung Mitteleuropas in der Bundesrepublik Deutschland. Mitt. Flor.-soz. Arbeitsgemeinschaft N.F. *15/16:* 14–21. 1973.
Haffer, J.: Speciation in Colombian Forest birds West of the Andes. Amer. Mus. Novitates *2294:* 1–57. 1967.
Haffer, J.: Zoogeographical notes on the "Nonforest" Lowland Bird Faunas of Northwestern South America. El. Hornero *10* (4): 315–333. 1967.
Haffer, J.: Speciation in Amazonian Forest Birds. Science *165:* 131–137. 1969.
Haffer, J.: Entstehung und Ausbreitung nord-andiner Bergvögel. Zool. Jb. Syst. *97:* 301–337. 1970.
Haffer, J.: Avian speciation in tropical South America. Pbl. Nuttall Ornithol. Club *14:* 1–390. 1974.
Hahn, J. and Aehnelt, E.: Die Fruchbarkeit der Tiere als biologischer Indikator für Umweltbelastungen. Verhdl. der Ges. f. Ökologie *I:* 49–54, Gießen. 1972.
Hahn, J. and Aehnelt, E.: Nachweis von Schädlichen Nahrungsfaktoren im Kaninchenversuch. Deutsch. Tieräztl. Wochenschr. *79:* 155–157. 1972.
Hahn, J., Günther, D., Maerckjlin, T. and Messow, C.: Befunde an Fortpflanzungsorganen und Nebennieren bei Kaninchen nach Futtergaben unterschiedlicher K-Konzentration. VII. Internat. Kongr. über tier. Fortpflanzung und Haustierbesamung. München. 1972.
Halbach, U.: Methoden der Populationsökologie. Verhdl. Ges. Ökologie, Erlangen, 1–24, Junk, The Hague. 1975.
Hall, C.A.S. and Day, J.W.: Ecosystems Modelling in Theory and Practice: An Introduction with Case Histories. Wiley & Sons, New York, Toronto. 1977.
Hammen, T. van der: The Pleistocene changes of vegetation and climate in tropical South America. J. Biogeography *1:* 3—26. 1974.

Hapke, H.-J.: Subklinische Bleivergiftung bei Schafen. Proc. Int. Symp. Environm. Health Aspects of Lead, 329–248, Amsterdam. 1972.
Hapke, H.-J.: Wirkungen und Schäden durch Blei, Cadmium und Zink bei Nutztieren. Staub-Reinhalt. Luft *34* (1): 8–10. 1974.
Hapke, H.-J. and Prigge, E.: Interactions of Lead and Glutathione with Delta-Aminolevulinic Acid Dehydratase. Arch. Toxikol. *31:* 153–161. 1973.
Hapke, H.-J. and Prigge, E.: Neue Aspekte der Bleivergiftung bei Wiederkäuern. Berl. Münch. Tierärztl. Wschr. *86:* 410–413. 1973.
Hardy, A.C.: The herring in relation to its animate environment. 1. The food and feeding habits of the herring with special reference to the east coast of England. Fish. Invest., London, ser. 2, vol. 7, no. 3. 1924.
Harrington, H.J.: Geology and Morphology of Antarctica. In: Biogeography and Ecology in Antarctica, 1–71, The Hague. 1965.
Harris, M.: A Field Guide to the Birds of Galapagos. Collins, London. 1974.
Harris, W.F. and Morris, G.: Ecologic significance of recurrent groups of Pollen and Spores in Quaternary Sequence from New Zealand. Palaeogeography, Palaeoclimatology, Palaeoecology *11:* 107–124. 1972.
Hartkamp, H.: Untersuchungen zur Immissionsstruktur einer Großstadt-Projekt "Großstadtluft". Schriftenreihe der LIB *83:* 30–38. 1975.
Hausen, H.: A Pre-canarian basement complex remains of an ancient african borderland. Acta del V Congr. Panafricano de Prehistoria y de estudio del cuaternario, 91–94, Santa Cruz de Tenerife. 1966.
Hawksworth, D.I. and Rose, F.: Lichens as Pollution Monitors. Studies in Biology *66:* 1–60, Camelot Press, Southampton. 1976.
Hays, H.R.: Das Abenteuer Biologie. Diederichs Verl., Düsseldorf. 1972.
Heath, J. and Leclercq, J.: Erfassung der europäischen Wirbellosen. Entomol. Zeit. *80* (19): 195–196. 1970.
Hedberg, O.: Adaptive evolution in a tropical-alpine environment. Taxonomy and Ecology, 71–92. London, New York. 1973.
Heilprin, A.: The geographical and geological distribution of animals. Kegan, Paul, London. 1887.
Heine, H.-H.: Beiträge zur Kenntnis der Adventiv- und Ruderalflora von Mannheim, Ludwigshafen und Umgebung. Jahresber. Ver. Naturkunde Mannheim *117/118:* 85–132. 1952.
Heine, K.: Bemerkungen zu neueren chronostratigraphischen Daten zum Verhältnis glazialer und pluvialer Klimabedingungen. Erdkunde *28:* 303–312. 1974.
Heine, K.: Schneegrenzdepressionen, Klimaentwicklung, Bodenerosion und Mensch im zentralmexikanischen Hochland im jüngeren Pleistozän und Holozän. Z. Geomorph. N.F. *24:* 160–176. 1976.
Hennig, I.: Geoökologie der Hawaii-Inseln, Franz Steiner Verl., Wiesbaden. 1974.
Hennig, W.: Grundzüge einer Theorie der phylogenetischen Systematik. Berlin. 1950.
Hennig, W.: Die Dipteren-Fauna von Neuseeland als systematisches und tiergeographisches Problem. Beitr. Ent. *10:* 221–329. 1960.
Hennig, W.: Die Stammesgeschichte der Insekten. Verl. Kramer, Frankfurt. 1969.
Heringer, E.P., Barroso, G.M., Rizzo, J.A. and Rizzini, C.T.: A Flora do Cerrado. IV. Simpósio sobre o Cerrado, 211–232, Ed. Universidade de São Paulo, São Paulo. 1976.
Herre, W. and Röhrs, M.: Haustiere — zoologisch gesehen. G. Fischer Verl., Stuttgart. 1973.
Herrmann, R.: Einführung in die Hydrologie. Teubner Verl., Stuttgart. 1977.

Hesse, R.: Tiergeographie auf ökologischer Grundlage. Verl. Fischer, Jena. 1924.
Hettche, O.: Zum Problem eines Immissionsgrenzwertes für Benzo (A) Pyren. Umwelthygiene *2:* 46–50. 1975.
Heydemann, B.: Die Biologische Grenze Land-Meer im Bereich der Salzwiesen. Verl. Steiner, Wiesbaden. 1967.
Heydemann, B.: Zum Aufbau semiterrestrischer Ökosysteme im Bereich der Salzwiesen der Nordseeküste, Faun.-ökolo. Mitt. *4:* 155–168. 1973.
Heyder, R.: Hundert Jahre Gartenamsel. Beitr. Vogelkde. *4:* 64–81. 1955.
Heyder, R.: Ein Fall des Gartenbrütens der Amsel, Turdus merula, 18. Jahrhundert. Beitr. Vogelkde. *15:* 87. 1969/70.
Heyer, R.W. and Berven, K.A.: Species diversities of Herpetofaunal samples from similar microhabitats at two tropical sites. Ecology *54:* 642-645. 1973.
Hildebrandt, G.: Biologische Rhythmen und Arbeit. Bausteine zur Chronobiologie und Chronohygiene der Arbeitsgestaltung. Verl. Springer, Vienna, New York. 1976.
Hill, D.S.: Agricultural Insect Pests of the Tropics and their Control. Cambridge Univ. Press London, New York, Melbourne. 1975.
Hirano, M.: Freshwater algae in the Antarctic Regions. In: Biogeography and Ecology in Antarctica, 127–193. The Hague. 1965.
Höhne, H. and Nebe, W.: Der Einfluß des Baumalters auf das Gewicht sowie den Mineral- und Stickstoffgehalt einjähriger Fichtennadeln. Arch. Forstwes. *13:* 153–61. 1964.
Holdgate, M.W.: Antarctic Ecology. Acad. Press, London and New York. 1970.
Holdhaus, K.: Die Spuren der Eiszeit in der Tierwelt Europas. Abhandl. Zool.-Bot. Ges. Vienna. 1954.
Holleman, D.F. and Luick, J.R.: Relationships between Potassium intake and Radiocesium retention in the Reindeer. Mineral Cycling in Southeastern Ecosystems, Springfield, Virginia. 1975.
Holm, E. and Edney, E.B.: Daily activity of Namib desert arthropods in relation to climate. Ecol. *54* (1): 45–56. 1973.
Holst, H. von: Sozialer Stress bei Tier und Mensch. Verhdl. Ges. Ökologie, Saarbrücken, p. 97–107, Junk, The Hague. 1974.
Holthuis, L.B.: Two new species of freshwater Shrimps (Crustacea, Decapoda) from the West Indies. Koninkl. Nederl. Akademie van Wetenschappen. *66:* 61–69, Amsterdam.
Holtmeier, F.-K.: Geoökologische Beobachtungen und Studien an der subarktischen und alpinen Waldgrenze in vergleichender Sicht. Erdw. Forschung *8,* Verl. Steiner, Wiesbaden. 1974.
Hopkins, D.M.: The Bering Land Bridge, Stanford. 1967.
Hopkins, D.M.: The palaeogeography and climatic history of Beringia during Late Cenozoic time. Inter-Nord *12.* 1972.
Horten, D.R.: The concept of Zoogeographic Subregions. Syst. Zool. *22:* 191–195. Lawrence 1973.
Howell, T.R.: Avian distribution in Central America. Auk *86:* 293–326. 1969.
Hower, J., Prinz, B., Gono, E. and Reusmann, G.: Untersuchungen zum Zusammenhang zwischen dem Blutspiegel bei Neugeborenen und der Bleiimmissionsbelastung der Mutter am Wohnort. Int. Symp. Environment and Health, Paris. 1974.
Hueck, H.: Die Wälder Südamerikas, Fischer Verl. Stuttgart.
Huggett, R.J., Cross, F.A. and Bender, M.E.: Distribution of Copper and Zinc in

Oysters and Sediments from three Coastal-Plain Estuaries. In: Mineral Cycling in Southeastern Ecoysystems. ERDA Distribution Category UC-11, Springfield, Virginia. 1975.
Hulten, E.: Flora of the Aleutian Islands. Cramer Verl., Weinheim. 1960.
Hulten, E.: The Circumpolar Plants. Almquist & Wiksell, Stockholm. 1962.
Hulten, E.: The circumpolar plants II. Kungl. Svensk Vetensk. Handl. *13* (1): 1–463. 1970.
Hupke, H.: Die Adventiv- und ruderalflora der Kölner Güterbahnhöfe, Hafenanlagen und Schuttplätze. 2 Ntr. Fed. Rep. Beih. *101:* 123–138. 1938.
Hurt, W.R.: Recent radiocarbon dates for Central and Southern Brazil. American Antiquity, *25:* 25–33. 1964.
Husmann, S.: Weitere Vorschläge für eine Klassifizierung subterraner Biotope und Biocoenosen der Süßwasserfauna. Int. Rev. Ges. Hydrobiol. *55:* 115–129. 1970.
Hutt, C. and Vaizey, M.J.: Group density in social behaviour. In: Neue Ergebnisse der Primatologie, 225–227, Verl. Fischer, Stuttgart. 1967.
Iglisch, I.: Potentielle Brutgewässer für Hausmücken (Arten aus der Culex-pipiens-Gruppe) im städtischen Bereich. Umwelthygiene *6:* 151–156. 1975.
Ihering, H. von: Die Geschichte des Atlantischen Ozeans. Jena. 1927.
Illies, J.: Der biologische Aspekt der limnologischen Fließwassertypisierung. Arch. f. Hydrobiol. *22* (3/4): 337–346. 1955.
Illies, J.: Die Barbenregion mitteleuropäischer Fließgewässer. Verh. internat. Ver. Limnol. *13:* 834–844. 1958.
Illies, J.: Die Wegenersche Kontinentalverschiebungstheorie im Lichte der modernen Biogeographie. Naturwissenschaften *52* (18): 505–51. 1965.
Illies, J.: Limnofauna Europas. Stuttgart. 1967.
Immelmann, K.: Einführung in die Verhaltensforschung. Verl. Parey, Berlin und Hamburg. 1976.
Irion, G.: Quaternary sediments of the upper Amazon Lowlands of Brazil. Biogeographica *7:* 163–167. 1976.
Irmler, U.: Inundation-Forest types in the Vicinity of Manaus, Biogeographica *8:* 17–29. 1977.
Jain, S.K.: Patterns of Survival and Microevolution in Plant Population. 1975.
Jakobi, H.: Ökosysteme Ostparans. Biogeographica *8:* 43–67, Junk, The Hague. 1977.
Jalas, J.: Hemerobe und hemerochore Pflanzenarten. Ein terminologischer Reformversuch. Acta Soc. Fauna Flora Fenn. *72:* 1–15. 1955.
Jalas, J.: Fälle von Introgression in der Flora Finnlands, hervorgerufen durch die Tätigkeit des Menschen. Fennia *85:* 58–81. 1961.
Janetschek, H.: Arthropod Ecology of South Victoria Land. Antarctic Res. Ser. *10:* 205–283, Washington. 1967.
Jefferies, D.J. and French, M.C.: Mercury, Cadmium, Zinc, Copper, and Organochlorine Insecticide Levels in Small Mammals Trapped in a Wheat Field. Environm. Pollut. *10:* 175–182. 1976.
Jefferson, G.T.: Cave Faunas. In: The science of Speleology. Acad. Press, London, New York. 1976.
Johannes, R.E.: Pollution and Degradation of Coral Reef Communities. Elsevier Oceanogr. *12:* 13–51. 1975.
Joiris, C.: La contamination par pesticides organochlores des oiseaux de proie en Belgique. Bull. Recherches Agronom. Gembloux. 479–483. 1974.
Jolly H. and Brown, J.M.: New Zealand Lakes. Auckland Univ. Press, New Zealand. 1975.

Jones, J.S.: Ecological Genetics and Natural Selection in Molluscs. Sciences *182:* 546–552. 1973.

Joos, H.-P.: Die Zusammensetzung der Bodenarthropodenfauna in der Umgebung einer Bundesautobahn. Thesis, Biogeography, Saarbrücken. 1975.

Jungbluth, J.H.: Über die Verbreitung des Edelkrebses Astacus (Astacus) astacus (L. 1758) im Vogelberg Oberhessen. (Decapoda: Astacidae). Pilippia *2:* 39–43. 1973.

Junk, H.: Verbreitung und Variabilität von Biston betularia. Thesis, Biogeography, Saarbrücken. 1975.

Jusatz, H.J.: Alte Seuchen auf neuen Wegen. Bild der Wissenschaft. 390–398. 1966.

Kalela, O.: Die geographische Verbreitung des Waldlemmings und seine Massenvorkommen in Finnland. Arch. Soc. Vanamo *18:* 9–16. 1963.

Kälin, K.: Populationsdichte und soziales Verhalten. Verl. Lang, Bern. 1972.

Kauffmann, E.G. and Hazel, J.E.: Concepts and Methods of Biostratigraphy. Dowden, Hutchinson & Ross, Pennsylvania. 1977.

Keast, A.: Vertebrate speciation in Australia: some comparisons between birds, marsupials and reptiles. Symp. Roy. Soc. Victoria, Melbourne Univ. Press. 1959.

Keast, A.: Bird speciation on the Australian continent. Bull. Mus. Comp. Zool. Harvard Coll. *123:* 305–495. 1961.

Keast, A.: Evolution of mammals on southern continents. IV Australian mammals: Zoogeography and Evolution. Quarterly Rev. Biol. *43* (4): 373–408. 1968.

Keil, U.: Hartes und weiches Trinkwasser und seine Beziehung zur Mortalität, besonders an kardiovaskulären Krankheiten. Öff. Gesundheits Wes. *35:* 255–263. 1973.

Keil, U., Pflanz, M. and Wolf, E.: Hartes und weiches Trinkiwasser und seine Beziehungen zur Mortalität, besonders an kardiovaskulären Krankheiten in der Stadt Hannover in den Jahren 1968 u. 1969. Umwelthygiene *4:* 110–117. 1975.

Keller, E.: Müll darf kein Müll mehr bleiben. Der Leitende Angestellte *1:* 12–13. 1976.

Keller, E.: Abfallwirtschaft und Recycling. Verl. Girardet, Essen. 1977.

Keller, Th.: Die Bedeutung des Waldes für den Umweltschutz. Schweiz. Z. Forstwesens *122* (12): 600–613. 1971.

Kenagy G.J.: Über die Rolle der Nahrungspezialisierung und des Wasserhaushaltes in der Ökologie und Evolution von einigen Wüstennagern. Verhdl. Ges. f. Ökologie, Saarbrücken, 89–95. The Hague. 1974.

Kendall, R.L.: An ecological history of the Lake Victoria Basin. Ecol. Monogr. *39:* 121–176. 1969.

Kettlewell, H.B.D.: Selection experiments on industrial melanism in the Lepidoptera. Heredity *9:* 323–342. 1955.

Kettlewell, H.B.D.: The evolution of melanism. Oxford. 1973.

Kikkawa, J. and Pearse, K.: Geographical distribution of Land Birds in Australia. Aust. J. Zool. *17:* 821–840. 1969.

Kjelvik, S. and Kärenlampi, L.: Plant Biomass and Primary Production of Fennoscandian Subarctic and Subalpine Forests and of Alpine Willow and Heath Ecosystems. In: Fennoscandian Tundra Ecosystems, *1* Verl. Springer, Berlin, New York. 1975.

Klein, H.: Zur Verwendung von Landkarten in der Biologie. Zool. Jb. Syst. *103:* 50–75. 1976.

Klemmer, K.: The Amphibia and Reptilia of the Canary Islands. In: Biogeographica

and Ecology in the Canary Islands. Junk, 433–456, The Hague. 1976.
Klinge, H.: Nährstoffe, Wasser und Durchwurzelung von Podsolen und Latosolen unter tropischem Regenwald bei Manaus-Amazonien. Biogeographica *7:* 45–58. 1976.
Klinge, H. and Rodrigues, W.A. : Litter Production in an Area of Amazonian terra firma Forest. Amazonia *1* (4): 287–310. 1968.
Kloft, W.J.: Ökologie der Tiere. UTB 729, Verlag E. Ulmer Stuttgart. 1978.
Kloft, W., Kunkel, H. and Ehrhard, P.: Weitere Beiträge zur Kenntnis der Fichtenröhrenlaus Elatobium abietinum (Walk.) unter besonderer Berücksichtigung ihrer Weltverbreitung. Z. angew. Entomol. *55:* 160–185. 1964.
Knabe, W., Brandt, C.S., Van Haut, H. and Brandt, C.J.: Nachweis photochemischer Luftverunreinigungen durch biologische Indikatoren in der Bundesrepublik Deutschland. Proc. of the Third. Int. Clean Air Congr., VDI-Verl., Düsseldorf. 1973.
Knapp, R.: Vegetation Dynamics. Junk, The Hague. 1974.
Knox, G.A.: Antarctic Marine Ecosystems. In: Antarctic Ecology, Acad. Press, London and New York. 1970.
Kobelt, W.: Die Verbreitung der Tierwelt. Jena. 1897.
Koeman, J.H.: The toxicological importance of chemical pollution for marine birds in the Netherlands. Die Vogelwarte *28:* 145–150. 1975.
Koepcke, H.-W.: Die Lebensformen (Grundlagen zu einer universell gültigen biologischen Theorie) Goecke & Evers, Krefeld. 1971.
Kohler, A.: Zur Ökologie submerser Gefäß-Makrophyten in Fließgewässern. Ber. Deutsch. Bot. Ges. Bd. *84* (11): 713–720. 1971.
Kohler, A.: Veränderung natürlicher submerser Fließgewässervegetationdurch organische Belastung. Daten und Dokumente zum Umweltschultz *14:* 59–66. 1975.
Kolkwitz, R. and Marsson, M.: Ökologie der pflanzlichen saprobien Ber. Deutsch. Bot. Ges. *26 (A):* 505–519. 1908.
Kolkwitz, R. and Marsson, M.: Ökologie der tierischen Saprobien. Int. Rev. Hydrobiol. *2:* 126–152. 1909.
Kosswig, C.: Zur Evolution der Höhlentiermerkmale. Rev. Fac. Sci. Instanbul (B) *9:* 285–287. 1944.
Kozhov, M.: Lake Baikal and its life. Junk, The Hague. 1963.
Kraus, O.: Taxonomische und tiergeographische Studien an Myriapoden und Araneen aus Zentralamerika. Dissertation, Frankfurt. 1955.
Krause, G.: Phytotoxische Wechselwirkung zwischen Schwefeldioxid und den Schwermetallen Zink und Cadmium, Schriftenr. LIB *34:* 86–91. 1975.
Krebs, C.: Ecology, the experimental analysis of distribution and abundance. Harper Int. Edit., New York, Evanston, San Francisco, London. 1972.
Kreh, W.: Verlust und Gewinn der Stuttgarter Flora im letzten Jahrhundert. Jh. Ver. vaterl. Naturkd. Württemberg *106:* 69–124. 1951.
Kreh, W.: Die Pflanzenwelt des Güterbahnhofs in ihrer Abhängigkeit von Technik und Verkehr. Mitt. florist.-soziol. Arbeitsgemeinschaft *3:* 86–109. Stolzenau. 1960.
Krolopp, E.: Faunengeschichtliche Untersuchungen im Karpatenbecken. Malacologia *9* (1): 111–119. 1969.
Kubiena, W.: Entwicklungslehre des Bodens. Vienna. 1948.
Kuhn, O.: Angewandte Chemie *84:* 838. 1972.
Kunick, W.: Veränderungen von Flora und Vegetation einer Großstadt dargestellt am Beispiel von Berlin (West). Dissertation, Techn. Univ. Berlin. 1974.

Kunkel, G.: Biogeography and Ecology in the Canary Islands. Junk, The Hague. 1976.
Kurten, B.: Transberingian Relationships of Ursus arctos LINNE (Brown and Grizzly Bears). Commentationes biologicae *65:* 3–10. 1973.
Kuss, E.: Abfolge und Alter der pleistozänen Säugetierfaunen der Insel Kreta. Ber. Naturfr. Ges. Freiburg *60:* 35–83. 1970.
Lack, D.: Darwin's Finches. Cambridge Univ. Press. 1947.
Lack, D.: Island Biology, illustrated by the land birds of Jamaica. Studies in Ecology, Vol. 3, pp. 445, Blackwell Scient. Publ., Oxford, London, Melbourne. 1976.
Lagerwerff, J.V. and Specht, A.W.: Contamination of Roadside Soil and Vegetation with Cadmium, Nickel, Lead and Zinc. Environ. Sci. & Tech. *4:* 583–586. 1970.
Lamb, H.H.: Climates and circulation regimes developed over the northern hemisphere during and since the last ice age. — Palaeogeography, Palaeoclimatol., Palaeoecol. *10* (1): 125–162. 1971.
Lamberti, F., Taylor, C.E. and Seinhorst, J.W.: Nematode Vectors of Plant Viruses. Plenum Press, London and New York. 1974.
Lamego, A.R.: Restingas na costa do Brasil, Min. Agr. Div. Geol. Min. *96:* 1–63. 1940.
Lamprecht, H.: Einige Strukturmerkmale natürlicher Tropenwaldtypen und ihre waldbauliche Bedeutung. Forstwissenschaftl. Centralbl. *91:* 270–277. 1972.
Lang, G.: Die Ufervegetation des Bodensees im farbigen Luftbild. Landeskdl. Luftbildauswertung im mitteleuropäischen Raum *8,* Bonn-Bad Godesberg. 1969.
Lange, O.L., Kappen, L. and Schultze, E.-D.: Water and Plant Life. Problems and Modern Approaches. Springer Verl., Berlin, Heidelberg, New York. 1976.
Lanphear, F.O.: Urban vegetation: values and stresses. Hortscience *6:* 332–334. 1971.
Larcher, W.: Produktionsökologie alpiner Zwergstrauchbestände auf dem Patscherkofel bei Innsbruck. Verhdl. Ges. Ökologie, Wien. *1975:* 3–7, Junk, The Hague. 1976.
Lattin, G. de: Höhlentiere und ihre Enstehuyng. Umschau *45:* 293–297. 1941.
Lattin, G. de: Postglaziale Disjunktionen und Rassenbildung bei europäischen Lepidopteren. Verhdl. Deutsch. Zool. Ges. Frankfurt. 1959.
Lattin, G. de: Grundriß der Zoogeographie. Jena. 1967.
Lauer, W.: Zur hygrischen Höhenstufung tropischer Gebirge. Biogeographica *7:* 169–182. Junk, The Hague. 1976.
Lauer, W. and Frankeberg, P.: Zum Problem der Tropengrenze in der Sahara. Erdkunde *31:* 1–15. 1977.
Lavrenko, E.M.: Die Gliederung des Schwarzmeer-Kasachstanischen Gebiets der Euroasiatischen Steppenzone in Provinzen. Bot. Z. *55:* 605–625. 1970.
Lazarus, R.S.: Psychological stress and the coping process. McGraw Hill, New York. 1966.
Lees, D.R., Creed, E.R. and Duckett, J.G.: Atmospheric pollution and industrial melanism. Heredity *30:* 227–232. 1973.
Lehn, H.: Veränderungen im Stoffhaushalt des Bodensees. Verhdl. Ges. Ökologie, Wien, Junk, The Hague. 1976.
Leithe, W.: Die Analyse der Luft und ihrer Verunreinigungen. In der freien Atmosphäre und am Arbeitsplatz. Wiss. Verl., Stuttgart. 1974.
Lenz, M.: Zum Problem der Erfassung von Brutvogelbeständen in Stadtbiotopen. Die Vogelwelt *92 (2):* 41–52. 1971.
Levi, L.: Stress and distress in response to psychological stimuli. Pergamon Press, Oxford. 1972.

Lewis, T.H.: Dark coloration in the reptiles of the Tularose Malpais, New Mexico. Copeia *3:* 181–184. 1949.

Lewontin, R.C.: The Meaning of Stability. In: Diversity and Stability in Ecological Systems. Symposia in Biology, 22 USAEC Report BNL-50175, Brookhaven, National Laboratory. 1970.

Li, H.L.: Urban botany: need for a new science, Bioscience *19:* 882–883. 1969.

Liebmann, H.: Der Wassergüteatlas. Seine Methodik und Anwendung. München. 1969.

Lieth, H. and Whittaker, R.H.: Primary productivity of the biosphere, Springer Verl., Berlin, Heidelberg, New York. 1975.

Lindroth, C.H.: The theory of glacial refugia in Scandinavia. Notulae Entomol. XLIX. 1969.

Lockwood, A.P.: Effects of pollutants on aquatic organisms. Cambridge Univ. Press, Cambridge, London, New York, Melbourne. 1976.

Löffler, E.: Evidence of Pleistocene glaciation in East Papua. Austral. Geograph. Stud. *8:* 16–26. 1970.

Lönnberg, E.: Einige Bemerkungen über den Einfluß der Klimaschwankungen auf die afrikanische Vogelwelt. J. Ornith. *74:* 259–273. 1926.

Lorenz, K. and Leyhausen, P.: Antriebe tierischen und menschlichen Verhaltens. Verl. Piper, München. 1968.

Löser, S.: Art und Ursachen der Verbreitung einiger Carabidenarten (Coleoptera) im Grenzraum Ebene-Mittelgebirge. Zool. Jb. Syst. Ökol. Geogr. *99:* 1972.

Lötschert, W. and Köhm, H.J.: Baumborke als Anzeiger von Luftverschmutzungen. Umschau *73:* 403–404. 1973.

Loub, W.: Umweltverschmutzung und Umweltschutz in naturwissenschaftlicher Sicht. Verl. Deuticke, Vienna. 1975.

Ludwig, G.: Probleme im Paläozoikum des Amazonas- und des Maranhão-Beckens in erdölgeologischer Sicht, Erdöl und Kohle, Erdgas, Petrochemie *19:* 798–807. 1966.

Ludwig, G.: Die geologische Entwicklung des Marajó-Beckens in nord Brasilien. Geol. Jb. *86:* 845–878. 1968.

Lüscher, M.: Phase and caste determination in insects. Endocrine Aspects. Pergamon Press, Oxford. 1976.

Lydekker, R.: A Geographical History of mammals. Cambridge. 1896.

MacArthur, R.H.: Fluctuations of Animal Populations, and a Measure of Community Stability. Ecology *36:* 533–536. 1955.

MacArthur, R.H.: Patterns of communities in the tropics. Biol. J. Linn. *1:* 19–30. 1969.

MacArthur, R.H.: Geographical Ecology, Patterns in the Distribution of Species, Harper & Row Publ., New York. 1972.

MacArthur, R.H. and Wilson, E.O.: An equilibrium theory of insular zoogeography. Evolution *17:* 373–387. 1963.

MacArthur, R.H. and Wilson, E.O.: The theory of island biogeography. Princeton Univ. Press, Princeton. 1967.

MacArthur, R.H. and Wilson, E.O.: Biogeographie der Inseln. Verl. Goldmann, München. 1971.

MacArthur, R.H., Diamond, J.M. and Karr, J.R.: Density compensation in island faunas. Ecology *53:* 330–342. 1972.

MacArthur, R.H., Recher, H. and Cody, M.: On the relationship between habitat selection and species diversity. Am. Nat. *100:* 319–332. 1966.

MacLean, S.F., Fitzgerald, B. and Pitelka, F.: Population cycles in arctic Lemmings Winter Reproduction and Predation by weasels. Arctic and Alpine Research *6* (1): 1–12. 1974.

Magaard, L. and Rheinheimer, G.: Meereskunde der Ostsee. Springer Verl. Berlin, Heidelberg, New York. 1974.

Magalhaes, G.M.: Contribuiçao para o conhecimento da flora dos campos alpinos de Minas Gerais. Anais V Reunia o Anual Soc. Bot. Brasil, 227–304, Porto Alegre. 1956.

Mägdefrau, K.: Die Geographie der Moose, ihre Begründung und Entwicklung. Acta Hist. Leopoldina *9:* 95–110, Halle. 1975.

Mägdefrau, K.: Die ersten Alpen-Botanniker. Jb. Verein zum Schutze der Alpenpflanzen und -tiere, *40:* 1–14. 1975.

Mägdefrau, K.: Das Alter der Drachenbäume auf Tenerife, Flora *164:* 347–357. 1975.

Malavolta, E., Sarruge, J.R. and Bittencourt, V.C.: Toxidez de Alumino e de Manganês. 4. Simp. Cerrado, 275–301, Ed. Univ. São Paulo, São Paulo. 1976.

Malicky, H.: Gebirgsbach und Gebirgsbachleben. Jb. Verein zum Schutze der Alpenpflanzen und -tiere. *38:* 1–13. 1973.

Malins, D.C. and Sargent, J.R.: Biochemical and Biophysical Perspectives in Marine Biology *1,2.* Acad. Press, London, New York, San Francisco. 1975.

Maloiy, G.M.O.: Comparative Physiology of Desert Animals. Zool. Soc.-London, New York and London. 1972.

Margalef, R.: La teoría de la información en ecologia. Mem. Real. Acad. Cienc. Artes Barcelona *32* (13): 373–436. 1957.

Margalef, R.: Information theory in ecology. Gen. Systems *3:* 37–71. 1958.

Margalef, R.: Perspectives in ecological theory. 4th edition, Chicago, London. 1975.

Marking, L.L. and Dawson, V.K.: Investigations in Fish control. 67. Method for Assessment of Toxicity or Efficacy of Mixtures of Chemicals. US Dep. Interior, Fish & Wildlife Serv., Report. La Crosse, Wisconsin USA. 1975.

Martin, M.H. and Coughtrey, P.J.: Preliminary Observations on the Levels of Cadmium in a Contaminated Environment. Chemosphere *3:* 155–160. 1975.

Masironi, R.: Cardiovascular mortality in relation to radioactivity and hardness of local water supplies in the USA. Bull. Wld. Hlth. Org. *43:* 687–697. 1970.

Mason, S.: Geschichte der Naturwissenschaft. Kröner Verl., Stuttgart. 1974.

Matsch, C. L.: North America and the Great Ice Age. McGraw Hill Book Comp., New York. 1976.

May, R.M.: Will a large complex system be stable? Nature *238:* 413–414. 1972.

May, R.M.: Stability and Complexity in Model Ecosystems. Princeton Univ. Press. 1973.

May, R.M.: Thresholds and breakpoints in ecosystems with a multiplicity of stable states. Nature *269:* 471–477. 1977.

Mayer, H.: Waldbau auf soziologisch-ökologischer Grundlage. Verl. G. Fischer, Stuttgart, New York. 1977.

Mayer, R.: Bioelementflüsse im Wurzelraum saurer Waldböden. Mitt. Deutsch. Bodenk. Ges. *16:* 136–145. 1972.

Mayer R., Ulrich, B. and Khanna, P.K.: Die Ausfilterung von Luftverunreinigungen durch Wälder-Einflüsse auf die Azidität der Niederschläge und deren Auswirkungen auf den Boden. Mitt. Deutsch. Bodenk. Ges. *22:* 339–348. 1975.

Mayr, E.: The zoogeographic position of the Hawaiian Islands. Condor *45:* 45–48. 1943.

Mayr, E.: Interferences concerning the Tertiary American bird faunas. Proc. Nat.

Acad. Sci. *51*. 1964.
Mayr, E.: Artbegriff und Evolution. Parey Verl., Hamburg und Berlin. 1967.
Mayr, E.: Grundlagen der Zoologischen Systematik. Verl. Parey, Hamburg and Berlin. 1975.
Mayr, E. and Diamond, J.M.: Birds on islands in the sky: Origin of the montane avifauna of Northern Melanesia. Proc. Nat. Acad. Sci. USA *73* (5): 1765–1769. 1976.
Mayr, E., Linsley, E.G. and Usinger, R.L.: Methods and principles of systematic Zoology. New York, Toronto, London. 1953.
McLusky, D.S.: Ecology of estuaries. Heinemann Books Ltd., London. 1971.
Mertens, R.: Die Insel-Reptilien, ihre Ausbreitung, Variation und Artbildung, Zoologica *84:* 1–2091. 1934.
Mertens, R.: Lacerta goliath n.sp. eine ausgestorbene Rieseneidechse von der Kanaren. Senckenbergiana *25:* 330–339. Frankfurt. 1942.
Mertens, R.: Die Amphibien und Reptilien von El Salvador. Abhdl. Senck. Nat. Ges. *487*. 1952.
Mertens, R.: Froschfang-Exkursionen in Brasilien. Die Aquarien- und Terrarienzeitschrift (DATZ) *10* (1): 22–25. 1957.
Meyer, H.: Wälder des Ostalpenraumes. Verl. Fischer, Stuttgart. 1974.
Miettinen, J.K.: Plutonium Foodchains. Helsinki. 1976.
Miller, M.W. and Stannard, J.N.: Environmental Toxicity of Aquatic Radionuclides. Models and Mechanisms. Ann Arbor Sciences. 1976.
Miller, StL. and Orgell, L.E.: The Origins of Life on the Earth. Prentice-Hall, Inc., Englewood Cliffs, New Jersey. 1974.
Milliman, D.J. and Emery, K.O.: Sea levels during the past 35,000 years. Science *162:* 1121–1123. 1968.
Milliman, D.J. and Summerhayes, C.P.: Upper continental margin sedimentation of Brazil. In: Contribution to Sedimentology *4,* E. Schwizerbart'sche Verlagsbuchhdl. Stuttgart. 1975.
Mitscherlich, G.: Wald, Wachstum und Umwelt *3*. Sauerländer's Verl., Frankfurt. 1975.
Miyawaki, A and Tüxen, R.: Vegetation Science and Environmental Protection. Tokyo. 1977.
Möbius, K.: Die Auster und die Austernwirtschaft. Berlin. 1877.
Montanucci, R.R.: Convergence, Polymorphism or Introgressive Hybridization? An Analysis of Interaction between Crotaphytus collaris and C. reticulatus (Sauria: Iguanidae). Copeia *1974* (1): 87–101. 1974.
Moore, V.: A pesticide monitoring system with special reference to the selection of indicator species. J. appl. Ecol. *3:* 261–269. 1966.
Moore, V.: Pesticide monitoring from the national and international point of view. Bull. Recherches Agronom. Gembloux, Vol. extraord. 464–478, Gembloux. 1974.
Morafka, D.J.: A Biogeographical Analysis of the Chihuahuan Desert through its Herpetofauna. Biogeographica *9,* Junk, The Hague. 1977.
Moreau, R.E.: Pleistocene climatic changes and the distribution of life in East Africa. J. Ecol. *21:* 415–435. 1933.
Moreau, R.E.: Vicissitudes of the African biomes in the Late Pleistocene. Proc. Zool. Soc. London *141:* 395–421. 1963.
Moreau, R.E.: The bird faunas of Africa and its islands. London and New York. 1966.
Moreau, R.E.: Climatic changes and the distribution of forest vertebrates in West

Africa. J. Zool. *158:* 39–61. 1969.
Morgan, C.J. and King, P.E.: British Tardigrades. Synopses of the British Fauna (New Series) *9:* 1–133;, Acad. Press, London, New York, San Francisco. 1976.
Morrow, P.A., Bellas, T.E. and Eisner, T.: Eucalyptus Oils in the Defensive Oral Discharge of Australian Sawfly Larvae (Hymenoptera: Pergidae). Oecologia *24:* 193–206, 1976.
Mückenhausen, E.: Entstehung, Eigenschaften und Systematik der Böden der Bundesrepublik Deutschland. Frankfurt. 1962.
Müller, D. and Nielsen, J.: Production brute, pertes par respiration et production nette dans la forêt ombrophile tropicale. Forst. Forsgv. in Danmark *29:* 69–160. 1965.
Müller, G.: Attersee, Vorläufige Ergebnisse des OECD-Seeneutrophie-rungs- und des MAB-Programms. Gmunden. 1976.
Müller, G.: Die Salmonellen im Lebensraum einer Grosßstadt. Beiträge zur Hygiene und Epidemiologie. Barth Verl., Leipzig. 1965.
Müller, H.: Zur Morphologie pleistozäner Seebecken im westlichen schleswig-holsteinischen Jungmoränengebiet. Z. Geomorph. N.F. *20* (3): 350–360. 1976.
Müller, J.: Palynological study of Holocene peat in Sarawak. Symp. Ecol. Res. Hum. Trop. Veg. Kuching 1963, Tokyo, 147–156. 1975.
Müller, J.: Palynological evidence for change in geomorphology, climate, and vegetation in the Mio-Pliocene of Malasia. Trans. 2nd Aberdeen-Hull Symp. Mal. Ecol., 6 16. 1972.
Müller, J.: Pollen analytical studies of peat and coal from Northwest Borneo. In: Barsta, H.R. van & Casparie, W.A. (eds.). Modern Quaternary research in Southeast Asia. Symp. Mod. Quat. Res. Indonesia, Rotterdam, 83–86. 1975.
Müller, P.: Einige Bemerkungen zur Verbreitung von Vipera aspis (Serpentes, Viperidae) in Spanien, Salamandra *5* (1/2): 57–62. 1969.
Müller, P.: Vertebratenfaunen brasilianischer Inseln als Indikatoren für glaziale aund postglaziale Vegetationsfluktuationen. Abhdl. Deutsch. Zool. Ges. Würzburg *1969:* 97–107. 1970.
Müller, P.: Die Ausbreitungszentren und Evolution in der Neotropis. Mitt. Biogeogr. Abt. Univers. Saarl. *1:* 1–20. 1971.
Müller, P.: Biogeographische Probleme des Saar-Mosel-Raumes, dargestellt am Hammelsberg bei Perl. Faun.-flor. Not. aus dem Saarl. *4* (1/2): 1–14, 1971.
Müller, P.: Die Bedeutung der Biogeographie für die ökologische Landschaftsforschung. Biogeographica *1.* 1972.
Müller, P.: Die Bedeutung der Ausbreitungszentren für die Evolution neotropischer Vertebraten. Zool. Anz. *189* (1/2). 1972.
Müller, P.: The Dispersal Centres of Terrestrial Vertebrates in the Neotropical Realm. Biogeographica *2:* 1–244, Junk, The Hague. 1973.
Müller, P.: Die Verbreitung der Tiere, Grzimeks-Tierleben *16,* Kindler-Verl. 1973.
Müller, P.: Historisch-biogeographische Probleme des Artenreichtums der Südamerikanischen Regenwälder. Amazoniana *4* (3): 229–242. 1973.
Müller, P.: Probleme des Ökosystems einer Industriestadt, dargestellt am Beispiel von Saarbrücken. In: Belastung und Belastbarkeit von Ökosystemen. Tagungsber. Ges. Ökol. Gießen *1972:* 123–132. 1973.
Müller, P.: Die Erfassung der europäischen Fauna als europäische Aufgabe. Mitt. Abt. Biogeogr. Saarbrücken *5:* 1–2. 1973.
Müller, P.: Was ist Ökologie? Geoforum *18.* 1974.
Müller, P.: Aspects of Zoogeography. Junk, The Hague. 1974.

Müller, P.: La structuration de l'environment naturel dans les regions de concentration urbaine. Bull. des Recherches agronomiques de Gembloux. Vol. extraordinaire édité à l'occasion de la semaine d'étude agriculture et environment, pp. 742–761, Gembloux. 1974.

Müller, P.: Beiträge der Biogeographie zur Geomedizin und Ökologie des Menschen, Fortschritte der Geomedizinischen Forschung, p. 88–109, Steiner Verl., Wiesbaden. 1974.

Müller, P.: Erfassung der westpaläarktischen Invertebraten. Fol. Ent. Hung. *27:* 405–430. 1974.

Müller, P.: Biogéographie et évolution en Amérique du Sud. C. R. Soc. Biogéogr., Séance *448:* 15–22. 1975.

Müller, P.: Ökologische Kriterien für die Ram- und Stadtplanung, Umwelt-Saar *1974:* 6–51, Homburg. 1975.

Müller, P.: Voraussetzungen für die Integration faunistischer Daten in die Landesplanung der Bundesrepublik Deutschland. Schriftenr. Vegetationskde., Vol. *10:*27–47, Bonn-Bad Godesberg. 1976.

Müller, P.: Fundortkataster der Bundesrepublik Deutschland, Erfassung der westpaläarktischen Tiergruppen, Teil 2: Lepidoptera, Bearb. H. Schreiber, Schwerpunkt Biogeographie, Universität des Saarlandes, Saarbrücken. 1976.

Müller, P.: Belastbarkeit von Ökosystemen. Mitt. *8,* Schwerp. Biogeogr., Saarbrücken. 1977.

Müller, P.: Biogeographie und Raumbewertung. Verl. Wiss. Buchgesellschaft, Darmstadt. 1977.

Müller, P.: Tiergeographie: Struktur, Funktion, Geschichte und Indikatorbedeutung von Arealen. Verl. Teubner, Stuttgart. 1977.

Müller, P.: Abfallwirtschaft als ökologisch-ökonomisches System. In: Abfallwirtschaft und Recycling, Verl. Girardet, Essen. 1977.

Müller, P.: Stand und Probleme der Erfassung der westplalaearktischen Tiergruppen in der BRD. Int. Entomol. Sympos. Lunz, Junk, The Hague. 1977.

Müller, P. and Blatt, G.: Die Mortalitästrate von Importschildkröten im Saarland. Salamandra. 1975.

Müller, P., Klomann, U., Nagel, P., Reis, H. and Schäfer, A: Indikatorwert unterschiedlicher biotischer Diversität im Verdichtungsraum von Saarbrücken. Verhdl. Gesellsch. Ökologie, Erlangen, 113–128. 1975.

Muller, P. and Schäfer, A.: Diversitätsuntersuchungen und Expositionstest in der mittleren Saar, Umwelt-Forum *2:* 43–46. 1976.

Müller, P. and Schmithüsen, J.: Probleme der Genese südamerikanischer Biota. Festschr. Gentz, Verl. Hirt, Kiel. 1970.

Müller, P. and Steiniger, H.: Evolutionsgeschwindigkeit und chorologische Verwandtschaft brasilianischer Erdleguane der Gattung Liolaemus (Sauria, Iguanidae). Mitt. *9,* Schwerpunkt für Biogeographie, Saarbrücken. 1977.

Müller-Hohenstein, K.: Die Landschaftsgürtel der Erde. Teubner, Stuttgart. 1979.

Mulsow, R.: Untersuchungen zur Siedlungsdichte der Hamburger Vogelwelt. Abhd. Verh. Naturwiss. Ver. Hamburg N.F. *12:* 123–188. 1968.

Murray, M.D.: Insect parasites of marine birds and mammals. In: Marine insects, North-Holland Publ. Comp., Amsterdam, Oxford, New York. 1976.

Murton, R. and Westwood, N.: Birds as Pests. In: Applied Biology *1:* 89–181, Acad. Press, London, New York. 1976.

Myers, J.H.: Distribution and Dispersal in Populations Capable of Resource Depletion. A Simulation Model. Oecologia *24:* 255–269. 1976.

Nadig, A.: Über die Bedeutung der Massifs de Réfuge am südlichen Alpenrand. Mitt. schweiz. entom. Ges. *41:* 341–358. 1968.

Nagel, P.: Studien zur Ökologie und Chorologie der Coleopteren (Insecta) xerothermer Standorte des Saar-Mosel-Raumes mit besonderer Berücksichtigung der die Bodenoberfläche besiedelnden Arten. Dissertation, Biogeographie, Saarbrücken. 1975.

Nagel, P.: Die Darstellung der Diversität von Biozönosen, Schriftenr. Vegetationskde., *10,* Bonn-Bad Godesberg. 1976.

Nelson, G.J.: The problem of Historical Biogeography. Syst. Zool. *18* (2): 243–246. 1969.

Neri, L.C., Hewitt, D. and Schreiber, G.B.: Can Epidemiology elucidate the water story? Am. J. Epidemiol. *99:* 75–88. 1974.

Neumann, D.: Die intraspezifische Variabilität der lunaren und täglichen Schlüpfzeiten von Clunio marinus (Diptera: Chironomidae). Verhdl. Ges. Zool., 223–233. 1965.

Neumann, D.: Zielsetzungen der Physiologischen Ökologie. Verhdl. Ges. Ökologie, Saarbrücken, 1–9, Junk, The Hague. 1974.

Nicolai, J.: Der Rotmaskenastrild (Pytilia hypogrammica) als Wirt der Togo-Paradieswitwe (Steganura togoensis). J. Orn. *188:* 175–188. 1977.

Nielsen, E.S.: Marine Photosynthesis with special emphasis on the ecological aspects. Elsevier Publ., Amsterdam, Oxford, New York. 1975.

Niethammer, G.: Die Rolle der Auslese bei Wüstenvögeln. Bonn. Zool. Beitr. *10:* 179–197. 1959.

Niethammer, G.: Die Einbürgerung von Säugetieren und Vögeln in Europa. Parey Verl., Hamburg, Berlin. 1963.

Niethammer, J.: Die Igel von Teneriffa. Zool. Beitr. *18* (2): 307–309. 1972.

Niklfeld, H.: Bericht über die Kartierung der Flora Mitteleuropas. Taxon *20* (4): 545–571. 1971.

Njogu, A.R. and Kinoti, G.K.: Observations in the breeding sites of mosquitoes in Lake Manyara, a saline lake in the East African Rift Valley. Bull. Ent. Ges. *60:* 473–479. 1971.

Nogales, J. and Schminke, H.-U.: El Pino enterrado de la Canada de las Arenas (Gran Canaria). Cuad. Bot. Canar. *5:* 23–25. 1969.

Noodt, W.: Syncarida. Biota Acuatica de Sudamerica Austral., San Diego State University, San Diego. 1977.

Nowak, E.: Die Ausbreitung der Tiere, Brehm-Bücherei, Ziemsen Verl. Wittenberg-Lutherstadt. 1975.

Nuorteva, P.: The synanthropy of birds as an expression of the ecological cycle disorder caused by urbanization. Ann. Zool. Fennici *8:* 547–553;. 1971.

Nylander, W.: Les lichens du Jardin du Luxembourg. Bull. Soc. bot. Fr. *13:* 364–372. 1866.

Olsson, I.U.: Radiocarbon Variations and Absolute Chronology. Nobel Symposium *12.* Almquist and Wiksell, Stockholm. 1971.

Omori, M.: The Biology of Pelagic Shrimps in the ocean. Adv. mar. Biol. *12:* 233–324. 1974.

Orians, G.H.: The number of the bird species in some tropical forests. Ecology *50:* 783–801. 1969.

Overbeck, J.: Über die Kompartimentierung der stehenden Gewässer. Ein Beitrag zur Struktur und Funktion des limnishcen Ökosystems. Verdhl. Ges. Ökol. Saarbrücken, Junk, The Hague. 1974.

Paijmans, K.: New Guinea Vegetation. Els. Publ. New York. 1976.

Palmén, E.: Die anemohydrochore Ausbreitung der Insekten als zoogeographischer Faktor, Ann. Zool. Soc. Zool. Bot. Fenn. Vanamo *10* (1): 1–262. 1944.

Patten, B.C.: The Zero State and Ecosystem Stability. Proc. First International Congr. of Ecology. Centre for Agricult. Publ. and Documentation, Wageningen. 1974.

Patten, B.C.: Systems Analysis and Simulation in Ecology. Acad. Press, New York, San Francisco, London. 1975.

Patton, J.L., Yang, S.Y. and Myers, P.: Genetic and Morphologic Divergence among introduced Rat Populations (Rattus rattus) of the Galapagos Archipelago. Ecuador. Syst. Zool. *24:* 296–310. 1975.

Patton, J.L.: Biosystematics of the rodent fauna of the Galapagos Archipelago. American Philos. Soc. Year Book *1975:* 352–353. 1975.

Paulian, R.: Some ecological and biogeographical problems of the Entomofauna of Madagascar. In: Biogeography and ecology in Madagascar. The Hague. 1972.

Peake, J.F.: The evolution of terrestrial faunas in the western Indian Ocean. Phil. Trans. Roy. Soc. London. B *260:* 581–610. 1971.

Pearson, O.P. and Patton, J.L.: Relationships among South American Phyllotine Rodents based on Chromosome analysis. J. Mammalogy *57* (2): 339–350. 1976.

Peary, J. and Castenholz, R.W.: Temperature strains of a thermophilic blue-green alga. Nature *202:* 720–721. 1964.

Perring, F.H.: Data-processing for the Atlas of the British Flora. Taxon *12:* 183–190. 1963.

Perry, R. and Young, R.: Handbook of Air Pollution Analysis. Chapman and Hall, London. 1977.

Peters, G.: Studien zur Taxonomie und Ökologie der Smaragdeidechsen. IV. Zur Ökologie und Geschichte der Populationen von Lacerta v. viridis (Laurenti) im mitteleuropäischen Flachland. Veröff. Bezirksheimatmuseums Potsdam. *21:* 49–119. 1970.

Peters, H.: Fliegen- und Rattenbekämpfung — wichtige Aufgaben der Stadthygiene. G. Ing. *9/10:* 160–169. 1949.

Peadenhauer, J.: Beziehungen zwischen Standortseinheiten, Klima, Stickstoffernährung und potentieller Wuchsleistung der Fichte im Bayerischen Flyschgebiet — dargestellt am Beispiel des Teisenbergs. Dissert. Botanicae *30:* 1–239, Verl. Cramer, Vaduz. 1975.

Pianka, E.R.: On r- and K-selection. Amer. Natur. *104:* 592–597. 1970.

Pianka, E.R.: Evolutionary Ecology. Harper & Row Publ., New York. 1974.

Pielou, E.C.: An introduction to mathematical ecology. New York, London. 1969.

Pielou, E.C.: Ecological Diversity. Wiley Publ., New York. 1975.

Pielou, E.C.: Mathematical Ecology. J. Wiley & Sons Publ., New York. 1977.

Pietzsch, W.: Straßenplanung. Werner Verl., Düsseldorf. 1976.

Pignatti, S.: Waldvegetation Japans und Westeuropas. Ein Vergleich. In: Vegetation Science and Environmental Protection, 495–500, Maruzen Ltd., Tokyo. 1977.

Pijl, L. van der: The dispersal of plants by bats. Acta Bot. Neerl. *6:* 291–315. 1957.

Pimienov, M.G.: The analysis of the distribution of species of Angelica occurring in the Soviet Far East. Bot. Z. Moscow *53:* 932–946. 1968.

Pimienta, J.: A feixa costeira meridional de Santa Catarina, Brasil. Bol. Div. Geol. Mineral Brasil *176:* 10–104. 1958.

Pitelka, F.A.: An Avifaunal review for the Barrow Region and North Slope of Arctic Alaska. Arctic and Alpine Res. *6* (2): 161–184. 1974.

Povolny, D.: Synanthropy. Flies and Disease. Princeton Univ. Press, Princeton, New Jersey. 1971.

Power, D.M.: Similarity among avifaunas of the Galápagos Islands. Ecology *56:* 616–626. 1975.

Po-Yung, L., Metcalf, R.L., Furman, R., Vogel, R. and Hasset, J.: Model Ecosystem Studies of Lead and Cadmium and of Urban Sewage Sludge Containing these Elements. J. Environ. Qual. *4.* 975.

Precht, H., Christophersen, J., Hensel, H. and Larcher, W.: Temperature and Life, Verl. Springer. Berlin, Heidelberg, New York. 1973.

Preston, F.W.: The canonical distribution of commonness and rarity. Ecology *43:* 185–215, 410–432. 1962.

Prestt, I. and Ratcliff, D.A.: Effects of organochlorine insecticides on European birdlife. Proc. Int. orn. Congr. 15, 486–513. 1972.

Prinz, B.: Immissionswirkungskataster in Nordrhein-Westfalen als Planungskriterium. Umwelt-Saar *1974.* 1975.

Prinz, B. and Scholl, G.: Erhebungen über die Aufnahme und Wirkung gas- und partikelförmiger Luftverunreinigungen im Rahmen eines Wirkungskatasters, Schriftenr. LIB *36:* 62–86. Essen. 1975.

Purchon, R.D.: The Biology of the Mollusca. Pergamon Press, Oxford, New York. 1977.

Pütter, A.: Altersbestimmungen an Drachenbäumen von Tenerife. Sitz. Ber. Heidelb. Akad. Wiss., math.-phys. Kl. *12.* 1925.

Rachid, M.: Transpiração sistemas subterrâneos da vegetação de verão dos campos cerrados de Emas. Bol. Fac. Fil. Ciénc. e Letr. USP. 80, Botânica *5:* 1–135. 1947.

Rambo, B.: Analyse historica da Flora de Pôrto Alegre. Sellowia *6:* 1–112. Itajai, Brazil. 1954.

Rambo, B.: História da Flora do Littoral Riograndense. Sellowia *6:* 113–172, Itajai, Brazil. 1954.

Rat der Sachverständigen in Umweltfragen: Umweltgutachten 1974. Kohlhammer Verl. Stuttgart und Mainz. 1974.

Ratter, J.A., Askew, G.P., Montgomery, R.F. and Gifford, D.R.: Observações adicionais sobre o cerrado de solos mesotrôficos no Brasil central. 4 Simpos. Cerrado, 303–316, Ed. Univ. São Paulo, São Paulo. 1976.

Rawitscher, F.K.: Algunas nocões sôbre a vegetação do Litoral brasiliero. Bol. Assoc. Geogr. Brasil *5:* 13–28. 1944.

Rawitscher, F.K., Ferri, M.G. and Rachid, M.: Profundidade dos solos e vegetação em campos cerrados do Brasil meridional. An. Acad. Brasil Ciênc. *15* (4): 267–294. 1943.

Reichenback-Klinke, H.H.: Der Süßwasserfisch als Nährstoffquelle und Umweltindikator. G. Fischer Verl., Stuttgart. 1974.

Reinig, W.F.: Die Holarktis. Jena. 1936.

Reinig, W.F.: Bastardierungszonen und Mischpopulationen bei Hummeln (Bombus) und Schmarotzerhummeln (Psithyrus). Mitt. Münchn. Entomol. Ges. *59:* 1–89. 1970.

Reinke, W.: Das Klima Amazoniens. Dissertation, Tübingen. 1962.

Remane, A.: Die Bedeutung der Lebensfortypen für die Ökologie. Biol. Generalis *17:* 164–182, 1943.

Remmert, H.: Über den Tagesrhythmus arktischer Tiere. Z. Morp.-ökol. Tiere *55:* 142–160. 1965.

Remmert, H.: Die Tundra Spitzbergens als terrestrisches Ökosystem. Umschau *72* (2): 41–44. 1972.

Reyment, R.A.: Minor ebb structures and shell orientation on a tidal beach (Bay of Arcachon, France). Palaeogeography, Palaeoclimatol., Palaeoecol. *10*. 1971.

Richards, P.W.: Speciation in the tropical rain forest and the concept of the niche. Biol. J. Linn. Soc. *1:* 149–153. 1969.

Ricker, W.E.: An ecological classification of certain Ontario streams. Univ. of Toronto Studies Biol. Ser. *37:* 1–114. Toronto. 1934.

Ridley, H.N.: The Dispersal of Plants throughout the World. Ashford, Kent. 1930.

Riley, J.P. and Skirrow, G.: Chemical Oceanography. Acad. Press, London, New York, San Francisco. 1975.

Rizzini, C.T.: Fitogeografia do Brasil. Ed. Univ. São Paulo, São Paulo. 1976.

Rizzini, C.T.: Tratado de Fitogeografia do Brasil. Aspectos ecológicos. Ed. Univ. São Paulo, São Paulo. 1976.

Rode, W.W.: Schutzeinrichtungen von Früchten und Samen gegen die Einwirkung Fließenden Meerwassers. Dissertation, Gleiwitz.

Rosmanith, J., Schröder, A., Einbrodt, H.J. and Ehm, W.: Untersuchungen an Kindern aus einem mit Blei und Zink belasteten Industriegebiet. Unwelthygiene *9:* 266–271. 1975.

Rump, H.-H., Symader, W. and Herrmann, R.: Mathematical modelling of Water Quality in small rivers (Nutrients, Pesticides and other chemical properties). Catena *3:* 1–16. 1976.

Runge, M.: Böden im Bereich der Bebauung, Umweltschutzforum *8,* Berlin. 1973.

Rutford, R.H., Craddock, C. and Bastien, T.W.: Late Tertiary glaciation and Sea-Level changes in Antarctica. Palaeogeography, Palaeoclimatol., Palaeoecol. *5:* 15–39. 1968.

Ruthsatz, B.: Pflanzengesellschaften und ihre Lebensbedingungen in den Andinen Halbwüsten Nordwest-Argentiniens. Verl. Cramer, Vaduz. 1977.

Rzoska, J.: The Nile, Biology of an ancient river. Monographiae Biologicae. Vol. *29:* 1–417, W. Junk, The Hague. 1976.

Saarisalo-Taubert, A.: Die Flora in ihrer Beziehung zur Siedlung und Siedlungsgeschichte in den südfinnischen Städten Provoo, Loviisa und Hamina. Ann. Bot. Soc. Vanamo *35* (1): 1–190. 1963.

Saemann, D.: Zur typisierung städtischer Lebensräume im Hinblick auf avifaunistische Untersuchungen. Mitt. IG Avifauna DDR *1:* 81–88. 1968.

Saemenn, D.: Der gegenwärtige Stand der Urbanisierung der Wacholderdrossel, Turdus pilaris L., in einer sächsischen Großstadt. Beitr. Vogelkd. *20:* 12–41. 1974.

Salomonsen, F.: Fuglene pa Grønland. Rhodos, Copenhagen. 1967.

Sauer, E.G. and Rothe, P.: Ratite Eggshells from Lanzarote, Canary Islands. Science *176:* 43–45. 1972.

Saunders: Insects Clocks. Pergamon Press, Oxford, New York. 1976.

Savage, J.M.: The origins and history of the Centrale American Herpetofauna. Copeia *1966* (4): 719–766. 1966.

Schauermann, J.: Zur Abundanz- und Biomassendynamik der Tiere in Buchenwäldern des Sollings. Verhdl. Ges. Ökologie *1976:* 113–124, Junk, The Hague. 1977.

Schenkel, W.: Abfallwirtschaft-Umwelt und Ressourcen. In: Abfallwirtschaft und Recycling. Verl. Girardet, Essen. 1977.

Scherner, E.R.: Möglichkeiten und Grenzen ornithologischer Beiträge zur Land-

eskunde und Umweltforschung am Beispiel der Avifauna des Solling. Dissertation, Göttingen. 1977.
Scheuermann, R.: Mittelmeerpflanzen der Güterbahnhöfe des rheinisch-westfälischen Industriegebietes. Fed. Rep. Beih. *76:* 65–99. 1934.
Scheuermann, R.: Die Pflanzen des Vogelfutters. Natur am Niederrhein *17* (1): 1–13. 1941.
Schimper, A.F.W.: Pflanzengeographie auf physiologischer Grundlage. Jena. 1935.
Schmid, J.A.: Urban vegetation. A review and Chicago case study. Univ. Chicago 161. 1975.
Schmidt, E.: Die Libellenfauna des Lübecker Raumes. Ber. Ver. H. Nat. Hist. Mus. Lübeck *13/14:* 25–43. 1975.
Schmidt, G.H.: Sozialpolymorphismus bei Insekten. Wissenschaftl. Verlagsgesellsch. Stuttgart. 1974.
Schmincke, K.H.: Mesozoic Intercontinental Relationships as evidenced by Bathynellia Crustacea (Syncarida: Malacostraca). Syst. Zool. *23:* 157–164. 1974.
Schmithüsen, J.: Der geistige Gehalt in der Kulturlandschaft. Wissenschaftl. Buchgesellsch., Darmstadt. 1967.
Schmithüsen, J.: Begriff und Inhaltsbestimmung der Landschaft als Forschungsobjekt vom geographischen und biologischen Standpunkt. Arch. Naturschutz. u. Landschaftsforsch. *8* (2): 101–112. 1968.
Schmithüsen, J.: Geschichte der Geographischen Wissenschaft. Bibliogr. Inst., Mannheim. 1970.
Schmithüsen, J.: Allgemeine Geosynergetik, Verl. de Gruyter, Berlin, New York. 1976.
Schmitt, W.: The species of Aegla, endemic South American Fresh-Water Crustaceans. Proc. US Nat. Mus. *91:* 431–519. Washington. 1942.
Schneider, S.: Umweltforschung an der Saar mit Hilfe von Fernerkundungsverfahren. Saarheimat *3:* 55–59. 1972.
Schönbeck, H.: Fine Methode zur Erfassung der biologischen Wirkung von Luftverunreinigungen durch transplantierte; Flechten. Staub *29:* 14–18. 1969.
Schönbeck, H.: Untersuchungen in Nordrhein-Westfalen über Flechten als Indikatoren für Luftverunreinigungen. Schriftenr. Landesanstalt Immissions- und Bodennutzungsschutz des Landes NRW in Essen *26:* 99–104. 1972.
Schönbeck, H. and Van Haut, H.: Messung von Luftverunreinigungen mit Hilfe pflanzlicher Organismen. Bioindicators of landscape deterioration. Prague. 1971.
Schönebeck, C.: Der Beitrag komplexer Stadtsimulationsmodelle (vom Forrester-Typ) zur Analyse und Prognose großstädtischer Systeme. Birkhäuser-Verl. Basel und Stuttgart. 1975.
Schoener, A.: Experimental zoogeography: colonization of marine mini-islands. Amer. Natur. *108:* 715–738. 1974.
Schoener, A.: Colonization curves for marine islands. Ecology *55:* 818–827. 1974.
Schrimpff, E.: Ein mathematisches Modell zur Vorhersage von Abflußereignissen im Bereich der Anden Kolumbiens/Südamerika. Dissertation, Cologne. 1975.
Schroeder, D.: Bodenkunde in Stichworten. Hirt Verl., Kiel. 1969.
Schroder, F.G.: Zur Klassifizierung der Anthropochoren. Vegetatio *16:* 225–238. 1969.
Schubart, H. and Beck, L.: Zur Coleopterenfauna amazonischer Böden. Amazonia *1* (4): 311–322. 1968.
Schulte, G.: Die Bindung von Landarthropoden an das Felslitoral der Meere und ihre

Ursachen. Math. Nat. Fak. Univ. Kiel. 1977.

Schulz, E.: Pollenanlytische Untersuchungen quartäre Sedimente des Nordwest-Tibesti. Pressedienst Wissenschaft FU Berlin *5:* 59–69. 1974.

Schulze, E.-D., Ziegler, H. and Stichler, W.: Der Säurestoffwechsel von Welwitschia mirabilis am natürlichen Standort in der Namib Wüste. Verhdl. Ges. Ökologie, Vienna, *1975:* 211–220, Junk, The Hague. 1976.

Schwarzbach, M.: Das Klima der Vorzeit. Enke Verl., Stuttgart. 1974.

Schwibach, J. and Gans, I.: Gefährdung durch radioaktive Stoffe in der Umwelt — Radioökologie. In: Kernenergie und Umwelt, 207–215. E. Schmidt Verl., Berlin. 1976.

Schwidetzky, I.: Die vorspanische Bevölkerung der Kanarischen Inseln. Göttingen. 1963.

Schwoerbel, J.: Einführung in die Limnologie. 3rd edition. UTB, Fischer, Stuttgart. 1977.

Sclater, P.L.: On the general geographical distribution of the members of the class Aves. J. Proc. Linn. Soc. London (Zool.) *2:* 130–145. 1858.

Scott, G.A.M. and Stone, I.G.: The Mosses of Southern Australia. Acad. Press, London, New York. 1976.

Seekamp, G. and Fassbender, H.: Zur Erfassung der Schwermetall-Belastung von industrie- und verkehrsfernen Waldökosystemen durch Niederschlagswasser. Mitt. Deutsch. Bodenkd. Ges. *20:* 493–499. 1974.

Segerstrale, S.G.: The freshwater amphipods Gammarus pulex (L.) and Gammarus lacustris (D.O. SARS) in Denmark and Fennoskandia: a contribution to the late- and postglacial immigration history of the aquatic fauna of Northern Europe. Soc. Sci. Fenn. Comment. Biol. *15.* 1954.

Segerstrale, S.G.: On the immigration of the glacial relicts of Northern Europe, with remarks on their prehistory. Soc. Sci. Fenn. Comment. Biol. *16.* 1957.

Sehnem, A.: As Filicineas do Sul do Brasil, sua distribução geográfica, sua ecologia e suas votas de migração. Pesquisas *31:* 1–108, São Leopoldo. 1977.

Seibert, P.: Übersichtskarte der natürlichen Vegetationsgebiete von Bayern 1:500 000 mit Erläuterungen. Schriftenr. Vegetationskd. *3.* 1968.

Selander, R. K.: Systematics and Speciation in Birds. In: Avian Biology. Acad. Press, London, New York. 1971.

Sells, S.B.: On the nature of stress. In: Social and psychological factors in stress. 134–139, Holt, Rinehart & Winston, New York. 1970.

Semmel, A: Grundzüge der Bodengeographie. Verl. Teubner, Stuttgart. 1977.

Sernander, R.: Studier öfer lafvarnes biologi. Svensk. Bot. Tidskr. Stockholm. 1926.

Servant, M. and Servant-Vildary, S.: Le Plio-Quaternaire du Bassin du Tschad. In: Le Quaternaire. Bull. l'Assoc. Franc. pour l'étude du Quaternaire *36:* 169–175. 1973.

Shannon, C.E. and Weaver, W.: The mathematical theory of communication. Urbana. 1949.

Simberloff, D.S.: Models in biogeography. In: T.J.M. Schopf (ed.) Models in paleobiology, Freeman, Cooper & Co., San Francisco. p. 1972.

Simberloff, D.S.: Trophic structure determination and equilibrium in an arthropod community. Ecology. *57:* 395–398. 1976.

Simmons, J.G.: The ecology of natural resources, Arnold, London. 1974.

Simon, D.E.: The growth of Dracaena draco. Journ. of the Arnold Arboretum *55:* 51–58. 1974.

Simpson G.G.: History of the fauna of Latin America. Amer. Scient. *38:* 361–389.

1950.
Sioli, H.: Zur Ökologie des Amazonas-Gebietes, Biogeography and Ecology in South America, Junk, The Hague, *1:* 137–170. 1968.
Smith, F.E.: Spatial Heterogeneity, Stability and Diversity in Ecosystems. Trans. Conn. Acad. Arts. Sci. *44:* 309–335. 1972.
Smith, G.N., Watson, B.S. and Fisher, F.S.: The metabolism of [C^{14}] o,o-diethyl o-(3, 4, 6-trichloro-2-pyridyl) phosphothioate (DURSBAN) in fish, J. econ. Ent. *59:* 1464–1475. 1966.
Smith, L.B.: Origins of the flora of southern Brazil. Contrib. US Nat. Herbarium Washington *35:* 215–249. 1964.
Södergren, A.: Transport, Distribution and Degradation of DDT and PCB in a South Swedish Lake Ecosystem. Vatten *2:* 90–107. 1973.
Solberg, J.J.: Principles of system modelling. Proc. Int. Sympos. Systems Eng. and Analysis, 67–74. 1972.
Sorauer, P.: Die makroskopische Analyse rauchgeschädigter Pflanzen. Samml. Abhdl. Abgase und Rauchschäden *7.* 1911.
Sournia, A.: Circadian periodicities in natural populations of marine phytoplankton. Adv. mar. Biol. *12:* 325–389. 1974.
Spassky, B. et al.: Geography of the Sibling Species related to Drosophila willistoni, and of the Semispecies of the Drosophila paulistorum complex. Evolution *25* (1): 129–140. 1971.
Stark, N.: Nutrient cycling II: Nutrient distribution in Amazonian vegetation. Trop. Ecol. *12* (2): 177–201. 1971.
Stearn, F.W.: Wildlife habitat in urban and suburban environments. Conference on North American Wildlife and Natural Resources Transactions. *32:* 61–69. 1967.
Stearn, F.W.: Urban botany: an essay on survival. University of Wisconsin at Milwaukee, Field Stations Bulletin *4* (1): 1–6. 1971.
Steele, J.H.: The structure of marine ecosystems. Harvard University Press, Cambridge. 1974.
Stein, N.: Coniferen im westlichen malayiischen Archip. Biogeographica *II,* Junk, The Hague.
Steiniger, H.: Genetische Variabilität bei Carabiden-Populationen inner- und Außerstädtischer Standorte (Coleoptera). Dissertation, Biogeographie, Saarbrücken. 1978.
Stevenson, F.J. and Cheng, C.N.: Amino acid levels in the Argentine Basin sediments: Correlation with Quaternary climatic changes. J. Sed. Petrol. *1969:* 345–349. 1969.
Stieglitz, W.: Bemerkenswerte Adventivarten aus der Umgebung von Mettmann. Göttinger Floristische Rundbriefe *11* (3): 45–49. 1977.
Stiehl, E.: Niederschlagsanalysen im Raum Marburg/Lahn unter besonderer Berücksichtigung der Witterungs- und Reliefeinflüsse. Dissertation, Marburg. 1970.
Stix, E.: Pollen- und Sporengehalt der Luft im Herbst über den Atlantik. Ocecologia *18:* 235–242. 1975.
Stocker, O.: Steppe, Wüste and Savanne. Festschr. Firbas. Veröff. Geobot. Inst. Rübel *37.* 1962.
Stöfen, D.: Bleiprobleme auf dem internationalen Symposium “Umwelt und Gesundheit” Paris 1974. Forum Umwelt Hygiene *8:* 233–235. 1975.
Stonehouse, B.: Tiere der Antarktis. BLV Verlagsges., Munich, Vienna. 1972.
Støp-Bowitz, C.: Did lumbricids survive the quaternary glaciations in Norway? Pedobiol. *9:* 93–98. 1969.

Stuart, E.B., Gal-Or, B. and Brainard, A.J.: A critical Review of Thermodynamics, Mono Book Corp., Baltimore. 1970.

Stugren, B.: Grundlagen der allgemeinen Ökologie. 2nd edition, Fischer, Jena. 1974.

Suddia, T.W.: Man, nature, city: the urban ecosystem. United States Department of the Interior, National Park Service, Urban Ecology *1*. 1972.

Sukopp, H.: Wandel von Flora und Vegetation in Mitteleuropa unter dem Einfluß des Menschen. Ber. Ldw. *50:* 112–139, Parey Verl., Hamburg and Berlin. 1972.

Sukopp, H.: Die Großstadt als Gegenstand ökologischer Forschung. Schr. Ver. zur Verbreit. naturwiss. Kenntnisse, Vienna. 1973.

Sukopp, H.: Dynamik und Konstanz in der Flora der Budesrepublik Deutschland. Vegetationskunde *10,* Bonn-Bad Godesberg. 1976.

Sukopp, H., Kunick, W., Runge M. and Zacharias, F.: Ökologische Charakteristik von Großstädten, dargestellt am Beispiel Berlins. Verhdl. Ges. für Ökologie *2:* 383–403. Junk, The Hague. 1974.

Sukopp, H. and Müller, P.: Symposium über Veränderungen von Flora und Fauna in der Bundesrepublik Deutschland — Ergebnisse und Konsequenzen. Schriftenr. Vegetationskde., *10:* 401–409, Bonn-Bad Godesberg. 1976.

Swink, F.A.: Plants of the Chicago region. Lisle, Illinois. 1974.

Tamayo, F.: Adaptaciones de la vegetatión pirofila. Bol. Soc. Venez. Cienc. Naturales *23* (101): 49–58. 1962.

Tanner, W.F.: Multiple influences on Sea-level changes in the Tertiary. Palaeogeogr., Palaeoclim., Palaeoecol. *5:* 165–171. 1968.

Tansley, A.G.: The use and abuse of vegetation concepts and terms. Ecology *16.* 1935.

Tenovuo, R.: Zur Urbanisierung der Vögel in Finnland. Acta Zool. Fennici *4:* 33–34. 1967.

Thaler, K.: Endemiten und Arktoalpine Arten in der Spinnenfauna der Ostalpen (Arachnida: Araneae). Entomol. Germanica *3* (1/2): 135–141. 1976.

Thannheiser, D.: Ufer- und Sumpfvegetation auf dem westlichen kanadischen Arktis-Archipel und Spitzbergen. Polarforschung *46* (2): 71–82. 1976.

Thiele, H.U.: Physiologisch-ökologische Studien an Laufkäfern zur Kausalanalyse ihrer Habitatbindung. Verdhl. Ges. für Ökologie *2:* 39–54. 1974.

Thiele, H.U.: Carabid Beetles in their Environments. Springer Verl., Heidelberg, New York. 1977.

Thienemann, A.: Lebensgemeinschaft und Lebensraum. Nat. Wschr. N.F. *17:* 281–290. 1918.

Thienemann, A.: Die Reliktenkrebse Mysis relicta, Pontoporeia affinis, Pallasea quadrispinosa und die von ihnen bewohnten norddeutschen Seen. Arch. f. Hydrobiol. *19:* 521–582. 1928.

Thienemann, A.: Verbreitungsgeschichte der Süßwassertierwelt Europas, Binnengewässer 18. Stuttgart. 1950.

Thomé, M.: Ökologische Kriterien zur Abgrenzung von Schadräumen in einen urbanen System. — Dargestellt am Beispiel der Stadt Saarbrücken. Dissertation Saarbrücken. 1976.

Thompson, A.W.: Über Wachstum und Form. Basel and Stuttgart. 1973.

Thower, N. and Bradsbury, D.: Chile — California Mediterranean scrub atlas: A comparative analysis, Dwoden, Hutchinson and Ross, Stroudsbourg. 1977.

Tischler, W.: Biozönotische Untersuchungen an Ruderalstellen. Zool. Jahrb. (Syst.) *81:* 122–174. 1952.

Topp, W.: Zur Ökologie der Müllhalden. Ann. Zool. Fennici *8:* 194–222. 1971.

Tricart, J., Santos, M., Cardoso Da Silva, T. and Dias Da Silva, A.: Estudos de Geografia da Bahia. Geografia e Planeamento. Publ. Univ. Bahia *4* (3): 1–243. 1958.

Tricart, J., Vogt, H. and Gomes, A.: Note préliminaire sur la morphologie du cordon littoral actuel entre Tramandai et Torres, Rio Grande do Sul, Brésil. Cah. océanogr. et Côtes, Paris *12:* 453–457. 1960.

Troll, C.: Über das Wesen der Hochgebirgsnatur. AV-Jahrbuch *80.* 1955.

Troll, C.: The relationships between the climate ecology and plant-geography of the southern cold temperature zone and the tropical high mountains. Proc. Royal Soc., B. *152:* 529–532.

Troll, C.: Landschaftökolige. Junk, The Hague. 1968.

Turner, J.R.G.: Two thousand generations of hybridisation in a Heliconius butterfly. Evolution *25:* 471–482. 1971.

Turner, J.R.G.: The genetics of some polymorphic forms of the butterflies Heliconius melpomene (Linnaeus) and H. erato (Linnaeus). II. The hybridization of subspecies of H. melpomene from Surinam and Trinidad. Zoologica, N.Y. *56:* 125–157. 1972.

Tüxen, R.: Grundriß einer Systematik der nitrophilen Unkrautgesellschaften in der Eurosibirischen Region Europas. Mitt. Flor.-soz. Arbeitsg. 2 Stolzenau/Weser. 1950.

Tüxen, R.: Die heutige potentielle Vegetation als Gegenstand der Vegetationskartierung. Pflanzensoz. *13.* 1956.

Tüxen, R.: Pflanzensoziologie als synthetische Wissenschaft. Misc. Pap. *5:* 141–159. 1970.

Udvardy, M.D.F.: Dynamic Zoogeography. New York. Cincinnati, Van Nostrand Reinhold. 1969.

Udvardy, M.D.F.: A classification of the Biogeographical Provinces of the World. IUCN Paper *18,* Morges. 1975.

Ule, E.: Die Vegetation von Cabo Frio an der Küste von Brasilien. Engl. Bot. Jahrbuch *28:* 511–528. 1901.

Vanzolini, P.E.: Problemas faunisticos do Cerrado. Simposio sôbre Cerrado. Univ. São Paulo. 1963.

Vanzolini, P.E.: Zoolgia systematica geografia e a origem das especies. Univ. São Paulo *3:* 1–56. 1970.

Vanzolini, P.E. and Ab'Saber, A.N.: Divergence rate in South American Lizards of the Genus Liolaemus (Sauria, Iguanidae). Pap. Avuls. Zool. *21:* 205–208. 1968.

Vanzolini, P.E. and Williams, E.E.: South American Anoles: The geographic differentiation and evolution of the Anolis chrysolepsis species group (Sauria, Iguanidae). Arq. Zool. *19:* 1–298. 1970.

Vareschi, V.: La quema como factor ecologico en los Llanos. Bol. Soc. Venz. Cienc. Naturales *23* (101): 9–26. 1962.

Varga, Z.: Geographische Isolation und Subspeziation der Lepidopteren in den Hochgebirgen des Balkans. Acta Entomol. Jugoslavica *11* (1–2): 5–39. 1975.

Vauk, G.: Die Vögel Helgolands. Hamburg and Berlin. 1972.

Vaurie, C.: A study of Asiatic larks. Bull. Amer. Mus. Nat. Hist. *97:* 431–526. 1951.

Vernberg, W.B. et al.: Effects of sublethal concentrations of cadmium on adult Palaemonetes pugio under static and flow-through conditions. Bull. Envir. Contamm. & Toxicology *17* (1): 16–24. 1977.

Vetter, H.: Belastungen und Schäden durch Schwermetalle in der Nähe einer Blei- und Zinkhütte in Niedersachsen. Staub-Reinhalt. Luft *34* (1): 10–14. 1974.

Vida, G.: Evolution in Plants, Symposia Biologica Hungarica, *12,* Budapest. 1972.
Visher, S.S.: Tropical cyclones and the dispersal of life from island to island in the Pacific. Amer. Natural. *59:* 70–78. 1925.
Vogl, R.J.: The effects of fire on the vegetational composition of bracken-grasslands. Trans. Wisconsin Acad. *53:* 67–81. 1964.
Vogel, S.: Chiropterophilie in der neotropischen Flora. Flora *158:* 289–323. 1969.
Vuilleumier, F.: Systematics and Evolution in Diglossa (Aves, Coerebidae). Am. Mus. Novitates *2382:* 1–44. 1969.
Vuilleumier, F.: Speciation in South American Birds: I Progress report, Act. IV Congr. Latin, Zool. *1:* 239–2551. 1970.
Vuilleumier, F.: Zoogeography. In: Avian biology *5:* 421–496. Acad. Press. New York, San Francisco, London. 1975.
Vuilleumier, F., Wada, O., Toyokawa, K., Urata, G., Yano, Y. and Nakao, K.: A simple method for the quantitative analysis of urinary delta-aminolevulinic acid to evaluate lead absorption. Brit. J. industr. Med. *26:* 240–243. 1969.
Wallace, A.R.: Geographical distribution of animals. London. 1876.
Wallace, A.R.: Island Life. 1880.
Walter, H.: Das Pampaproblem in vergl. ökologischer Betrachtung und seine Lösung. Erdkunde *21:* 181–203. 1967.
Walter, H.: Die Vegetation der Erde in Öko-physiologischer Betrachtung. Verl. Fischer, Stuttgart. 1968.
Walter, H.: War die Pampa von Natur aus baumfrei? Umschau *16:* 508–509. 1969.
Walter, H.: Biosphäre, Produktion der Pflanzendecke und Stoffkreislauf in ökologisch-geographischer Sicht. G.Z. *59* (2): 116–130. 1971.
Warming, G.E.: Plantesamfund. Copenhagen. 1895.
Warnecke, G.: Gibt es xerothermische Relikte unter den Makrolepidopteren des Oberrheingebietes von Basel bis Mainz? Arch. Insektenkd. Oberrheingeb. u. angrenz. Ld. *2* (3): 81–119. 1927.
Washburn, A.L.: Periglacial Processes and Environments. London. 1973.
Weber, H.: Die Paramos von Costa Rica und ihre pflanzengeographische Verkettung mit den Hochanden Südamerikas. Abhdl. Akad. Wiss. Lit. Mainz, *3.* 1958.
Weber-Oldecop, D.W.: Fließgewässertypologie in Niedersachsen auf floristisch-soziologischer Grundlage. Göttinger Flor. Rdbr. *10* (4): 73–79. 1977.
Webster, J.R., Waide, J.B. and Patten, B.C.: Nutrient recycling and the stability of Ecosystems. Mineral Cycling in Southeastern Ecosystems. ERDA Symp. Series, U.S. Energy Research and Development Administration, Springfield. 1975.
Wegener, A.: Die Entstehung der Kontinente und Ozeane. Vieweg, Braunschweig. 1929.
Weidner, H.: Schwebfliegen auf hoher See. Entomol. Z. *13:* 152–153. Kernen Verl., Stuttgart. 1958.
Weidner, H.: Schädlinge an Arzneidrogen und Gewürzen in Hamburg. Beitr. zur Entomologie *13* (3/4): 527–545. 1963.
Weidner, H.: Eingeschleppte und eingebürgerte Vorratsschädlinge in Hamburg. Z. ang. Entomol. *54* (1/2): 163–177. 1964.
Weischet, W.: Klimatologische Regeln zur Vertikalverteilung der Niederschlägen in Tropengebirgen. Die Erde *100:* 287–306. 1969.
Weischet, W.: Einführung in die Allgemeine Klimatologie. Verl. Teubner, Stuttgart. 1977.
Wendland, V.: Die Wirbeltiere Westberlins. Berlin. 1971.

Went, F.W. and Stark, N.: The biological and mechanical role of soil fungi. Proc. nat. Acad. Sci. USA *60:* 497–504. 1968.

Werner, G. et al.: Umweltbelastungsmodell einer Großstadtregion. Verl. E. Schmidt, Berlin. 1975.

Weyl, R.: Die Physica Sacra des J.J. Scheuchzer. Konstanzer Blätter für Hochschulfragen, Universitätsverlag, Konstanz. 1966.

Whitaker, J.H.: Submarine Canyons and Deep-Sea Fans. Dowden, Hutchinson and Ross, Inc. Stroundsburg, Pennsylvania. 1976.

Whittaker, R.H.: Communities and Ecosystems. MacMillan Publ., New York. 1975.

Whitton, B.A.: River Ecology, Blackwell Scient. Publ., Oxford, London, Edinburgh, Melbourne. 1975.

Whitton, B.A.: River Ecology. Blackwell Publ., Oxford. 1975.

Wiedenmayer, F.: Shallow-water sponges of the western Bahamas. Birkhäuser Verl., Basel and Stuttgart. 1977.

Wiener, N.: Cybernetics. New York. 1948.

Wiggins, I.L. and Porter, D.M.: Flora of the Galapagos Islands. Stanford Univ. Press, Stanford. 1971.

Wilkens, H.: Über das phylogenetische Alter von Höhlentieren. Z. f. zool. Systematik und Evolutionsforschung *II* (I): 49–60. 1973.

Williams, S.: An Introduction to Physical oceanography. Addison-Wesley. Publ., London. 1962.

Williams, C.B.: Patterns in the Balance of Nature. Acad. Press, London and New York. 1964.

Williams, C.H. and David, D.J.: The Accumulation in Soil of Cadmium Residues from Phospate Fertilizers and their Effect on the Cadmium Content of Plants. Soil Science *121:* 86–93. 1976.

Williams, W.D.: Biogeography and Ecology in Tasmania. W. Junk, The Hague. 1974.

Willis, J.C.: Age and Area. A study in geographical distribution and origin of species. Cambridge Univ. Press, Cambridge. 1922.

Winge, H.: Races of Drosophila willistoni sibling species: Probable origin in quaternary forest refuges of South America. Genetics *74:* 297–298. 1973.

Winkler, S.: Moose als Indikatoren bei SO_2- und Bleibelastung. Daten und Dokumente zum Umweltschutz, *19:* 43–55. Hohenheim. 1976.

Winterbourn, M.J. and Lewis, M.H.: Littoral Fauna. In: New Zealand Lakes. Auckland Univ. Press, New Zealand. 1975.

Witley, G.P.: The Freshwater Fishes of Australia. In: Biogeography and Ecology in Australia. The Hague. 1959.

Woltereck, R.: Über die Spezifität des Lebensraumes, der Nahrung und der Körperformen bei pelagischen Cladoceren und über "Ökologische Gestalt-Systeme". Biol. Zentralbl. *28:* 521–551. 1928.

Yordanov, D.: A statistical analysis of the Air Pollution data in connection with the metereological conditions. In: Benarie, Atmospheric Pollution, 60–70, Elsevier Scient. Publ., Amsterdam and Oxford, 1976.

Zaret, T.M. and Paine, R.T.: Species Introduction in a Tropical Lake, Science *182:* 449–455. 1973.

Ziegler, H.: Zur Physiologie austrocknungsfähiger Kormophyten. In: P. Müller (ed.) Verhdl. Gesellsch. Ökol. Saarbrücken *1973:* 65–73, Junk, The Hague. 1974.

Zimmerman, E.C.: Insects of Hawaii. Univ. Hawaii Press, Honolulu. 1948.

Zimmerman, K.: Die Randformen der mitteleuropäischen Wühlmäuse. — Syllegomena Biologica, 454–471, Verl. Ziemsen, Wittenberg. 1950.

Index

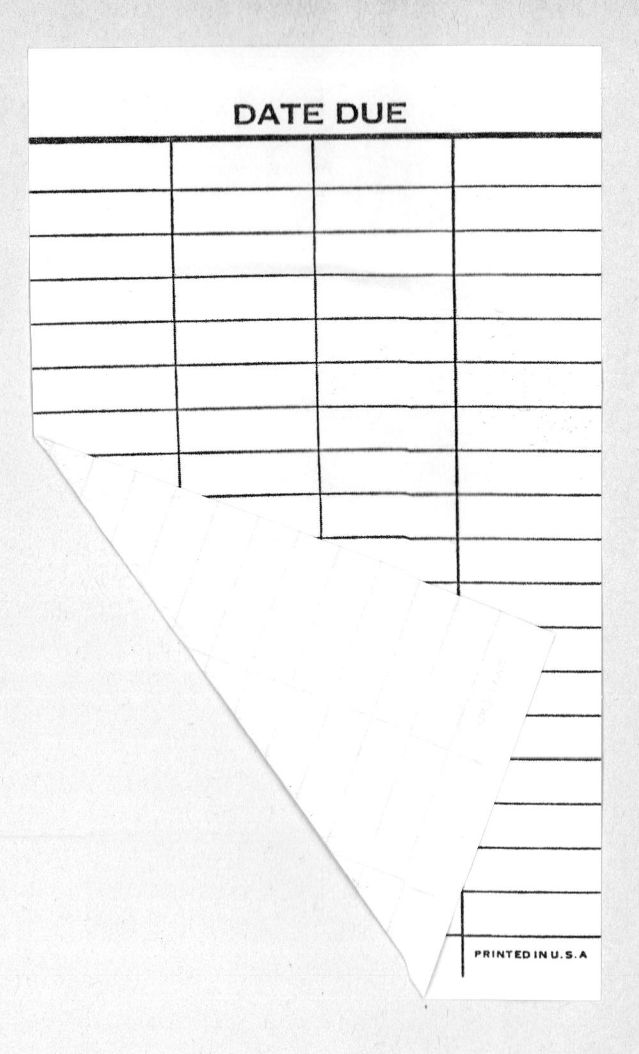
DATE DUE
PRINTED IN U.S.A